Impact of Environmental Factors on Aquatic Ecosystem

Impact of Environmental Factors on Aquatic Ecosystem

Guest Editors

Xuan Ban

Wenxian Guo

Yicheng Fu

MDPI

Basel • Beijing • Wuhan • Barcelona • Belgrade • Novi Sad • Cluj • Manchester

Guest Editors

Xuan Ban
Innovation Academy for Precision Measurement Science and Technology
Chinese Academy of Sciences
Wuhan
China

Wenxian Guo
School of Water Conservancy
North China University of Water Resources and Electric Power
Zhengzhou
China

Yicheng Fu
China Institute of Water Resources and Hydropower Research
Beijing
China

Editorial Office
MDPI AG
Grosspeteranlage 5
4052 Basel, Switzerland

This is a reprint of the Special Issue, published open access by the journal *Water* (ISSN 2073-4441), freely accessible at: https://www.mdpi.com/journal/water/special_issues/TAS387C6SH.

For citation purposes, cite each article independently as indicated on the article page online and as indicated below:

Lastname, A.A.; Lastname, B.B. Article Title. *Journal Name* **Year**, *Volume Number*, Page Range.

ISBN 978-3-7258-7555-9 (Hbk)
ISBN 978-3-7258-7556-6 (PDF)
https://doi.org/10.3390/books978-3-7258-7556-6

Contents

About the Editors

Xuan Ban

Xuan Ban is a research professor at the Innovation Academy for Precision Measurement Science and Technology, Chinese Academy of Sciences, in Wuhan, China. Her research primarily focuses on the theory and application of eco-hydraulics, with an emphasis on ecological conservation in the Yangtze River Basin. Her work encompasses habitat simulation, environmental flow assessment, and ecological restoration techniques. She has developed an innovative hydrological–ecological coupling model that has advanced restoration strategies for endangered species such as the Chinese sturgeon, the Yangtze sturgeon. She serves as a Guest Editor for the *Water* Special Issue 'Impact of Environmental Factors on Aquatic Ecosystem.' Dr. Ban has led 22 research projects, including grants from the National Natural Science Foundation of China and the Chinese Academy of Sciences. She has authored over 60 papers, contributed to four monographs, and holds four invention patents. Her accolades include a First Prize for Science and Technology from the Chinese Academy of Fishery Sciences, a Second Prize for Scientific and Technological Progress from the Hubei Provincial Government, and the Yangtze River Science and Technology Award, among others.

Wenxian Guo

Wenxian Guo is a Professor of North China University of Water Resources and Electric Power (NCWU). He was selected as the Innovative Talent of Higher Education Institutions in Henan Province, and Outstanding Master's Thesis Supervisor of Henan Province. His research interests cover river eco-hydrology and eco-hydraulics, ecological impact assessment of hydropower projects, reservoir ecological operation, eco-hydraulics engineering, river and lake ecological restoration, and water resources and water environment planning and management. He has presided over and participated in more than 30 national and provincial research projects, including the National Natural Science Foundation of China, National Major Water Science and Technology Special Project, and provincial key R&D programs. He has published 170 academic papers (80 indexed in SCI/EI) in domestic and international authoritative journals, published 8 academic monographs, and holds 12 authorized invention patents. He has won two provincial science and technology awards as the lead investigator, and multiple provincial and ministerial science and technology awards as a core participant.

Yicheng Fu

Yicheng Fu is a senior engineer at the China Institute of Water Resources and Hydropower Research, based in Beijing, China. His main research focuses on ecological impact assessment of regional coal mining, coordination of coal–water conflicts in watersheds, ecological restoration of rivers and lakes, and resource utilization of mine water. He has developed key technologies for multi-dimensional regulation of mine water quantity and quality in coal-rich watersheds and ecological restoration, which have been successfully applied to the ecological governance and restoration of near-river mining areas such as the Kuyi River and Morgele River. The groundwater level decline management plans for coal mining areas and water resource demonstration reports for mining area planning that he prepared have received approval at the provincial and ministerial levels. He has led seven national-level research projects and eight provincial/ministerial-level projects. To date, he has obtained 14 invention patents, including intelligent risk prediction for ecological utilization of mine water, dynamic zoning for ecological restoration in coal subsidence areas, and integrated watershed water-ecology management methods. He has also presided over the compilation and publication of three standards related to the utilization and quantitative calculation of mine water resources, published 126 academic papers (56 indexed by SCIE/EI), and authored nine monographs. His research achievements have won five provincial- and ministerial-level second prizes and three third prizes.

Preface

Aquatic ecosystems worldwide are under increasing pressure due to climate change, altered hydrology, pollution and habitat degradation. It is essential to understand how these environmental factors interact and influence the structure and function of ecosystems in order to develop effective conservation and management strategies. This Reprint brings together a selection of studies originally published in the Special Issue of *Water*, 'Impact of Environmental Factors on Aquatic Ecosystems', which we had the privilege of editing. The collection presents cutting-edge research from a variety of geographical regions and ecological settings, ranging from rivers and reservoirs to lakes and lagoons. The contributions employ a variety of approaches, including field investigations, advanced biomonitoring, remote sensing and numerical modelling, to address important themes such as the influence of hydrology and morphology, the impact of pollution and its remediation, the assessment of ecological health and the dynamics of eutrophication. This Reprint is intended for researchers, water resource managers and environmental practitioners seeking interdisciplinary insights into the resilience and vulnerability of aquatic systems. By showcasing innovative methodologies and evidence-based findings, we aim to promote science-based decision-making and encourage further research into the sustainable management of freshwater ecosystems.

Xuan Ban, Wenxian Guo, and Yicheng Fu
Guest Editors

water

MDPI

Editorial

Impact of Environmental Factors on Aquatic Ecosystem

Xuan Ban [1,*], Wenxian Guo [2] and Yicheng Fu [3]

1 Key Laboratory for Environment and Disaster Monitoring and Evaluation, Innovation Academy for Precision Measurement Science and Technology, Chinese Academy of Sciences, Wuhan 430071, China
2 Water Resources College, North China University of Water Resources and Electric Power, Zhengzhou 450046, China; guowenxian@ncwu.edu.cn
3 State Key Laboratory of Simulation and Regulation of River Basin Water Cycle, China Institute of Water Resources and Hydropower Research, Beijing 100038, China; swfyc@126.com
* Correspondence: banxuan@apm.ac.cn

1. Introduction

Aquatic ecosystems are facing unprecedented challenges due to anthropogenic activities and climate change [1]. Environmental factors such as hydrological alterations, pollution, eutrophication, and habitat degradation significantly affect biodiversity, water quality, and ecosystem services [2]. Understanding these impacts is critical for developing effective conservation and management strategies [3]. This Special Issue explores the multifaceted effects of environmental stressors on aquatic ecosystems through innovative methodologies, case studies, and interdisciplinary approaches. Its contributions highlight advancements in monitoring, modeling, and remediation techniques, offering insights into the resilience and vulnerability of aquatic environments.

This Special Issue of *Water* addresses these challenges by synthesizing cutting-edge research on the multifaceted impacts of environmental stressors across rivers, lakes, wetlands, and reservoirs. Its contributions highlight innovative methodologies—from advanced biomonitoring techniques to hybrid remediation strategies—that bridge gaps between theoretical understanding and practical conservation. By integrating interdisciplinary approaches (e.g., hydrology, ecotoxicology, remote sensing, and microbial ecology), this collection not only diagnoses ecosystem degradation, but also proposes scalable solutions tailored to diverse geographical and socioecological contexts.

Aquatic ecosystems face tipping points, where cumulative stressors may trigger irreversible shifts in function [4]. For instance, eutrophication-driven algal blooms, heavy metal bioaccumulation, and habitat fragmentation increasingly compromise water security and human health [5]. This issue underscores the need for science-based policy interventions, emphasizing resilience-building through adaptive management, community engagement, and technological innovation.

2. An Overview of the Contributions to This Special Issue

The 13 articles in this Special Issue address diverse aspects of aquatic ecosystem dynamics, and can be grouped into the following five thematic categories.

2.1. Hydrological and Morphological Influences on Aquatic Ecosystems

Hydrological and geomorphological changes profoundly affect riverine and lacustrine ecosystems. Yu et al. (Contribution 1) investigated how river sinuosity impacts the hydrodynamic conditions and ecological flow requirements in the Nansha River, China. Using the River2D model, they demonstrated that moderate sinuosity enhanced depth and velocity diversity, optimizing habitats for target species like *Cyprinus carpio*. Li et al. (Contribution

https://doi.org/10.3390/w17101453

2) examined the morphological traits of *Cynodon dactylon* in the Three Gorges Reservoir's riparian zone, linking plant adaptations to fluctuating water levels and sediment properties. Their findings revealed that soil moisture and nutrient gradients drive root and stem plasticity, enabling survival under prolonged submersion. Hong et al. (Contribution 3) combined LSTM and InVEST models to assess hydrological evolution and habitat quality in the Xiangjiang River Basin, identifying climate change and human activities as primary drivers of habitat fragmentation. Lastly, Firsova et al. (Contribution 4) explored how Lake Baikal's cold waters influence phytoplankton dynamics in the Irkutsk Reservoir, showing seasonal shifts in community structure driven by temperature and nutrient inputs.

2.2. Pollution Sources, Impacts, and Remediation Strategies

Pollution mitigation and remediation are central to restoring aquatic health. Sun et al. (Contribution 5) developed biochar/clay composite particles immobilized with *Flavobacterium* and *Aquamicrobium* species, achieving >80% removal of ammonia and petroleum hydrocarbons in contaminated wetlands. Sobiecka et al. (Contribution 6) demonstrated the efficacy of macrophytes (e.g., *Elodea canadensis*) in removing chlorpyrifos, with enzymatic responses (e.g., glutathione-S-transferase) indicating adaptive oxidative stress mechanisms. Papakonstantinou et al. (Contribution 7) assessed heavy metal contamination in Greece's Gialova Lagoon, identifying localized hotspots for Mo and Pb and emphasizing the role of sediment geochemistry in pollutant retention. Vallese et al. (Contribution 8) employed electrochemical detection to quantify Cd and Pb bioaccumulation in *Cyprinus carpio*, revealing tissue-specific metal accumulation patterns in the Colorado River. Vivien and Ferrari (Contribution 9) used oligochaete communities to evaluate wastewater treatment plant effluents, showing improved post-upgrade stream health through reduced pollutant loads.

2.3. Biomonitoring and Ecological Health Assessment

Biomonitoring tools provide critical insights into ecosystem stress. Kim et al. (Contribution 10) analyzed ferritin gene expression in *Litopenaeus vannamei* postlarvae, linking thermal and salinity stress to oxidative responses. Melkonyan et al. (Contribution 11) integrated acetylcholinesterase activity and oxidative stress biomarkers to assess whitefish health in Lake Sevan, Armenia, identifying eutrophication and hypoxia as key stressors.

2.4. Eutrophication and Algal Community Dynamics

Eutrophication drives shifts in primary producer communities. Ban et al. (Contribution 12) utilized satellite-derived ocean color data to estimate phytoplankton productivity in Qinghai Lake, revealing oligotrophic conditions with seasonal peaks in chlorophyll-*a*. Yang et al. (Contribution 13) studied filamentous algae (*Cladophora, Spirogyra*) in the Taihang catchment, showing nutrient-dependent succession and interactions with macrobenthos that influence habitat heterogeneity.

2.5. Advanced Technologies and Modeling Applications

Innovative technologies enhance monitoring and predictive capabilities. Hong et al. (Contribution 3) and Ban et al. (Contribution 12) exemplified the integration of hydrological (LSTM) and ecological (InVEST) models with remote sensing to track habitat and productivity changes. Vallese et al. (Contribution 8) advanced automated electrochemical detection for real-time heavy metal monitoring, validated against ICP-AES.

3. Conclusions

The studies in this Special Issue underscore the complexity of environmental impacts on aquatic ecosystems and the need for multidisciplinary solutions. The key findings include the following:

1. Hydrological and morphological alterations require balanced management to sustain biodiversity and ecological flows;
2. Pollution remediation benefits from hybrid approaches (e.g., biochar–microbe composites, phytoremediation) tailored to local conditions;
3. Biomonitoring tools (e.g., enzymatic biomarkers, oligochaete indices) are vital for early stress detection and policy formulation;
4. Remote sensing and modeling bridge spatial–temporal gaps in ecosystem assessment.

Future research should prioritize long-term monitoring, community engagement, and adaptive management to address emerging challenges like climate change and microplastic pollution. This collection advances our understanding of aquatic ecosystem resilience, providing a foundation for sustainable water resource management.

Acknowledgments: As guest editors of this Special Issue, the authors acknowledge the journal editors and all authors who submitted manuscripts to this Special Issue. Special thanks are extended to the referees who diligently reviewed all of the submissions, which greatly improved the quality of the published papers.

Conflicts of Interest: The authors declare no conflicts of interest.

List of Contributions

1. Yu, Z.; Fu, Y.; Zhang, Y.; Liu, Z.; Liu, Y. Quantifying the Impact of Changes in Sinuosity on River Ecosystems. *Water* **2023**, *15*, 2751. https://doi.org/10.3390/w15152751.
2. Li, X.; Li, S.; Xie, Y.; Wei, Z.; Li, Z. What Drives the Morphological Traits of Stress-Tolerant Plant *Cynodon dactylon* in a Riparian Zone of the Three Gorges Reservoir, China. *Water* **2023**, *15*, 3284. https://doi.org/10.3390/w15183183.
3. Hong, F.; Guo, W.; Wang, H. A Comprehensive Assessment of the Hydrological Evolution and Habitat Quality of the Xiangjiang River Basin. *Water* **2023**, *15*, 3626. https://doi.org/10.3390/w15203626.
4. Firsova, A.; Galachyants, Y.; Bessudova, A.; Hilkhanova, D.; Titova, L.; Nalimova, M.; Buzevich, V.; Marchenkov, A.; Sakirko, M.; Likhoshway, Y. The Influence of Lake Baikal on Phytoplankton in the Irkutsk Reservoir. *Water* **2024**, *16*, 3284. https://doi.org/10.3390/w16223284.
5. Sun, P.; Wei, J.; Gao, Y.; Zhu, Z.; Huang, X. Biochar/Clay Composite Particle Immobilized Compound Bacteria: Preparation, Collaborative Degradation Performance and Environmental Tolerance. *Water* **2023**, *15*, 2959. https://doi.org/10.3390/w15162959.
6. Sobiecka, E.; Mroczkowska, M.; Olejnik, T.; Nowak, A. The Use of Macrophytes for the Removal of Chlorpyrifos from the Aquatic Environment. *Water* **2024**, *16*, 724. https://doi.org/10.3390/w16050724.
7. Papakonstantinou, M.; Sergiou, S.; Geraga, M.; Prandekou, A.; Dimas, X.; Fakiris, E.; Christodoulou, D.; Papatheodorou, G. Sedimentological, Geochemical, and Environmental Assessment in an Eastern Mediterranean Coastal Setting: The Gialova Lagoon, SW Peloponnese, Greece. *Water* **2024**, *16*, 2312. https://doi.org/10.3390/w16162312.
8. Vallese, F.D.; Stupniki, S.; Trillini, M.; Belén, F.; Di Nezio, M.S.; Juan, A.; Pistonesi, M.F. Bioaccumulation Study of Cadmium and Lead in Cyprinus carpio Using Electrochemical Detection. *Water* **2025**, *17*, 77. https://doi.org/10.3390/w17010077.
9. Vivien, R.; Ferrari, B.J.D. New Data on the Use of Oligochaete Communities for Assessing WWTP Effluents Impacts. *Water* **2025**, *17*, 724. https://doi.org/10.3390/w17050724.
10. Kim, C.-W.; Lee, J.-W.; Kang, S.-W.; Kang, H.-S. Study on Ferritin Gene Expression to Evaluate the Health of White Leg Shrimp (*Litopenaeus vannamei*) Postlarvae Due

to Changes in Water Temperature, Salinity, and pH. *Water* **2024**, *16*, 1477. https://doi.org/10.3390/w16111477.

11. Melkonyan, H.; Chuiko, G.; Barseghyan, N.; Vardanyan, T.; Ghukasyan, E.; Kobelyan, H.; Gabrielyan, B. Assessment of the Health Status of Whitefish (*Coregonus lavaretus* Linnaeus, 1758) and the Quality of Its Habitat in Lake Sevan (Armenia) Using a Multi-Biomarker Approach. *Water* **2024**, *16*, 2789. https://doi.org/10.3390/w16192789.
12. Ban, X.; Dang, Y.; Shu, P.; Qi, H.; Luo, Y.; Xiao, F.; Feng, Q.; Zhou, Y. Estimation of Phytoplankton Primary Productivity in Qinghai Lake Using Ocean Color Satellite Data: Seasonal and Interannual Variations. *Water* **2024**, *16*, 1433. https://doi.org/10.3390/w16101433.
13. Yang, B.; Zhang, Y.; Zhang, M.; Lv, X.; Li, Y.; Zhang, J.; Wang, X.; Gao, X.; Zhao, X.; Wang, X. The Distribution and Succession of Filamentous Algae in the Southern Taihang Catchment. *Water* **2024**, *16*, 2453. https://doi.org/10.3390/w16172453.

References

1. Zolfagharpour, F.; Saghafian, B.; Delavar, M.J. Hydrological alteration and biodiversity change along the river network caused by anthropogenic activities and climate variability. *Ecol. Process.* **2022**, *11*, 19. [CrossRef]
2. Ban, X.; Qi, T.; Wang, H.Z.; Du, H.; Diplas, P.; Xiao, F.; Du, Y.; Gao, W.J.; Guo, W.X.; Shi, X.; et al. Comprehensive Environmental Flows Assessment for Multi-Guilds in the Riparian Habitats of the Yangtze River. *Water Resour. Res.* **2022**, *58*, e2021WR030408. [CrossRef]
3. Du, H.; Ban, X.; Shih, W.; Diplas, P.; Wu, J.; Li, J.; Cheng, P.; Li, P.; Liu, W.; He, F. The crucial role of ecohydraulic factors in triggering sturgeon reproduction: Implications for active habitat restoration strategies in the Yangtze River. *J. Appl. Ecol.* **2025**, *62*, 1052–1062. [CrossRef]
4. Davis, J.; Sim, L.; Chambers, J.J.F.B. Multiple stressors and regime shifts in shallow aquatic ecosystems in antipodean landscapes. *Freshw. Biol.* **2010**, *55* (Suppl. S1), 5–18. [CrossRef]
5. Gulati, R.D.; Donk, E.V.J.S.N. Lakes in the Netherlands, their origin, eutrophication and restoration: State-of-the-art review. *Hydrobiologia* **2002**, *478*, 73–106. [CrossRef]

water

MDPI

Article

New Data on the Use of Oligochaete Communities for Assessing the Impacts of Wastewater Treatment Plant Effluents on Receiving Streams

Régis Vivien * and Benoît J. D. Ferrari

Swiss Centre for Applied Ecotoxicology (Ecotox Centre), EPFL ENAC IIE-GE, 1015 Lausanne, Switzerland; benoit.ferrari@centreecotox.ch
* Correspondence: regis.vivien@centreecotox.ch

Abstract: Negative impacts of human activities on stream ecosystems include the reduction/modification of the connectivity between surface water and groundwater and the contamination of these resources. Vertical hydrological exchanges principally occur through the coarse surface sediments and the hyporheic zone (porous matrix) and these compartments have the property to store pollutants. Such hydrological exchanges participate in the self-purification of the stream and infiltration of polluted surface water can lead to the contamination of groundwater. A complete environmental monitoring program should therefore include the assessment of the biological quality of the porous matrix and of the dynamics of vertical hydrological exchanges. The Functional trait (FTR) method based on the study of oligochaete communities in the coarse surface sediments and the hyporheic zone, allows simultaneous assessment of the effects of pollutants present in these compartments and the dynamics of vertical hydrological exchanges. Here, we applied the FTR method upstream and downstream of the effluents of three different wastewater treatment plants (WWTPs) whose discharges were significantly polluted, and for one of them (Oberglatt), before and after its upgrading. We could clearly observe negative effects of the effluents of each of these WWTPs on oligochaete communities and the Oberglatt WWTP upgrading resulted, compared to the state before the upgrading, in a significant reduction of the polluted sludge effect at the downstream site of the effluent. In addition, the method allowed us to identify several sites where the stream had a high capacity to self-purify (through exfiltration of groundwater) and other sites where groundwater was vulnerable to pollution by surface water.

Keywords: Clitellata; bioindication; hyporheic zone; coarse surface sediments; stream functioning; chemical pollution

1. Introduction

Negative effects of human activities on aquatic ecosystems include the reduction or modification of the surface water and groundwater connectivity, as well as the contamination of these resources [1,2]. The dynamics of vertical hydrological exchange depend on the geomorphological context but can be significantly affected by anthropogenic activities [3,4]. Infiltration of surface water into groundwater can be caused by a number of factors, including excessive groundwater pumping, high discharges from combined sewer overflows and flash floods, and the presence of impervious surfaces in the catchment area [5]. Conversely, exfiltration may occur when the flow rate is low (in such condition, the groundwater can feed the stream) [6], or in case of the raise of the level of subterranean

 https://doi.org/10.3390/w17050724

waters relative to that of surface waters (for example upstream of a dam). The functioning of the watercourse ecosystem can be defined as the result of the interactions between the dynamics of vertical hydrological exchanges and the inputs of chemical substances (i.e., nutrients and pollutants) [5,7–9].

The coarse surface sediments and the hyporheic zone, which are here collectively referred to as the "porous matrix", are generally predominant habitats in the streams and play a significant ecological role [1,2,10–12]. These compartments possess the capacity to store pollutants and serve as the substrate through which vertical hydrological exchanges occur. Such exchanges have the potential to facilitate the self-purification of the stream [13]. However, vertical hydrological exchanges have also the potential to impair the quality of surface water, the porous matrix and groundwater [6]. Indeed, infiltration of polluted surface water may result in the contamination of groundwater, while polluted groundwater exfiltration may lead to a degradation of the quality of the porous matrix and surface water [14]. It is therefore recommended that the environmental monitoring programs include an assessment of the biological quality of the porous matrix and of the dynamics of vertical hydrological exchanges.

Oligochaetes are a common component of the interstitial biota of porous matrices in aquatic ecosystems, comprising species with a wide range of pollution tolerance and characteristic of surface sediments or groundwater [5,13,15]. The Functional Trait (FTR) method, which is based on the study of oligochaete communities in the coarse surface sediments and the hyporheic zone, allows for the simultaneous assessment of the effects of pollutants present in these compartments and the dynamics of hydrological exchanges between surface water and groundwater [5,13].

The FTR method was applied in France (in the Yseron, Chaudanne, Moselle, Loire, Azergues and Rhône rivers) and Switzerland (in the Roseg and Seymaz rivers) at sites showing a strong gradient of impairment, from pristine to highly impaired [5,16,17]. In these studies, oligochaete communities proved good indicators of the different degrees of stream alteration: sensitive taxa to chemical pollution and taxa indicator of groundwater exfiltration largely dominated at the unimpacted sites (e.g., in a mountain river), while at the most impaired sites, where chemical pollution was associated with man-induced infiltration (e.g., downstream of combined sewer overflows), resistant taxa to chemical pollution largely dominated, with an absence or very low percentages of taxa indicator of groundwater exfiltration. More recently, the FTR method was applied in Hochdorf and Buttisholz (Switzerland) upstream and downstream of the effluents of two outdated wastewater treatment plants (WWTPs) (at two different periods) to determine if it was suited to detect, in already impacted areas by human activities (agricultural/industrial areas), the effects of such effluents on the receiving streams [18]. Oligochaete results showed at the two different periods more altered biological quality and stream functioning downstream than upstream of the effluents of these WWTPs, in both the hyporheic zone and the coarse surface sediments. To confirm the suitability of the FTR method for assessing the effects of WWTP effluents on the receiving streams, more data were needed.

The objective of the present study was twofold: firstly, to obtain new data on the efficacy of the FTR method for assessing the impact of WWTP effluents on the biological quality and functioning of receiving streams; and secondly, to evaluate the potential of this method for monitoring the effects of a WWTP upgrade on the receiving stream. The FTR method was applied at upstream and downstream sites of three WWTPs (in Oberglatt, Muri and Vallorbe) whose discharges were found to be significantly contaminated. Two WWTPs (Oberglatt and Muri) were studied in detail, while the Vallorbe WWTP was only investigated with regard to the coarse surface sediments. Furthermore, the FTR method was applied following the upgrading of one of the WWTPs (Oberglatt) over the course of two sampling campaigns, with studies conducted on both compartments or only on coarse surface sediments.

2. Materials and Methods

2.1. Wastewater Treatment Plants and Studied Sites

Three WWTPs in Switzerland (in Oberglatt in the Canton of St. Gallen, Muri in the Canton of Aargau and Vallorbe in the Canton of Vaud), identified as significant sources of pollution for the receiving streams based on chemical measurements performed by the cantonal water protection services, were selected for our study. The Oberglatt WWTP was upgraded during our study, while the upgrading of the Muri and Vallorbe WWTPs was planned. In Oberglatt, one site upstream (UPS) and two sites downstream (Dws1–2) of the WWTP were selected (Glatt river) (Table 1). In Muri, two different discharges from the WWTP could be distinguished, a WWTP overflow and the WWTP effluent. Two sites were selected upstream of both discharges, UPSOa (upstream of the confluence with the Sörikerbach river) and UPS1 (downstream of this confluence), one site (UPS2) between the WWTP overflow and the WWTP effluent and three sites (Dws1–3) downstream of the WWTP effluent (Bünz river). In Vallorbe, three different discharges from the WWTP could be distinguished: two WWTP overflows and the WWTP effluent (from up- to downstream). We selected one site upstream (UPS1) of the three discharges, one site (UPS2) between the WWTP overflows and the WWTP effluent and two sites (Dws1–2) downstream of the WWTP effluent (Orbe river). Site Dws1 was located close to the outfall of the effluent (a few meters) and strongly impacted by this effluent, while site Dws2 was located approximately 200 m downstream of the effluent. For each of these WWTPs, the downstream sampling sites were chosen so that the WWTP effluents were completely mixed with the stream water across the stream channel.

Table 1. Detail concerning each study site: location, stream, geographical coordinates, studied compartments.

Location	Stream	Site	Coordinates	Studied Compartment(s)
Oberglatt	Glatt	UPS	47.413301° N, 9.206362° E	Hyporheic zone and coarse surface sediments except in November 2022 (only coarse surface sediments)
Oberglatt	Glatt	Dws1	47.414825° N, 9.198461° E	Hyporheic zone and coarse surface sediments except in November 2022 (only coarse surface sediments)
Oberglatt	Glatt	Dws2	47.419157° N, 9.195899° E	Hyporheic zone and coarse surface sediments except in November 2022 (only coarse surface sediments)
Muri	Bünz	UPSOa	47.279933° N, 8.341582° E	Hyporheic zone and coarse surface sediments
Muri	Bünz	UPS1	47.280316° N, 8.341538° E	Hyporheic zone and coarse surface sediments
Muri	Bünz	UPS2	47.281290° N, 8.341669° E	Hyporheic zone and coarse surface sediments
Muri	Bünz	Dws1	47.282608° N, 8.342234° E	Hyporheic zone and coarse surface sediments
Muri	Bünz	Dws2	47.291800° N, 8.339151° E	Hyporheic zone and coarse surface sediments
Muri	Bünz	Dws3	47.305217°N, 8.327193° E	Hyporheic zone and coarse surface sediments
Vallorbe	Orbe	UPS1	46.711611° N, 6.382140° E	Coarse surface sediments
Vallorbe	Orbe	UPS2	46.713835° N, 6.385747° E	Coarse surface sediments
Vallorbe	Orbe	Dws1	46.714113° N, 6.385774° E	Coarse surface sediments
Vallorbe	Orbe	Dws2	46.714958° N, 6.388094° E	Coarse surface sediments

We conducted three sampling campaigns at Oberglatt, one before the WWTP upgrade and two after the upgrade, while we conducted only one sampling campaign at the other two WWTPs (before their upgrade). In Oberglatt, the samplings took place on 27 and 28 October 2020 (before the WWTP upgrade) and on 31 May 2022 (after the WWTP upgrade) and on 9 November 2022 (after the WWTP upgrade). The November 2022 campaign was added to sample at the same period as before the WWTP upgrade, to ensure that eventual differences in community composition observed in 2022 could not be solely explained by seasonal variations in community composition. Sampling took place on 25 and 26 May 2021 in Muri and on 7 March 2022 in Vallorbe. Oligochaete communities were studied in the coarse surface sediments and in the hyporheic zone in Oberglatt in October 2020 and May 2022 and in Muri, and in only the coarse surface sediments in Oberglatt in November 2022 and in Vallorbe.

The watershed of the Glatt and Bünz rivers upstream of the sampling sites comprises agricultural, industrial and urban areas, so some chemical pollution was expected to be present at the upstream site(s) of the WWTPs. The Orbe river has its source near Rousses (France), forms three successive lakes (Rousses, Joux and Brenet) and then flows through karstic soils to emerge at the surface in the Vallorbe caves. Its biological and chemical quality is very good in the Vallorbe source [19]. Between the Vallorbe caves and the sampling sites, the stream is impacted by human activities (agricultural, industrial and urban areas) but over only 2 km (as the crow flies). For this reason, and also because the flow rate of the Orbe river is high, pollution at the upstream site was expected to be limited. We note the presence of two dams (or reservoirs) in Vallorbe, one upstream of site UPS1 (1–1.5 km upstream) and the other between sites UPS1 and UPS2. Finally, the stream ecomorphology at all sites in Oberglatt and Vallorbe was natural or rather natural, while it could be considered as semi-natural (straight canal with natural shores) at all sites in Muri.

For the Oberglatt and Muri WWTPs, we provide the following as supplementary information: the concentrations of nutrients (nitrate, ammonium, nitrite, phosphate (o-P), total phosphorus (T-P), dissolved organic carbon and total organic carbon), metals and micropollutants measured by the Swiss Federal Institute of aquatic science and technology Eawag (Switzerland) in the stream water directly impacted by the effluent of the WWTPs during the same months as our sampling of oligochaetes (on 28 October 2020 for Oberglatt before the WWTP upgrade, on 18 May 2022 for Oberglatt after the WWTP upgrade (first campaign) and on 1 November 2022 for Oberglatt after the WWTP upgrade (second campaign); on 18 May 2021 for Muri). The description of the measurement methods can be found in [20] (in "Box 5"). For the Vallorbe WWTP, we provide as supplemental information the concentrations of nutrients (nitrate, ammonium, nitrite, phosphate, total phosphorus, dissolved organic carbon and total organic carbon) measured by the Direction générale de l'environnement, Protection des eaux (State of Vaud, Switzerland) in the treated water before its release into the river, in the same month as our sampling of oligochaetes (on 14 March 2022) and in the two previous months (on 18 January and 14 February 2022). The nutrient and metal concentrations were compared to the Swiss water quality criteria provided in Liechti [21] and Conseil fédéral Suisse [22], respectively. The concentrations of at least three parameters (nutrients) in Oberglatt (before WWTP upgrading), Muri and Vallorbe exceeded the Swiss quality criteria (Tables S1 and S2). Concerning metals, concentrations of Cu and Zn in Oberglatt (before WWTP upgrading) and of Zn in Muri exceeded the Swiss quality criteria (Table S3). These results confirmed that the effluents of the three WWTPs represented a source of chemical pollution. The upgrade of the Oberglatt WWTP consisted of the addition of a treatment stage using powdered activated carbon. This treatment process aims primarily to reduce the concentrations of micropollutants but also has the effect of reducing the concentrations of nutrients and metals. This new treatment process was operative in autumn 2021. We observed that the concentrations of most micropollutants and of Cu and Zn were lower after the (Oberglatt) WWTP upgrade than before the WWTP upgrade (Tables S3 and S4). Furthermore, in Oberglatt, the concentrations of only one nutrient in May 2022 and three nutrients in November 2022 exceeded the Swiss quality criteria, while the concentrations of six nutrients exceeded these quality criteria before the WWTP upgrading (Table S1). The efficacy of the upgrade of this WWTP was therefore confirmed by these results.

2.2. Oligochaete Community Analysis

2.2.1. Sampling and Laboratory Procedures

At each site, four subsamples (spaced 10–20 m apart) of coarse surface sediments and hyporheic material were collected and then pooled (one bucket contained the coarse surface sediments and another, the hyporheic material).

Coarse surface sediments (5–10 cm depth) were collected using a shovel in a 5 L plastic bucket. The grain size of the coarse sediment samples was similar between sites (mostly gravel, with sand and pebbles). Samples from the hyporheic zone were collected in a 5 L plastic bucket using a probe inserted into the coarse surface sediment to a depth of 20–30 cm and a Bou-Rouch pump [23] fitted to the probe (Uwitec, Mondsee, Austria). The supernatant water was sieved in the field using a 0.2 mm mesh size sieve. The material retained in the sieve was transferred to the bucket (one per site). Neutral buffered formalin was then added to the buckets to a final concentration of 4% formaldehyde. The samples were then transported to the laboratory at ambient temperature and preserved in the laboratory at 4 °C.

In the laboratory, the coarse surface sediment and hyporheic zone samples were sieved on a column of 5 mm and 0.2 mm mesh size sieves (Fisherbrand, Fisher Scientific, Reinach, Switzerland) within 1 to 7 days after the sampling. The material retained on the 0.2 mm mesh size sieve was preserved in absolute ethanol at −20 °C (in a plastic box). For the extraction of oligochaetes, each sieved sample was transferred into a square subsampling box (5 × 5 cells). The contents of randomly selected cells were transferred to a Petri dish and examined under a stereomicroscope (Olympus, model SZ51, Basel, Switzerland). Successive cells were examined until 100 identifiable oligochaetes were obtained. Oligochaete specimens were mounted on slides in a semi-permanent coating solution consisting of lactic acid, glycerol and polyvinyl alcohol (Mowiol 4-88) [24] and identified using a compound microscope (Olympus, model BX43, Basel, Switzerland) to the lowest practical level (species if possible or genus or family), following the literature [25,26].

Ideally, the number of specimens identified per site should be 100. However, the results of the community analysis can be interpreted when the number of specimens obtained per site is less than 100, which is common in hyporheic zone samples. However, if the number of specimens is <20, the results should be interpreted with caution.

2.2.2. Functional Traits

We used the classification of oligochaete taxa of the porous matrix into 5 FTR categories proposed by Vivier [13], Lafont and Vivier [4] and Lafont et al. [5] (Table 2). Analysis of the percentages of these 5 FTRs obtained per site allows an assessment of the degree of chemical pollution as well as of the dynamics of hydrological exchanges between surface water and groundwater. FTR1 includes taxa indicating presence of groundwater exfiltration. They are either subterranean or stygophilic, i.e., living both in surface and in groundwater. FTR2, FTRi and FTR3 include sensitive, moderately resistant and resistant taxa to chemical pollution, respectively. FTR4 includes taxa that indicate the presence of polluted sludge within the interstitial spaces of the porous matrix. These taxa are highly tolerant to pollution.

A significant percentage of FTR1 taxa in the porous matrix indicates the exfiltration of groundwater. The intensity of exfiltration can be roughly assessed as follows: FTR1 <15%: low intensity; 16–30%: moderate intensity; 31–50%: high intensity; >50%: very high intensity. Conversely, the predominance of taxa belonging to a FTR other than FTR1 in the hyporheic zone indicates surface water infiltration. Surface water infiltration is low if the percentage of these taxa in the hyporheic zone is <15%; moderate (16–30%); high (31–50%); very high (>50%). FTR1 taxa may belong to FTR2 (sensitive taxa to pollution) or to FTR3 (resistant taxa to pollution) or to no other FTR (no pollution resistance status assigned). For example, *Marionina argentea* belongs to FTR1 and FTR2, *Pristina jenkinae* to FTR1 and FTR3 and *Stylodrilus heringianus* to FTR1 only. Calculating the percentages of the resistant and sensitive taxa included in FTR1 allows an assessment of the pollution status in groundwater.

Table 2. Description of the five different functional traits (FTRs) and examples of characteristic taxa of each FTR.

Functional Traits	Examples of Taxa
FTR1: Includes taxa which are characteristic of groundwater. Includes sensitive and resistant species/taxa to chemical pollution	All Lumbriculidae except *Lumbriculus variegatus; Haber* spp., *Pristina* spp., *Cernosvitoviella* spp., *Achaeta* spp., *Marionina argentea, Haplotaxis gordioides, Propappus volki, Chaetogaster parvus*
FTR2: Includes taxa which are sensitive to chemical pollution	*Cernosvitoviella* spp., *Marionina argentea, Eiseniella tetraedra, Nais alpina, Vejdovskyella comata, Rhyacodrilus falciformis, Propappus volki, Haplotaxis gordioides*
FTR3: Includes taxa which are resistant to chemical pollution	*Nais elinguis, Pristina jenkinae, Dero digitata, Globulidrilus riparius, Lumbriculus variegatus*
FTR4: Includes taxa which are very resistant to chemical pollution. Indicative of the presence of polluted sludge within sediment interstices ("polluted sludge effect")	All Tubificinae with and without hair setae except *Embolocephalus velutinus* and *Spirosperma ferox; Bothrioneurum* sp., *Lumbricillus* spp.
FTRi: Includes taxa which are moderately resistant to chemical pollution	*Chaetogaster diastrophus, C. diaphanus, Nais communis, N. christinae, N. barbata, N. pardalis, Slavina appendiculata*

2.2.3. Ecological Potential

The ecological potential (EP) describes the state of the functioning of the porous matrix (each compartment separately) and is calculated using the following equation [5]:

$$EP = \text{Log}_2\ [(\%FTR1 + \%FTR2) + 1]\ /\ [(\%FTR3 + \%FTR4) + 1]$$

The EP corresponds to the ratio between the sum of the percentages of FTRs considered as representative of a preserved functioning and the sum of the percentages of FTRs representative of an impaired functioning. The EP makes it possible to distinguish between different functioning states, from preserved to very altered, as shown in Table 3.

Table 3. Classes of functioning state according to ecological potential (EP) values.

EP	State of Functioning
≥7	Preserved
4–6.9	Slightly altered
2–3.9	Moderately altered
0.1–1.9	Altered
≤0	Very altered

3. Results

3.1. *Oligochaete Diversity*

In total, 44 oligochaete taxa were found: 29 taxa of Naididae (11 Tubificinae, 14 Naidinae and 4 Pristininae), 9 taxa of Enchytraeidae, 4 taxa of Lumbriculidae and 2 taxa of Lumbricidae (Tables S5–S7). All the functional traits (1–4, i) were well represented.

3.2. *WWTP of Oberglatt*

(1) Before WWTP upgrading

The percentages of FTR3 (which include resistant taxa to chemical pollution) were high at the three sites in both the coarse surface sediments and the hyporheic zone (Figure 1). High percentages of FTR4 (which includes very resistant taxa to chemical pollution, indicative of the presence of polluted sludge) at site Dws1 in both the coarse surface sediments and the hyporheic zone and low percentages of this functional trait at sites UPS and Dws2 in both compartments were observed. At site Dws1, the percentage of FTR4 was particularly high in the hyporheic zone (41%). The percentages of FTR1 (which includes taxa indicator of groundwater exfiltration) and FTR2 (which includes sensitive taxa to chemical pollution) were higher at site UPS than at site Dws1 in both compartments (Table 4). In the hyporheic zone, the EP value at site Dws1 (very altered functioning) was lower than at sites UPS and Dws2 (altered functioning). In the coarse surface sediments, the EP values at sites Dws1 and Dws2 were lower than at site UPS, but all three EP values were very low and indicated a very altered functioning. At the three sites, the percentages of FTR1 and FTR2 were moderate to quite high in the hyporheic zone and low in the coarse surface sediments.

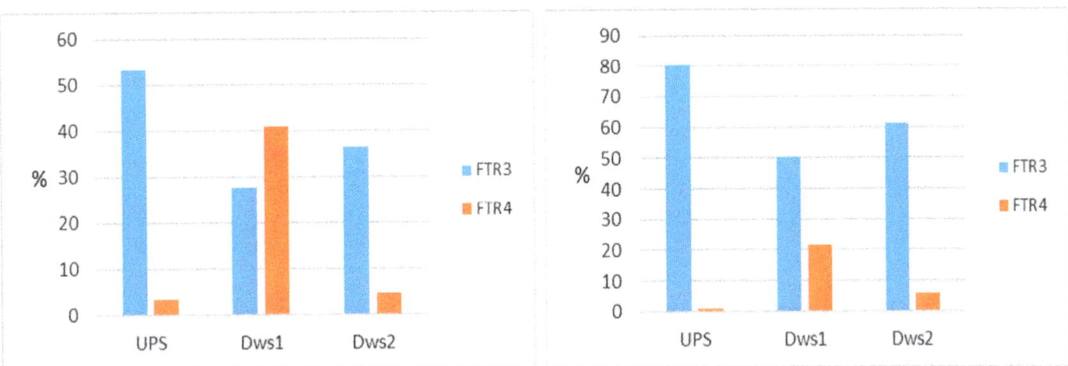

Figure 1. Percentages of FTR3 and FTR4 in the hyporheic zone (**left**) and in the coarse surface sediments (**right**) obtained in Oberglatt before the WWTP upgrading (October 2020).

Table 4. Percentages of the functional traits (FTR1–4; FTRi) and ecological potential (EP) values obtained in Oberglatt in the hyporheic zone and the coarse surface sediments before the WWTP upgrading (October 2020). For EP values, red color = very altered functioning; orange color = altered functioning.

		FTR1	FTR2	FTR3	FTR4	FTRi	EP
Hyporheic zone	UPS	40.7	30.20	53.5	3.5	5.8	0.31
	Dws1	22.9	16.2	27.6	41	12.4	−0.80
	Dws2	29.4	27.1	36.5	4.7	30.6	0.45
Coarse surface sediments	UPS	4.9	10.8	80.4	1	3.9	−2.30
	Dws1	3	3	50.5	21.8	22.8	−3.39
	Dws2	1	2.9	61.5	5.8	28.8	−3.80

(2) After WWTP upgrading

(2a) In May 2022

Compared to the state before the WWTP upgrade (October 2020), we observed in May 2022 a significant decrease in the percentages of FTR3 at the three sites in the coarse surface sediments and hyporheic zone, and a decrease in the percentages of FTR4 at site Dws1 in both compartments (Figure 2 and Table 5). EP values at sites Dws1 and Dws2 were higher than before the WWTP upgrading in both the coarse surface sediments and the hyporheic zone (Figure 3). Site UPS showed high percentages of FTR4 in both compartments. The percentages of FTR4 at sites Dws1 and Dws2 were much lower than at site UPS in both compartments, yet remained quite high. The percentages of FTR1 were particularly high at sites Dws1 and Dws2 in both compartments, indicating strong groundwater exfiltration. However, the percentages of FTR2 (sensitive taxa) were low at these sites, the taxa of FTR1 mainly belonging to Lumbriculidae sp. and *Stylodrilus heringianus*, to which no pollution resistance status is attributed (only FTR1) (Table S5). In both compartments, EP values at sites Dws1 and Dws2 were higher than at site UPS, due to the lower percentages of FTR4 and higher percentages of FTR1 at sites Dws1 and Dws2 than at site UPS.

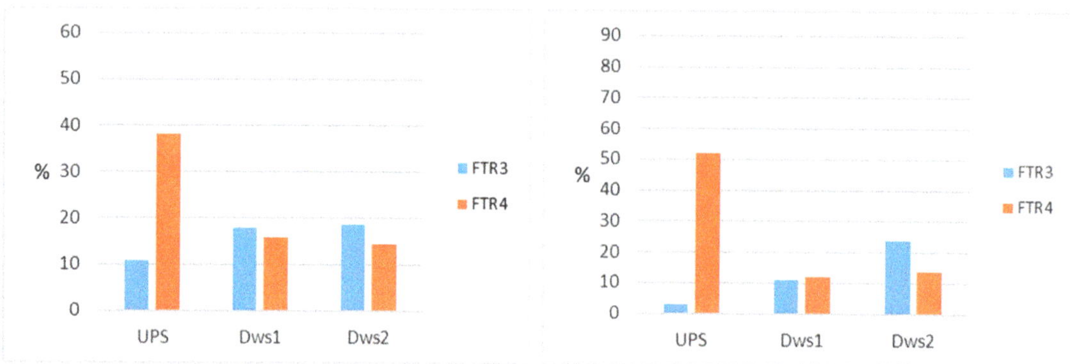

Figure 2. Percentages of FTR3 and FTR4 in the hyporheic zone (**left**) and in the coarse surface sediments (**right**) obtained in Oberglatt in May 2022 (after the WWTP upgrading).

Table 5. Percentages of the functional traits (FTR1–4; FTRi) and ecological potential (EP) values obtained in Oberglatt in May 2022 (hyporheic zone and coarse surface sediments) and November 2022 (coarse surface sediments). For EP values, red color = very altered functioning; orange color = altered functioning.

		FTR1	FTR2	FTR3	FTR4	FTRi	EP
Hyporheic zone in May 2022	UPS	29.1	5.5	10.9	38.2	21.8	−0.49
	Dws1	76	10	18	16	8	1.31
	Dws2	59.4	14.5	18.8	14.5	21.7	1.13
Coarse surface sediments in May 2022	UPS	34.7	6.1	3.1	52	11.2	−0.42
	Dws1	83.8	7.1	11.1	12.1	3	1.93
	Dws2	60.4	5	23.8	13.9	23.8	0.78
Coarse surface sediments in November 2022	UPS	23.8	16.8	14.9	0	44.6	1.39
	Dws1	9	5	9	6	72	−0.09
	Dws2	4.9	2.9	2.9	1.9	88.3	0.60

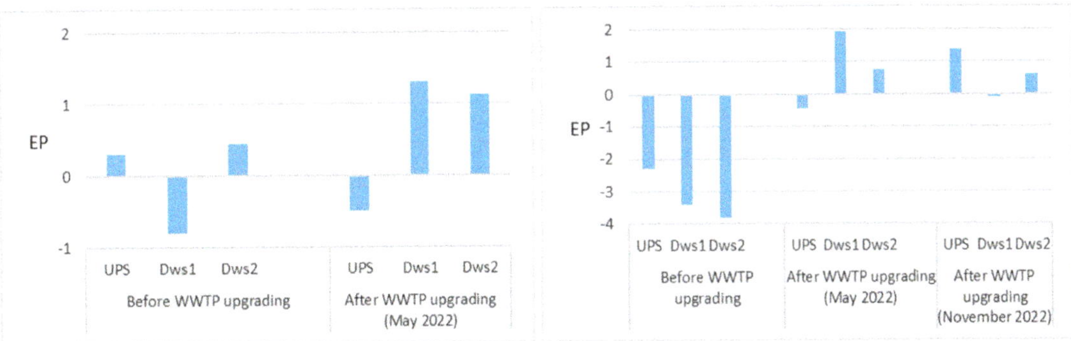

Figure 3. Ecological potential (EP) values at sites UPS, Dws1 and Dws2 in Oberglatt before (October 2020) and after the WWTP upgrading (May and November 2022); (**left**) in the hyporheic zone; (**right**) in the coarse surface sediments.

(2b) In November 2022

Compared to the state before the WWTP upgrade (October 2020), we observed in November 2022 (in the coarse surface sediments) markedly lower percentages of FTR3 at the three sites, and a lower percentage of FTR4 at site Dws1 (Figure 4). Furthermore, EP values at the three sites were higher than those obtained prior to the WWTP upgrade (Figure 3). At sites Dws1 and Dws2, particularly high percentages of FTRi (moderately resistant taxa to chemical pollution) and low percentages of FTR1 and FTR2 were observed (Table 5). The percentages of Naidinae were, like in October 2020 and unlike in May 2022, very high at the three sites (≥75%) (Table S5). At sites Dws1 and Dws2, the percentages of Naidinae species belonging to FTR3 (11% and 3% of Naidinae specimens, respectively) were significantly lower than in October 2020 (67% at both sites). Conversely, the percentages of Naidinae species belonging to FTRi (85% and 95% of Naidinae specimens, respectively) were notably higher than in October 2020 (29% and 31%, respectively). The EP values at sites Dws1 and Dws2 were found to be lower than at site UPS, which was explained by the lower percentages of FTR1 and FTR2 at sites Dws1 and Dws2 than at site UPS. EP values at sites Dws1 and Dws2 were lower than in May 2022, due to the much lower percentages of FTR1 at these sites in November 2022 than in May 2022.

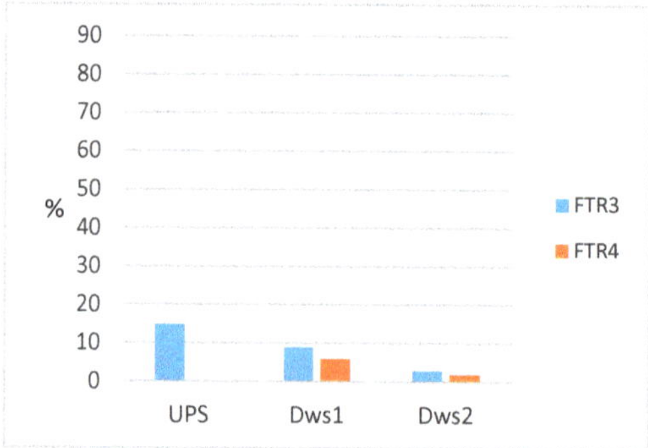

Figure 4. Percentages of FTR3 and FTR4 (in coarse surface sediments) obtained in Oberglatt in November 2022 (after the WWTP upgrading).

3.3. *WWTP of Muri*

In both compartments, an increase in the percentages of FTR3 was observed from the two upstream sites to the downstream sites, according to the following trend: UPSOa–UPS1 < UPS2 < Dws1–3 (Figure 5). All samples exhibited no or a slight to moderate polluted sludge effect (low percentages of FTR4). In both compartments, the extent of groundwater exfiltrations as indicated by the percentage of FTR1 decreased from upstream to the downstream sites: UPSOa > UPS1 > UPS2 > Dws1–3 (Table 6). In both compartments, the percentages of FTR2 were lower at sites Dws1–3 than at sites UPSOa, UPS1 and UPS2, with the exception of site UPSOa in the hyporheic zone. Finally, in both compartments, EP values were higher at sites UPSOa and UPS1 than at sites downstream of the WWTP overflow and WWTP effluents (Figure 6). There was a progressive decrease in EP values from upstream to the downstream sites: UPSOa–UPS1 (moderately altered or altered functioning) > UPS2 (altered functioning) > Dws1–3 (very altered functioning).

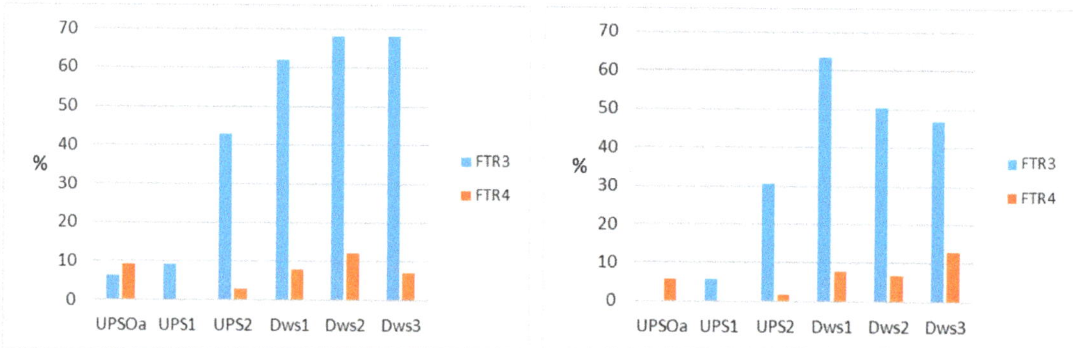

Figure 5. Percentages of FTR3 and FTR4 in the hyporheic zone (**left**) and in the coarse surface sediments (**right**) obtained in Muri.

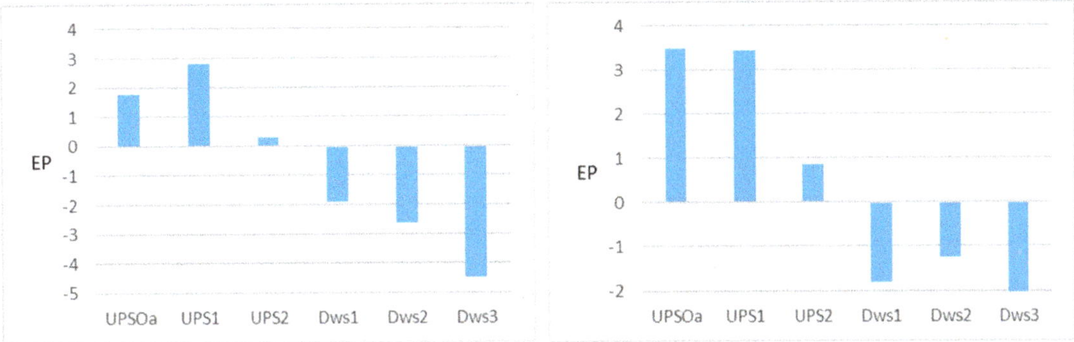

Figure 6. Ecological potential (EP) values obtained in Muri in the hyporheic zone (**left**) and in the coarse surface sediments (**right**).

Table 6. Percentages of the functional traits (FTR1–4; FTRi) and ecological potential (EP) values obtained in Muri in the hyporheic zone and in the coarse surface sediments. For EP values, red color = very altered functioning; orange color = altered functioning; yellow color = moderately altered functioning.

		FTR1	FTR2	FTR3	FTR4	FTRi	EP
Hyporheic zone	UPSOa	46.9	9.4	6.3	9.4	31.3	1.78
	UPS1	35.8	37.7	9.4	0	20.8	2.84
	UPS2	22.9	34.3	42.9	2.9	0	0.31
	Dws1	8	10	62	8	16	−1.90
	Dws2	4.9	7.3	68.3	12.2	7.3	−2.63
	Dws3	2.4	0	68.3	7.3	22	−4.49
Coarse surface sediments	UPSOa	34	41.7	0	5.8	24.3	3.50
	UPS1	22.5	52	5.9	0	24.5	3.45
	UPS2	17.3	43.3	30.8	1.9	6.7	0.87
	Dws1	5	14.9	63.4	7.9	9.9	−1.79
	Dws2	8.9	14.9	50.5	6.9	21.8	−1.24
	Dws3	9	5	47	13	29	−2.02

3.4. WWTP of Vallorbe

The coarse surface sediments at sites UPS1, UPS2 and Dws2 contained high percentages of FTR1 and FTR2 and low percentages of resistant taxa (FTR3–4) (Figure 7 and Table 7). At site Dws1, the percentage of FTR2 was lower than at the other sites, and a high percentage (33%) of FTR4 was observed. The percentage of FTR1 at site Dws1 was lower than at the other sites, yet remained high, indicating a significant groundwater exfiltration. The EP values indicated that the functioning was slightly altered at sites UPS1 and UPS2, moderately altered (but close to the slightly altered class) at site Dws2 and altered at site Dws1.

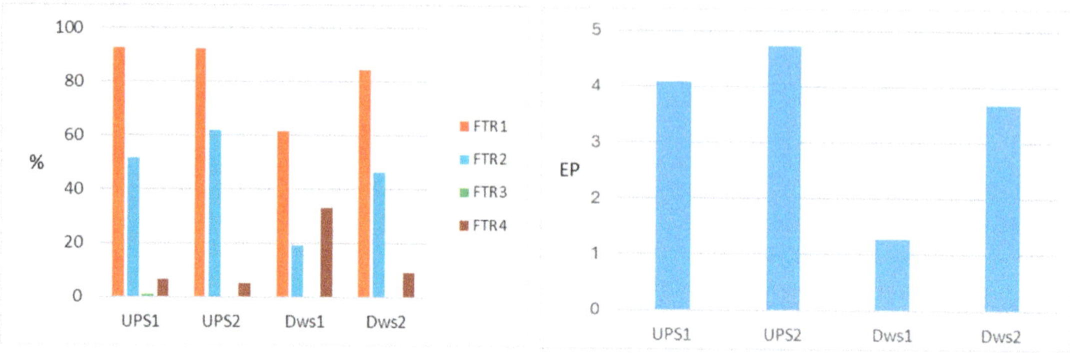

Figure 7. Percentages of the functional traits FTR1–4 (**left**) and ecological potential (EP) values (**right**) obtained in Vallorbe (in coarse surface sediments).

Table 7. Percentages of the functional traits (FTR1–4; FTRi) and ecological potentials (EP) values obtained in Vallorbe (in coarse surface sediments). For EP values, orange color = altered functioning; yellow color = moderately altered functioning; green color = slightly altered functioning.

	FTR1	FTR2	FTR3	FTR4	FTRi	EP
UPS1	92.5	51.6	1.1	6.5	1.1	4.08
UPS2	92.2	62.1	0	4.9	1	4.72
Dws1	61.7	19.1	0	33	5.2	1.27
Dws2	84.5	46.4	0	9.3	6.2	3.68

4. Discussion

This study confirmed that the FTR method was effective in detecting the effects of WWTP effluents on the biological quality and functioning of the receiving streams, even in the context of globally impacted areas (agricultural/industrial/urban areas). It also provided valuable information on the self-purification capacity of the stream environment and on the vulnerability of groundwater to surface water pollution.

The composition of oligochaete communities upstream and downstream of the effluent and overflow(s) of each WWTP was diverse:

- In Oberglatt (state before the WWTP upgrade), even if site UPS already showed significant alteration of the biological quality (high percentages of FTR3), site Dws1 differed from site UPS in higher percentages of FTR4, indicative of strong pollution (polluted sludge effect), in both compartments and in lower percentages of FTR1 and FTR2 in both compartments. The percentage of FTR4 was particularly high in the hyporheic zone (site Dws1), suggesting accumulation of polluted sludge in this compartment. The EP values were lower at site Dws1 than at site UPS, but the differences were not important given that oligochaete communities already indicated the presence of significant chemical pollution at site UPS. The absence of polluted sludge effect at site Dws2 showed that the stream was capable of self-purification (natural ecomorphology). The higher percentages of FTR1 and FTR2 in the hyporheic zone than in the coarse surface sediments (at the three sites) indicate that exfiltration reached the hyporheic zone and not or much less the coarse surface sediments. This could be explained by clogging in the coarse surface sediments or by more accused pollution in the coarse surface sediments than in the hyporheic zone (FTR1 partly includes sensitive taxa).

- In Muri, the effects of the WWTP overflow and effluent were observed in both compartments, with an increase in the percentages of FTR3 both downstream of the overflow (UPS2, first increase) and downstream of the effluents (Dws1, second increase). In addition, the percentages of FTR2 decreased significantly in both compartments downstream of the WWTP effluent. Thus, the specific effects of the effluents could be observed even when the oligochaete communities downstream of the overflow already showed an altered biological quality (site UPS2). Results also indicated that the increase in flow rates due to the confluence with the Sörikerbach river (UPS1) and to the discharges from the WWTP overflow (UPS2) and WWTP effluents (Dws1–3) might have induced enhanced infiltrations of surface water (decrease in the percentages of FTR1 from site UPSOa to sites Dws1–3). The percentages of FTR1 were particularly low at sites Dws1–3, suggesting vulnerability of groundwater to surface water pollution at these sites. At sites Dws2 and Dws3, the percentages of FTR3 remained high, showing that the stream was unable to recover the quality and functioning it had upstream of the WWTP overflow and effluents, possibly due to an insufficient capacity of the stream to self-purify (straight canal, weak exfiltration) and/or to the presence of other sources of pollution downstream of the WWTP. The percentages of FTR4 were low at all sites, possibly due to the cleaning of the porous matrix by the high flows in the weeks prior to sampling. Contrary to Oberglatt, the upstream sites (UPSOa, UPS1) were only moderately impacted by human activities, and therefore, results of the functioning state obtained upstream and downstream of the WWTP overflow and effluent were very contrasted.
- In Vallorbe, the effects of WWTP effluents were visible (at site Dws1) by an increase in the percentage of FTR4 and a decrease in the percentages of FTR1 and FTR2. The polluted sludge effect was local, as it was much lower about two hundred meters downstream (site Dws2). The high percentage of FTR1 at site Dws1 indicates a high capacity of the stream to self-purify. Contrary to Muri, the effects of the WWTP overflows were not visible on oligochaete communities (site UPS2). This lack of effects at sites UPS2 and Dws2 could be explained by the high flow rates leading to dilution of the local pollution and cleaning of the coarse surface sediments by the current, and by the strong groundwater exfiltration. At site UPS1, the percentage of sensitive taxa (FTR2) was high, showing, as expected, limited impacts from human activities in Vallorbe. Results of the functioning state obtained at sites UPS1 and Dws1 were thus very contrasted. It is acknowledged that a dam leads downstream to significant hydrological disturbances [27]. Lafont et al. [5] mention that rapid increases in water release from a dam tend to induce infiltration dynamics downstream. Oligochaete results at the three sites (UPS2, Dws1 and Dws2) located immediately downstream of the dam between sites UPS1 and UPS2 (c.f. Section 2.1) did not indicate significant infiltration of surface water. The sampling was performed when the flow rate was low, and outside a strong water release event from the reservoir. The indication of exfiltration observed in Vallorbe should, however, be interpreted with caution as it was shown that oligochaetes could be less pertinent indicators of vertical exchanges in rivers subjected to hydrological disturbance and hydropeaking than in rivers with no hydrological disturbance [15].

The results of the present work were globally concordant with those obtained upstream and downstream of the effluents of the two WWTPs in Hochdorf and Buttisholz [18] (see introduction), in the sense that in both studies, negative effects of the WWTP effluents were clearly observed on oligochaete communities. Vivien et al. [18] also observed various responses of oligochaete communities to the WWTP effluents: while the sum of percentages of FTR1 and FTR2 was much lower at the downstream site than at the upstream site in all samples, either only the percentage of FTR3 (two samples) or only the percentage of FTR4 (two samples), or the percentages of FTR3 and FTR4 (three samples) were higher at the downstream sites than at the upstream sites. For one of the samples, the percentages of FTR3 and FTR4 were not higher at the downstream site than at the upstream site.

The results obtained at Oberglatt suggest that the FTR method could be suited for monitoring the effects of a WWTP upgrade on the receiving stream. The WWTP upgrading resulted, compared to the state before the upgrading, in a significant reduction in the percentages of FTR4 at site Dws1 (reduction of the polluted sludge effect). Such results were expected given the improvement of the quality (described in Section 2.1) of the WWTP effluents following the (WWTP) upgrade. Although a potential seasonal variation in oligochaete communities was observed in this stream, with Naidinae being more abundant in autumn than in spring, the results in May and November 2022 were consistent, indicating that the WWTP upgrade was beneficial to the receiving stream. These results suggest that when evaluating the impact of a WWTP upgrade on oligochaete communities, it could not be necessary to conduct pre- and post-upgrade sampling during the same period. The quite high percentages of FTR4 observed at sites Dws1 and Dws2 in May 2022 in both compartments could be explained more by the pollution upstream of the WWTP effluent than by the WWTP effluent. The very high percentage of FTR4 obtained in May 2022 at site UPS (both compartments) could be explained by an accidental release of organic matter into the stream from agricultural activities. The dilution of this pollution by the WWTP effluent could explain that the percentages of FTR4 were much lower at sites Dws1 and Dws2 (both compartments) than at site UPS. After the WWTP upgrade, only oligochaete results obtained in November (2022) showed a negative impact of the WWTP effluents on the receiving stream (EP value lower at site Dws1 than at site UPS). However, this impact was moderate, the sum of the percentages of resistant taxa (FTR3) and very resistant taxa (FTR4) at site Dws1 being quite low and not higher than at site UPS.

The WWTP upgrading in Oberglatt did not result in a higher state of stream functioning than altered at sites Dws1 and Dws2. In May 2022, despite the strong exfiltration of groundwater, the percentages of sensitive taxa were low in both the coarse surface sediments and the hyporheic zone. This can be attributed to a very low abundance of sensitive species in the groundwater, which would suggest that the biological quality in this compartment may have been altered to some extent. Another explanation is that the groundwater could contain a significant abundance of sensitive taxa that are unable to maintain a viable population in the hyporheic zone and in coarse surface sediments due to the pollution present in these compartments. In November 2022, the percentages of FTR1 and FTR2 were very low, with oligochaete communities largely dominated by moderately resistant taxa. The study sites are situated in an agricultural area and downstream of industrial and urban areas. The altered or very altered functioning obtained at site UPS shows that it cannot be expected to achieve a preserved or slightly altered functioning at sites Dws1 and Dws2. Instead, a reasonable objective for these sites would be to constantly obtain a moderately altered functioning (EP values between 2 and 3.9) or, at the very least, an altered functioning (EP values between 0.1 and 1.9).

The rationale for including the hyporheic zone into the FTR method is that this compartment tends to store pollutants in greater concentrations than the coarse surface sediments and provides insights into the dynamics of surface water infiltrations into the hyporheic zone and of groundwater exfiltration into the hyporheic zone. While there was a general similarity, in the present study, between community composition in the hyporheic zone and coarse surface sediments, results obtained in the hyporheic zone provided some relevant ecological information (see results in Oberglatt in October 2020). In Vivien et al. [16,18], oligochaete results obtained in the hyporheic zone and coarse surface sediments (six different sites at two different periods) were also globally concordant, with EP values indicating in both compartments the same class of functioning state. Even if it would be preferable to systematically consider the hyporheic zone, the analysis of oligochaete communities in only coarse surface sediments already provides valuable information on the biological quality and functioning of the stream, while significantly reducing the logistical effort and cost analyses per site.

As stated by Vivien et al. [18], the FTR method should be continuously refined as new data becomes available. This can be achieved by incorporating additional information on functional traits and adapting the attribution of functional traits to each oligochaete species. However, it is important to note that the FTR method is already validated and ready to be applied in monitoring programs.

The implementation of the FTR method requires expertise in oligochaete taxonomy, which would certainly restrict the utilization of this methodology in biomonitoring programs. Advances in the use of DNA barcodes for the identification of oligochaete species [28] may allow to facilitate the application of the FTR method. A method based on high-throughput sequencing of genetically tagged oligochaete specimens is currently being developed for assessing the biological quality in fine/sandy sediments [29]. This could be transferred to the FTR method. An alternative approach would be to propose a simplified morphological FTR method, whereby specimens would be assigned directly to FTRs, without the attribution of a species name to each specimen. This morphological method would require identifying families and subfamilies, as well as a limited number of species.

5. Conclusions

The results of the present study confirmed that the Functional trait (FTR) method was suited to detect the effects of WWTP effluents on the biological quality and functioning of the receiving stream, even in areas already impacted by other sources of pollution. In addition, the results obtained in Oberglatt suggested that this method could be adequate to monitor the effects of a WWTP upgrade on the receiving stream. It is anticipated that new FTR data will be collected at sites exhibiting a range of chemical pollution levels and different vertical hydrological exchange dynamics (upstream and downstream of point sources of pollution, before and after the implementation of restoration measures). In the near future, we intend to confirm on new FTR data that the upgrade of the Oberglatt WWTP was beneficial to the receiving stream and to monitor the effects of the current/future upgrade of the Muri and Vallorbe WWTPs. Furthermore, FTR methods will be developed, as mentioned above, in a manner that renders them accessible to non-experts in the systematics of oligochaetes.

Supplementary Materials: The following supporting information can be downloaded at: https://www.mdpi.com/article/10.3390/w17050724/s1, Table S1: Concentrations of nitrate, ammonium, nitrite, phosphate (o-P), total phosphor (T-P), dissolved organic carbon (DOC) and total organic carbon (TOC) obtained in Oberglatt and Muri (effluents). Classification according to [21]: in blue: very good quality; in green: good quality; in yellow: medium quality; in orange: poor quality; in red: bad quality. Table S2: Concentrations of nitrate, ammonium, nitrite, phosphate (o-P), total phosphor (T-P), dissolved organic carbon (DOC) and total organic carbon (TOC) obtained in Vallorbe (effluents). Classification according to [21]: in blue: very good quality; in green: good quality; in yellow: medium quality; in orange: poor quality; in red: bad quality. Table S3: Concentrations of metals in µg/L obtained in Oberglatt and Muri (effluents). In bold: concentrations higher than the Swiss quality criteria [22]. Table S4: Concentrations of micropollutants in ng/L obtained in Oberglatt (campaigns of 2020 and May 2022) and Muri (effluents). Table S5: Oligochaete communities per sample during the campaigns of October 2020, May 2022 and November 2022 in Oberglatt. The values correspond to the numbers of specimens per taxon. The functional trait(s) FTR(s) of each taxon is(are) indicated in brackets. Table S6: Oligochaete communities per sample in Muri. The values correspond to the numbers of specimens per taxon. The functional trait(s) FTR(s) of each taxon is(are) indicated in brackets. Table S7: Oligochaete communities per sample in Vallorbe. The values correspond to the numbers of specimens per taxon. The functional trait(s) FTR(s) of each taxon is(are) indicated in brackets.

Author Contributions: Conceptualization, R.V. and B.J.D.F.; Sampling, R.V. and B.J.D.F.; methodology, R.V. and B.J.D.F.; formal analysis, R.V.; writing—original draft preparation, R.V.; writing—review and editing, R.V. and B.J.D.F. All authors have read and agreed to the published version of the manuscript.

Funding: The research was supported by Eawag Discretionary Funds and by the Ecotox Centre.

Data Availability Statement: Data are contained within the article.

Acknowledgments: We would like to thank Christian Stamm, Louis Carles, Simon Wullschleger, Baptiste Clerc, Denise Freudemann and the AuA lab (Eawag), Adriano Joss and René Schönenberger for the organization of the EcoImpact 2 project of which the present work is part and for providing us with the results of the chemical analyses. We also thank Mathieu Renaud for his assistance in the field.

Conflicts of Interest: The authors have no conflicts of interest.

References

1. Brunke, M.; Gonser, T. The ecological significance of exchange processes between rivers and groundwater. *Freshw. Biol.* **1997**, *37*, 1–33. [CrossRef]
2. Boulton, A.J. Hyporheic rehabilitation in rivers: Restoring vertical connectivity. *Freshw. Biol.* **2007**, *52*, 632–650. [CrossRef]
3. Breil, P.; Grimm, N.; Vervier, P. Surface water–ground water exchange processes and fluvial ecosystem function: An analysis of temporal and spatial scale dependency. In *Hydroecology and Ecohydrology: Past, Present and Future*; Wood, P.J., Hannah, D.M., Sadler, P.J., Eds.; John Wiley and Sons Ltd.: New York, NY, USA, 2007; pp. 93–111.
4. Lafont, M.; Vivier, A. Oligochaete assemblages in the hyporheic zone and coarse surface sediments: Their importance for understanding of ecological functioning of water courses. *Hydrobiologia* **2006**, *564*, 171–181. [CrossRef]
5. Lafont, M.; Jézéquel, C.; Vivier, A.; Breil, P.; Schmitt, L.; Bernoud, S. Refinement of biomonitoring of urban water courses by combining descriptive and ecohydrological approaches. *Ecohydrol. Hydrobiol.* **2010**, *10*, 3–11. [CrossRef]
6. Paran, F.; Arthaud, F.; Novel, M.; Graillot, D.; Piscart, C.; Bornette, G.; Marmonier, P.; Lavastre, V.; Travi, Y.; Cadilhac, L. *Caractérisation des Echanges Nappes/Rivières en Milieu Alluvionnaire—Guide Méthodologique*; Zone Atelier Bassin du Rhône (ZABR) et Agence de l'eau Rhône Méditerranée Corse: Lyon, France, 2015; 180p.
7. Lafont, M. A conceptual approach to the biomonitoring of freshwater: The Ecological Ambience System. *J. Limnol.* **2001**, *60*, 17–24. [CrossRef]
8. Schmitt, L.; Lafont, M.; Trémolières, M.; Jézéquel, C.; Vivier, A.; Breil, P.; Namour, P.; Valin, K.; Valette, L. Use hydrogeomorphological typologies in functional ecology: Preliminary results results in contrasted hydrosystems. *Phys. Chem. Earth* **2011**, *36*, 539–548. [CrossRef]
9. Schmitt, L.; Grosprêtre, L.; Breil, P.; Namour, P.; Lafont, M.; Delile, H.; Eschbach, D.; Jacob-Rousseau, N.; Cournoyer, B. L'hydromorphologie, une dimension-clé pour l'étude interdisciplinaire des petits hydrosystèmes périurbains (bassin de l'Yzeron, France). *Bull. Soc. Géogr. Liège* **2016**, *67*, 161–179.
10. Hynes, H.B.N. Groundwater and stream ecology. *Hydrobiologia* **1983**, *100*, 93–99. [CrossRef]
11. Malard, F.; Tockner, K.; Dole-Olivier, M.J.; Ward, J.V. A landscape perspective of surface-subsurface hydrological exchanges in river corridors. *Freshw. Biol.* **2002**, *47*, 621–640. [CrossRef]
12. Boulton, A.J. River ecosystem health down under: Assessing ecological conditions in riverine groundwater zones in Australia. *Ecosyst. Health* **2000**, *6*, 108–118. [CrossRef]
13. Vivier, A. Effets Ecologiques de Rejets Urbains de Temps de Pluie sur Deux Cours d'eau Périurbains de l'ouest Lyonnais et un Ruisseau Phréatique en Plaine d'Alsace. Ph.D. Thesis, L.P. University, Strasbourg, France, 2006; 208p.
14. Lafont, M.; Vivier, A.; Nogueira, S.; Namour, P.; Breil, P. Surface and hyporheic Oligochaete assemblages in a French suburban stream. *Hydrobiologia* **2006**, *564*, 183–193. [CrossRef]
15. Creuzé des Châtelliers, M.; Dolédec, S.; Lafont, M.; Dole-Olivier, M.J.; Konecny, L.; Marmonier, P. Are hyporheic oligochaetes efficient indicators of hydrological exchanges in river bed sediment? A test in a semi-natural and a regulated river. *River Res. Appl.* **2021**, *37*, 399–407. [CrossRef]

16. Vivien, R.; Lafont, M.; Ferrari, B.J.D. Utilisation des communautés d'oligochètes pour l'évaluation de la qualité biologique et du fonctionnement des cours d'eau: Un bilan à partir de données genevoises (Suisse). *Arch. Sci.* **2015**, *68*, 105–116.
17. Vivien, R.; Jézéquel, C.; Lafont, M. Evaluation des effets de l'augmentation des débits réservés à l'aval de deux barrages du Haut-Rhône français à l'aide de l'examen des communautés d'oligochètes. *Bull. Soc. Vaud. Sci. Nat.* **2021**, *100*, 103–117.
18. Vivien, R.; Lafont, M.; Werner, I.; Laluc, M.; Ferrari, B.J.D. Assessment of the effects of wastewater treatment plant effluents on receiving streams using oligochaete communities of the porous matrix. *Knowl. Manag. Aquat. Ecosyst.* **2019**, *420*, 18. [CrossRef]
19. De Source sûre. La Qualité des Cours d'eau Vaudois. Etat mars 2018. *Direction Générale de L'environnement (DGE), DIREV-Protection des Eaux;* Etat de Vaud: Lausanne, Switzerland, 2018.
20. Stamm, C.; Räsänen, K.; Burdon, F.J.; Altermatt, F.; Jokela, J.; Joss, A.; Ackermann, M.; Eggen, R.I.L. Unravelling the impacts of micropollutants in aquatic ecosystems: Interdisciplinary studies at the interface of large-scale ecology. In *Advances in Ecological Research: Vol. 55. Large-Scale Ecology: Model Systems to Global Perspectives;* Dumbrell, A.J., Kordas, R.L., Woodward, G., Eds.; Academic Press: Oxford, England, 2016; pp. 183–223. [CrossRef]
21. Liechti, P. Méthodes d'analyse et d'appréciation des cours d'eau. Analyses physico-chimiques, nutriments. In *L'environnement Pratique;* Office fédéral de l'environnement OFEV: Bern, Switzerland, 2010; 44p.
22. Conseil fédéral suisse. *Ordonnance sur la Protection des eaux du 28 octobre 1998;* Bern, Switzerland, 1998.
23. Bou, C.; Rouch, R. Un nouveau champ de recherches sur la faune aquatique souterraine. *C. R. Acad. Sci.* **1967**, *265*, 369–370.
24. Reymond, O. Préparations microscopiques permanentes d'oligochètes: Une méthode simple. *Bull. Soc. Vaud. Sci. Nat.* **1994**, *83*, 1–3.
25. Timm, T.; Veldhuizen van Zanten, H.H. *Freshwater Oligochaeta of North-West Europe;* Expert Center for Taxonomic Identification, World Biodiversity Database, University of Amsterdam: Amsterdam, The Netherlands, 2002; CD-ROM.
26. Timm, T. A guide to the freshwater Oligochaeta and Polychaeta of Northern and Central Europe. *Lauterbornia* **2009**, *66*, 1–235.
27. Hancock, P.J. Human impacts on the stream-groundwater exchange zone. *Environ. Manag.* **2002**, *29*, 763–781. [CrossRef]
28. Lefrançois, E.; Apothéloz-Perret-Gentil, L.; Blancher, P.; Botreau, S.; Chardon, C.; Crepin, L.; Cordier, T.; Cordonier, A.; Domaizon, I.; Ferrari, B.J.D.; et al. Development and implementation of eco-genomic tools for aquatic ecosystem biomonitoring: The SYNAQUA French-Swiss program. *Environ. Sci. Pollut. Res.* **2018**, *25*, 33858–33866. [CrossRef]
29. Vivien, R.; Apothéloz-Perret-Gentil, L.; Pawlowski, P.; Werner, I.; Ferrari, B.J.D. High-throughput DNA barcoding of oligochaetes for abundance-based indices to assess the biological quality of sediments in streams and lakes. *Sci. Rep.* **2020**, *10*, 2041. [CrossRef] [PubMed]

water

MDPI

Article

Bioaccumulation Study of Cadmium and Lead in *Cyprinus carpio* from the Colorado River, Using Automated Electrochemical Detection

Federico Danilo Vallese [1,*], Sofia Stupniki [2], Mariano Trillini [1], Federico Belén [1], María Susana Di Nezio [1], Alfredo Juan [3] and Marcelo Fabian Pistonesi [1]

1 Departamento de Química, Universidad Nacional del Sur, Instituto de Química del Sur (INQUISUR), Consejo Nacional de Investigaciones Científicas y Técnicas (CONICET), Avenida Alem 1253, Bahía Blanca 8000, Argentina; mariano.trillini@uns.edu.ar (M.T.); federico.belen@uns.edu.ar (F.B.); sdinezio@criba.edu.ar (M.S.D.N.); mpistone@criba.edu.ar (M.F.P.)

2 Departamento de Biología, Bioquímica y Farmacia, Universidad Nacional del Sur, San Juan 670, Bahía Blanca 8000, Argentina; sofistupniki@gmail.com

3 Departamento de Física, Universidad Nacional del Sur, Instituto de Física del Sur (IFISUR), Consejo Nacional de Investigaciones Científicas y Técnicas (CONICET), Avenida Alem 1253, Bahía Blanca 8000, Argentina; cajuan@criba.edu.ar

* Correspondence: federico.vallese@uns.edu.ar

Abstract: The monitoring of heavy metals in aquatic ecosystems is of critical importance due to the toxic effects that these elements can have on wildlife and the potential risks that they pose to human health. Rivers situated in close proximity to agricultural regions are particularly susceptible to contamination from a combination of natural and anthropogenic sources. The study of bioaccumulation is of great importance for the early detection of environmental stressors. The combination of electrochemical techniques, such as square-wave anodic stripping voltammetry (SWASV), with automated flow-batch systems represents an efficient and cost-effective approach for the detection of trace metals in environmental samples. This study examines the bioaccumulation of cadmium and lead in *Cyprinus carpio*, a bioindicator of contamination in the Colorado River, Argentina. The fish were exposed to sublethal metal concentrations for 24, 48, and 96 h. Metal quantification was conducted using a novel automatic flow-batch system with SWASV and a bismuth film electrode. To the best of our knowledge, this constitutes the first application of this methodology on aquatic bioindicators for the assessment of metal accumulation in a natural environment. The technique demonstrated enhanced sensitivity and selectivity for the detection of trace metals. The bioaccumulation results demonstrated an increase in cadmium and lead concentrations in fish liver tissue after 96 h, reaching 10.5 $\mu g\ g^{-1}$ and 11.9 $\mu g\ g^{-1}$, respectively. Validation with inductively coupled plasma–atomic emission spectrometry (ICP-AES) demonstrated a satisfactory correlation, confirming the reliability of the method. This novel electrochemical approach offers enhanced accuracy and efficiency, making it a promising tool for environmental monitoring. The results indicate that Colorado River water is within safe levels for aquatic life regarding these metals. However, continuous monitoring is recommended to detect changes in contamination levels and protect ecosystem health, especially during water crises and under climate change.

Keywords: environmental toxicology; heavy metal contamination; electrochemical analysis; aquatic bioindicators; aquatic pollution indicators; bismuth film electrode

 https://doi.org/10.3390/w17010077

1. Introduction

Environmental pollution represents a substantial global challenge, mainly in the context of water contamination. This phenomenon has been identified as a crucial driver of deterioration in aquatic ecosystems, with adverse implications for water quality and biota [1–3]. Freshwaters are susceptible to contamination as a consequence of human activity. The introduction of waste and industrial waters, which are not always adequately subject to purification treatments, into discharge systems with a final destination of rivers and seas results in the deterioration of ecosystems [4]. The contamination of the environment can result in the poisoning, disease, and even death of fish, representing a significant environmental concern due to the prolonged presence of these pollutants and their high toxicity, even at low levels [5]. The presence of heavy metals in water can have severe and far-reaching consequences for the ecological balance of the environment, with the potential for biomagnification up the food chain. This phenomenon amplifies their impact, not only within the ecosystem and its wildlife but also on individuals who consume contaminated organisms. Consequently, this results in significant adverse effects on human health, whether through direct exposure or bioaccumulation [6–8].

The importance of bioindicators in biovigilance programs has grown significantly in recent years. They constitute a valuable, specific tool for the assessment of the physiological state of and stress in natural systems, thereby promoting an enhanced understanding of the mechanisms by which organisms respond to environmental changes [9–12]. Such methods provide information on the exposure and effects of substances, as well as on interactive processes occurring under natural conditions. As the apex of the food chain, fish are susceptible to direct and indirect pollution effects, rendering them a suitable bioindicator.

The common carp (*Cyprinus carpio*) is a well-documented fish species that has been observed in a variety of aquatic habitats over an extended period of time. Originally from Eurasia, the common carp was introduced to Argentina and was first recorded in Buenos Aires province, where it now coexists with local ichthyofauna in various aquatic settings, such as rivers and lakes [13]. Its adaptability has allowed it to establish itself in the Colorado River ecosystem, where its reproductive rate and feeding habits raise concerns regarding its status as an invasive species.

The Colorado River is a critical water resource for agriculture, livestock, and human consumption in the region. Its ecological importance lies in its role as a habitat for a diverse range of species and its contribution to the regional hydrological balance. However, the river faces increasing vulnerability to anthropogenic impacts, including pollution, land use changes, and water extraction for irrigation. In recent years, prolonged droughts and reduced flow rates have heightened the risk of salinization and the accumulation of pollutants, further endangering the ecosystem.

Due to its easy identification and capture, as well as the extensive knowledge of its life cycle, the carp serves as an ideal sentinel bioindicator in environmental studies. Consequently, it is of interest in studies on bioaccumulative metals of environmental and toxicological significance, such as cadmium (Cd) and lead (Pb) [14–17]. These metals are highly toxic heavy metals that accumulate in various tissues, including the liver, kidneys, and gills of fish [18]. Cadmium, recognized as a carcinogen, ranks among the most hazardous substances according to the United States Agency for Toxic Substances and Disease Registry (ATSDR), and it frequently contaminates aquatic environments. Studies have shown that Cd impairs essential functions in fish, such as hematological indices, iron metabolism, and antioxidant enzyme activity, while disrupting their reproductive capacities. Similarly, lead accumulates across multiple organs, where it impairs enzymatic activities, cellular structures, and physiological functions, adversely affecting growth, reproduction, and feeding efficiency. The 48 h median lethal concentration (LC_{50}/48 h) for the combined effect of these metals is 50 $\mu g\ mL^{-1}$ [19,20]. Given these harmful effects, the bioaccumulation study of Cd and Pb in *Cyprinus carpio* presents valuable insights into contamination dynamics and early warnings for ecosystem health, especially in the Colorado River watershed.

Cadmium and lead determination in a variety of matrices is typically conducted using techniques such as atomic absorption spectroscopy (AAS), inductively coupled plasma mass spectrometry (ICP-MS), or atomic fluorescence spectroscopy (AFS). These techniques necessitate the use of sophisticated instrumentation, entail the consumption of considerable quantities of reagents, and are typically conducted by an experienced analyst. In certain instances, the extraction procedures are lengthy and complex [21–24]. Conversely, electrochemical techniques, particularly those based on pulsed voltammetry and redissolution, are widely employed for the quantification of metals in samples of environmental interest. This is due to their high sensitivity, selectivity, rapidity of measurement and relatively inexpensive instrumentation. With regard to the indicator electrodes employed in these techniques, bismuth film electrodes (BiFEs) are widely utilized, given that they offer a number of advantages, including undistorted signals (well-defined peaks) and a highly reproducible, close peak resolution and a wide linear range [25,26]. Bismuth has the potential to form multicomponent alloys with the metals under investigation, which enables their accurate detection. Such alloys can be deposited on a variety of supports, including vitreous carbon electrodes, carbon paste, graphite, pencil leads, and screen-printed electrodes. The generation of the bismuth film on different supports can be achieved through the "in situ" modality, which allows for the simultaneous generation of the bismuth film and the presence of the metals under study [27,28]. Pierini et al. employed these systems for cadmium and lead determination in samples of bee products originating from the province of Buenos Aires, Argentina. The automation of electrochemical systems has the potential to enhance the efficiency and precision of analytical processes, facilitating a notable increase in the frequency, accuracy, and reliability of analyses. These systems permit the handling of unstable and toxic substances and facilitate continuous monitoring in any process, thereby reducing costs for large volumes of analysis and the control of equipment with minimal operator intervention. In this regard, Eggly et al. put forth a prototype comprising a potentiostat with an automated flow-batch system, which incorporates a voltammetric cell with a BiFE for the analysis of propolis samples [29]. The combination of flow-batch systems with electrochemical detection makes them an attractive option for routine environmental analysis laboratories.

In this work, the bioaccumulation of cadmium and lead in the livers of *Cyprinus carpio* specimens captured in the Colorado River (Argentina) was studied. These organisms are used as bioindicators of contamination. To assess the effectiveness of this approach, specimens of fish were subjected to controlled exposure to sublethal concentrations of the aforementioned metals for a period of 24, 48 and 96 h. The study was based on the hypothesis that this species exhibits a quantifiable and proportional accumulation of these metals in liver tissue, which could be effectively quantified by a novel electrochemical methodology. To verify this hypothesis, an automated flow-batch system was designed and developed using square-wave anodic stripping voltammetry with a bismuth film electrode as the analytical method. The results are intended to contribute to the development of sensitive tools for monitoring heavy metal contamination in freshwater ecosystems.

2. Materials and Methods

2.1. Chemicals and Reagents

All reagents were of analytical grade, and solutions were prepared employing ultrapure water (18 MΩ). The Bi (III) standard stock solution was prepared through the dissolution of 0.0534 g of $Bi(NO_3)_3 \cdot 5\ H_2O$ (99.999%, Sigma-Aldrich, Buenos Aires, Argentina) in 5 mL of 20% ($v\ v^{-1}$) nitric acid, which was then brought up to 25.0 mL with water. Pb (II) and Cd (II) working solutions were obtained by diluting the corresponding standard solution (1000 mg L^{-1}, Merck, Buenos Aires, Argentina). The supporting electrolyte was 0.1 M acetate buffer (Cicarelli, Santa Fé, Argentina) with a pH of 4.50. Tris-HCl buffer with a concentration of 50 mM and a pH of 8.10 (Fluka, Buchs, Switzerland) was employed.

2.2. Study Area

The Colorado River basin is situated in the northern semi-arid region of Argentina's Patagonia. The river extends from its source in the Cordillera de los Andes to its point of discharge into the Atlantic Ocean, covering a total area of 47,459 km^2. The mean annual flow of the basin is 143.5 $m^3\ s^{-1}$, which is largely generated by snowmelt in the Cordillera de los Andes. This flows downstream through the basin's two major tributaries, the Rio Grande and the Barrancas. The Colorado River basin is characterized by the presence of three distinct climatic–terrestrial regions, which are distinguished by variations in climate, land cover, and land use. The aforementioned zones can be further subdivided as follows: upper, middle, and lower. The upper zone, situated to the northwest, encompasses a portion of the Cordillera de los Andes and plays a significant role in precipitation (mainly through snowfall) and runoff (mainly through snowmelt) within the basin. The middle region extends in a southeasterly direction to the Casa de Piedra Dam. The dam was constructed in 1990 with the objective of supplying water to local populations, generating hydroelectric energy, mitigating the impact of floods, and regulating the flow of water within the basin. It is the only basin to have a flow distribution agreement in place. The lower area, commonly designated the Valle Bonaerense del Río Colorado (VBRC), has been primarily developed as a significant irrigation zone for agricultural production. The irrigation water concession system is monitored and regulated by the Corporación de Fomento del Valle Bonaerense del Río Colorado (CORFO).

Since August 2015, a collaboration and assistance agreement has been in place between four institutions (Universidad Nacional del Sur (UNS), Instituto Nacional de Tecnología Agropecuaria (INTA Hilario Ascasubi), Comisión de Investigaciones Científicas de la provincia de Buenos Aires (CIC), and CORFO) with the objective of promoting the development of research and carrying out studies for the preservation of the VBRC's water resources. The aforementioned agreement designated the Paso Alsina station (39°22′2.60″ S 63°14′16.26″ W) as a sampling station (Figure 1).

2.3. Water Samples

Water samples were collected on a monthly basis over the course of the period between August 2015 and February 2021. The procedure was conducted in accordance with the established guidelines, employing the standard methods for the examination of water and wastewater [30]. The water samples were obtained directly from the Paso Alsina station and stored in high-density polyethylene bottles with a nominal volume of 500 milliliters. In the laboratory, all samples were stored at a temperature of 4 °C until analysis, which was conducted within a seven-day period. The following parameters were analyzed: total dissolved solids (TDS), pH, electrical conductivity (EC), calcium (Ca^{2+}), magnesium (Mg^{2+}), sodium (Na^+), potassium (K^+), carbonates (CO_3^{2-}), bicarbonates (HCO_3^-), chlorides (Cl^-), sulfates (SO_4^{2-}), cadmium (Cd), and lead (Pb). The pH, temperature, EC, and TDS were determined in situ.

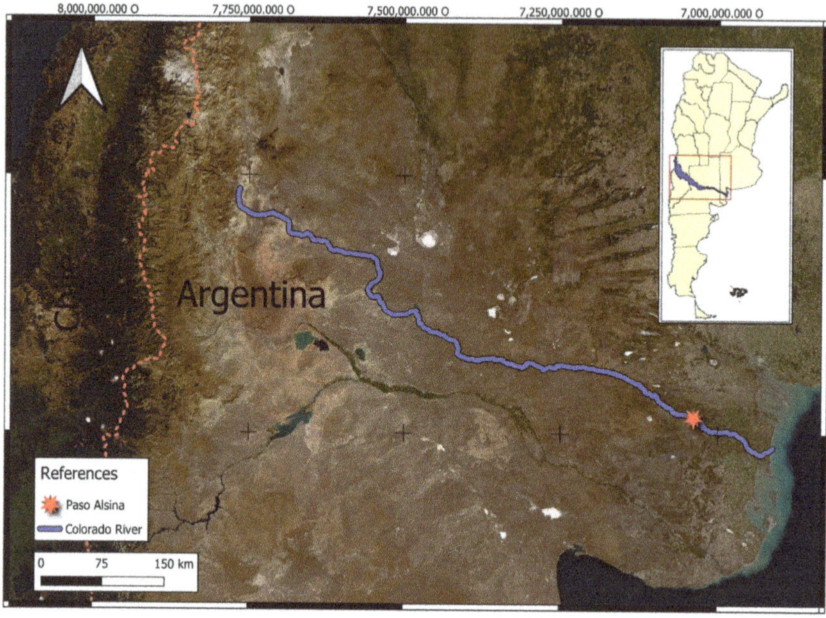

Figure 1. Study area, with the Paso Alsina station as the monitoring site.

2.4. Fish Samples and Pre-Treatment

Cyprinus carpio experiments were carried out in accordance with the international laws (EU Directive 2010/63/EU for animal experiments) and institutional guidelines for the protection and welfare of research animals. Following these criteria, the volume of water, the size of the container, the temperature, and the lighting of the fish were appropriate for the species [31].

The freshwater *Cyprinus carpio* were caught exclusively at the Paso Alsina station using beam trawls. This site was selected as it represents a typical habitat for local population of *Cyprinus carpio* and is frequently used for fishing activities. Local fishermen were consulted to identify the most suitable area for capture within this station, thus ensuring controlled conditions with minimal environmental variability. *Cyprinus carpio* specimens were identified using standard ichthyological keys and morphological characteristics, such as body shape, fin structure, the presence of two pairs of barbels, and large, thick scales, as outlined in Kottelat and Freyhof's *Handbook of European Freshwater Fishes* [32]. Additionally, local fishermen with extensive experience in catching *Cyprinus carpio* in the region contributed to confirming the species identification, ensuring accuracy through their practical knowledge. The fish were subsequently transported in ice-cold water to the laboratory, where they were acclimated for a period of seven days in a 500 L aquarium [33]. A total of 36 specimens were employed in the course of the experiments. The fish were provided with standard powdered food and were fasted for a period of 24 h prior to the commencement of experimentation. A 12 h photoperiod was maintained throughout the course of the experimental work. A continuous airflow was maintained, and the fish were provided with artificial dry food on a daily basis. During the course of the experiment, the pH of the water was maintained at 8.00 ± 0.30, with a temperature of 20 ± 1 °C. It is noteworthy that the water used for this experiment was sourced directly from the Paso Alsina station in the Colorado River.

The fish were separated into four distinct groups: the first served as the control group, while the other three were designated as experimental groups. The fish were randomly exposed to a solution of cadmium and lead, with a combined sublethal concentration of 5.0 μg mL^{-1} (1/10th of LC_{50}/48 h) [19]. The exposure period was 24, 48, and 96 h, respectively, in experimental tanks designated as T1, T2, and T3 (Table 1) [19]. Each experimental group consisted of nine fish, with three fish assigned per exposure period, ensuring three replicates for each condition. The water was maintained at a consistent level of aeration and refreshed at 24 h intervals. Prior to anesthesia and dissection of the organs, the weights and lengths of the fish were determined (Table 1).

Table 1. Experimental data for bioaccumulation of cadmium and lead.

	Experimental Tank 0 (Control)	**Experimental Tank 1 (T1)**	**Experimental Tank 2 (T2)**	**Experimental Tank 3 (T3)**
Average length	38.7 ± 1.8 cm			
Average weight	810 ± 35 g			
Cd	-	2.5 mg L^{-1}	1.0 mg L^{-1}	4.0 mg L^{-1}
Pb	-	2.5 mg L^{-1}	4.0 mg L^{-1}	1.0 mg L^{-1}
Total	-	5.0 mg L^{-1}	5.0 mg L^{-1}	5.0 mg L^{-1}

Subsequently, their livers were separated and weighed for analysis. An appropriate quantity of tissue was homogenized using ultrasound in a 50 mM Tris-HCl solution (pH 8.10), employing a piezoelectric ultrasonic device for 4 min, followed by a 10 min centrifugation at 2500 rpm. The resulting supernatant was heated to 95 °C for a period of 5 min on a hot plate and then filtered. The resulting homogenate was subjected to analysis using the voltammetric technique, as previously described by Fan et al. [34]. Figure 2 shows a flow chart of liver sample collection.

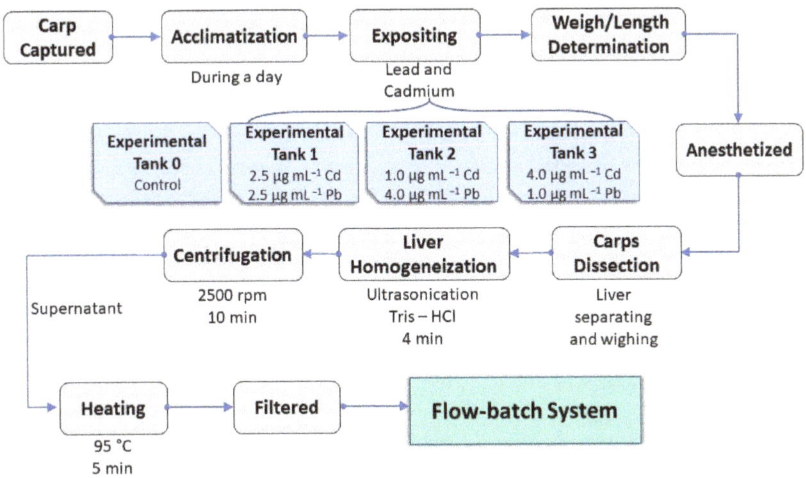

Figure 2. Flow diagram of the experimental procedure: fish acclimatization, metal exposure, liver collection, and sample preparation for analysis.

2.5. Flow-Batch System with Voltammetric Detection

The SWASV was performed using a lab-made prototype (Figure 3) consisting of a lab-made embedded flow-batch system with a potentiostat for electrochemical detection [29]. This device employs a low-cost microcontroller-based platform on an open-hardware Arduino kit. The electrochemical reaction takes place in a 25.0 mL electrochemical cell

(EC), which contains a PTFE cap with three holes for the attachment of a three-electrode configuration. This configuration includes a bismuth film prepared in situ on a glassy carbon as the working electrode, an Ag/AgCl (3 M KCl) reference electrode, and a platinum wire counter electrode.

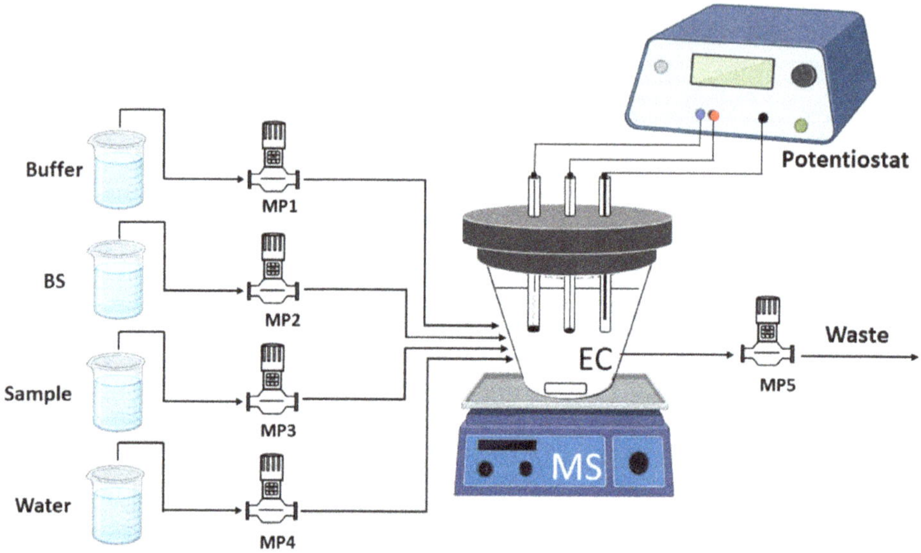

Figure 3. Embedded flow-batch system with electrochemical detection. MP: metering pump; EC: electrochemical cell; MS: magnetic stirrer; Buffer: buffer solution of 0.1 M of sodium acetate; BS: bismuth solution acidified with 5 mL of nitric acid; Sample: cadmium and lead standard solution or sample; Water: ultrapure water (18 MΩ).

2.6. Procedure

The experiments were conducted in the EC. The glassy carbon electrode (GCE) was polished with 0.05 μm alumina and subsequently rinsed with deionized water. In order to stabilize the surface of the working electrode, cyclic voltammograms were performed between −1.4 and −0.2 V using the supporting electrolyte solution (buffer acetic acid–sodium acetate, pH 4.50), at a scan rate of 100 mV s^{-1}, with the objective of stabilizing the surface of the working electrode. SWASV measurements were conducted by immersing the three electrodes in the EC, which contained the supporting electrolyte, a solution of 1000 μg L^{-1} of Bi^{3+}, and the sample. This process allows the bismuth film to be generated in situ on the surface of the GCE. The deposition potential was set at −1.4 V for a period of 120 s, with a rotation rate of 6000 rpm for the working electrode. Subsequently, the stirring was concluded, and following a 10 s interval, the voltammogram was recorded in square-wave voltammetry mode between −1.4 and 0.2 V. The SWASV parameters were as follows: step potential (ΔEs) 2 mV, amplitude of the square wave (ΔESW) 40 mV, and frequency (f) 25 Hz [27]. These parameters allowed for the effective resolution of the lead and cadmium peaks. Subsequently, in order to clean the surface of the working electrode, a potential of 0.2 V was applied for 30 s under stirring conditions of 6000 rpm. All experiments were conducted at room temperature and without the removal of oxygen.

Table 2 illustrates the flow-batch parameters employed for the cadmium and lead determination in the samples. Prior to the initiation of the analytical process, the metering pumps (MP1, MP2, MP3, MP4) were activated to facilitate the filling of the channels with the requisite solutions. Subsequently, the solution within the EC was discarded to the waste by activating the MP5. Afterward, MP1 was activated for 75 s to introduce the buffer, to stabilize the pH; MP2 for 25 s to add the bismuth solution, which serves as a film-forming agent on the electrode surface; and MP3 for 5 s to incorporate the sample into the EC. These steps were conducted with the continuous agitation of a magnetic stirrer.

Table 2. Flow-batch procedure steps for sample preparation and metal detection using SWASV.

Steps	Events	Time (s)	Pump Setup (μL) @ 1 Hz	Volume (mL)
1	Buffer (V1)	75	200	15.0
2	Bismuth (V2)	25	200	5.0
3	Sample (V3)	5	20	0.1
4	Deposition	120	-	-
5	Quiet time	10	-	-
6	Detection (SWASV)	32	-	-
7	Cleaning (V4)	480	200	20.0

Once all the reagents were introduced into the EC, the deposition step was performed at −1.4 V for 120 s, allowing cadmium and lead to accumulate on the bismuth film electrode. A resting time of 10 s followed to stabilize the system. Finally, the detection step was carried out for 32 s, recording the voltammograms between −1.4 and 0.2 V. After completing this stage, the EC underwent a cleaning process. First, MP5 was activated to remove waste solutions, followed by MP4 to introduce washing water, and MP5 again to eliminate the water. The total cleaning time was 480 s, preparing the EC for subsequent analyses.

3. Results and Discussions

3.1. Water Quality

Table 3 presents the results of the hydrosaline analysis of the Colorado River water, evaluated at the Paso Alsina station. The data set comprises the concentrations of the ions sodium (Na^+), potassium (K^+), calcium (Ca^{2+}), magnesium (Mg^{2+}), chloride (Cl^-), carbonate (CO_3^{2-} + HCO_3^-), and sulfate (SO_4^{2-}), the total hardness (TH), electrical conductivity (EC), pH, and total dissolved solids (TDS). The common carp were collected in July 2021, a period with particular implications in terms of water quality. This was due to the high levels of dissolved salts and metals resulting from drought and the flow regulation by the Casa de Piedra Dam [35,36]. The effects of these extreme conditions on the health of aquatic species, particularly carp, are of particular interest. Additionally, water samples were collected at the fishing site the same day and analyzed using ICP-AES. Metal concentrations (Cd and Pb) in the water were below the detection limit.

With regard to heavy metals, the Pb levels observed during the study period exhibited a range between 0.6 and 28.5 μg L^{-1}, while the Cd levels ranged between 0.22 and 1.3 μg L^{-1}. In some cases, these concentrations exceeded the guideline values for the protection of aquatic life established by the Canadian Environmental Quality Guidelines, which specify 7 μg L^{-1} for lead and 0.37 μg L^{-1} for cadmium under conditions of total water hardness above 180 mg L^{-1} and 280 mg L^{-1}, respectively [37]. From a biological perspective, increased salinity levels necessitate greater energy expenditure by fish to maintain osmotic balance, which subsequently reduces their capacity to effectively process toxic metals [38]. This additional effort may result in saline stress, which in turn affects the bioaccumulation of metals such as Cd and Pb, primarily affecting organs such as the liver, where the majority of accumulation occurs [39].

Table 3. Physicochemical characteristics of Colorado River basin in Paso Alsina station.

Parameters	Units	Min	Max	Median	Mean	Std	Method [30]	Instruments and Equipment
Na^+	mg L^{-1}	119	227	158	164	28	3500-Na^+ B	Photometer Metrolab 315. Manufacturer: Metrolab. Buenos Aires, Argentina
K^+	mg L^{-1}	3	6	4	4	0.6	3500-K^+ B	Photometer Metrolab 315. Manufacturer: Metrolab. Buenos Aires, Argentina
Ca^{2+}	mg L^{-1}	73	174	134	135	19	3500-Ca^{2+} B	Sartorius Model Biotrate 50 mL. Manufacturer: Sartorius. Göttingen, Germany
Mg^{2+}	mg L^{-1}	3	42	19	20	7.5	3500-Mg^{2+} B	Sartorius Model Biotrate 50 mL. Manufacturer: Sartorius. Göttingen, Germany
Cl^-	mg L^{-1}	171	400	238	248	48	4500-Cl^- B	Sartorius Model Biotrate 50 mL. Manufacturer: Sartorius. Göttingen, Germany
CO_3^{2-} + HCO_3^-	mg L^{-1}	91	161	121	121	15	2320-B	Sartorius Model Biotrate 50 mL. Manufacturer: Sartorius. Göttingen, Germany
SO_4^{2-}	mg L^{-1}	83	455	312	293	83	4500-SO_4^{2-} E	Spectrophotometer Lambda 35 UV–Vis Perkin Elmer. Manufacturer: Perkin Elmer. Waltham, MA, USA
TH	mg L^{-1}	330	517	418	419	46		
EC	µs cm^{-1}	870	1940	1400	1454	240	2520 B	Hanna 991301, Manufacturer: Hanna Instruments. Woonsocket, RI, USA,
pH	-	7.7	8.6	8.2	8.2	0.2	4500-H^+ B	Hanna 991301. Manufacturer: Hanna Instruments. Woonsocket, RI, USA,
Temperature	°C	16	26	21	20	3	2550 A	Hanna 991301. Manufacturer: Hanna Instruments. Woonsocket, RI, USA,
TDS	mg L^{-1}	600	1300	1000	983	166	2540 B	Gravimetrics 321 LX 220A. Manufacturer: Precisa. Buenos Aires, Argentina

Furthermore, the joint impact of salinity and heavy metals can result in synergistic interactions, whereby the toxicity of both stressors is amplified [40,41]. It is possible that the toxic effects of heavy metals on aquatic organisms may be intensified by drought stress and water regulation. Consequently, this period represents a significant context for the evaluation of bioaccumulation risks under conditions of environmental adversity, which could intensify metal accumulation and impact the health of fauna.

3.2. Analytical Parameters

Experimental calibration curves were constructed for the cadmium and lead quantification in the range between 1.00 and 30.00 µg L^{-1}. The resulting calibration curves were as follows: Ipn = (0.244 ± 0.009) C* Cd + (0.192 ± 0.072) (R^2 = 0.996) and Ipn = (0.211 ± 0.012 C* Pb) + (0.151 ± 0.065) (R^2 = 0.998), where Ipn represents the maximum current peak and C* the concentration (µg L^{-1}). The estimated detection limits, calculated as three times the standard deviation of the signal divided by the slope, were 0.92 μg L^{-1} and 0.88 μg L^{-1} for cadmium and lead, respectively. The reproducibility was evaluated as the percentage of variation relative to standard deviation of six independent measurements of 15.00 μg L^{-1} of cadmium and lead solutions, respectively. The resulting values were 6.2% for cadmium and 5.3% for lead.

3.3. Bioaccumulation Study

Cadmium and lead concentrations in liver samples of *Cyprinus carpio* were quantified using SWASV with a BiFE, a method known for its ability to detect trace concentrations of metals. In contrast to traditional techniques, SWASV offers high sensitivity and is particularly effective for distinguishing between metals in complex biological samples, making it a promising alternative to conventional methods like atomic absorption spectrometry or ICP-AES. This method is especially useful in bioaccumulation studies of metals in fish, where cadmium and lead levels can be low but still pose significant toxicological risks.

The initial analysis of the control group demonstrated the absence of quantifiable metal concentrations, as illustrated in Figure 4A (the signal at −0.11 V corresponds to Bi^{3+} ions), thereby confirming the initial conditions of the experiment. The subsequent analysis of samples exposed to metals at varying time points revealed a notable increase in lead and cadmium levels, with discernible signals for both elements in samples collected after 48 h (Pb at −0.52 V and Cd at −0.74 V). This is illustrated in Figure 4B. These findings demonstrate a clear accumulation of the metals in the liver tissues following exposure.

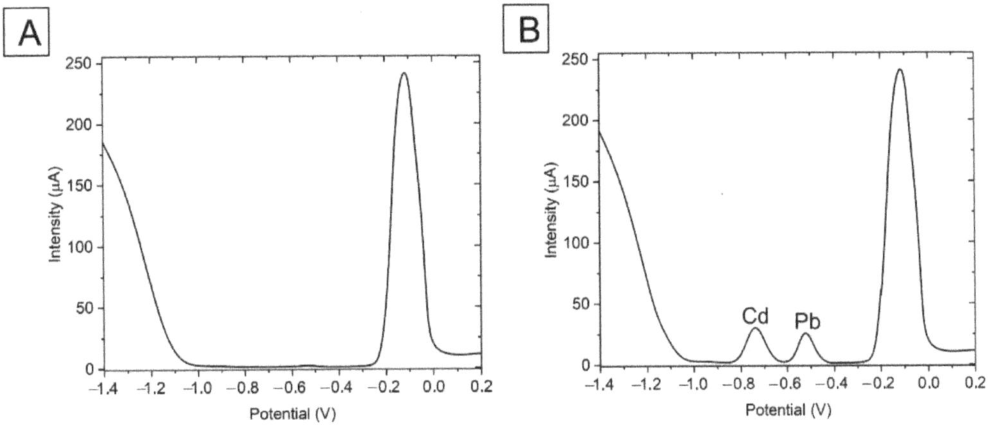

Figure 4. (**A**) Voltagram for a resulting homogenate of *Cyprinus carpio* taken from the Colorado River belonging to the control group. (**B**) Voltagram for a resulting homogenate of *Cyprinus carpio* after 48 h of exposure to metals.

The experimental tanks were subjected to different treatments, as was detailed in Table 1:

- T1: Equal concentrations of Cd and Pb at sublethal levels (1/10th of LC50/48 h for both metals) for exposure periods of 24, 48, and 96 h;
- T2: A higher concentration of Pb four times greater than Cd, with the same sublethal exposure times of 24, 48, and 96 h;
- T3: A higher concentration of Cd four times greater than Pb, with the same sublethal exposure times of 24, 48, and 96 h.

The results of the bioaccumulation study of these analytes are presented in Table 4. With regard to the temporal progression of bioaccumulation, a considerable increase in the concentration of metals in the liver was observed during the initial 24 h period, followed by a period of stabilization in the subsequent periods. This pattern indicates an initial phase of rapid uptake and accumulation in tissues, which is likely associated with the high permeability of liver membranes to metal ions under conditions of stress. Prior research has substantiated that the bioaccumulation of metals in fish exhibits considerable variability contingent on the specific metal in question. Additionally, factors such as the age, size, and physiological state of the fish also exert a significant influence [42,43].

Table 4. Bioaccumulation of metals in each experimental tank for the different times measured (n = 3).

	Experimental Tank 0 (Control)		Experimental Tank 1 (T1)		Experimental Tank 2 (T2)		Experimental Tank 3 (T3)	
	Cd	Pb	Cd	Pb	Cd	Pb	Cd	Pb
				$\mu g\ g^{-1}$ Dry wt.				
24 h	<LD	<LD	8.5 ± 0.3	9.9 ± 0.1	2.0 ± 0.3	14.0 ± 0.6	12.9 ± 0.3	2.5 ± 0.2
48 h	<LD	<LD	10.4 ± 0.8	11.6 ± 0.6	2.0 ± 0.1	14.7 ± 0.5	13.2 ± 0.1	2.5 ± 0.1
96 h	<LD	<LD	10.5 ± 0.6	11.9 ± 0.7	2.4 ± 0.2	15.1 ± 0.7	13.4 ± 0.2	2.6 ± 0.1

The results of the study demonstrated that the concentration of lead was higher than that of cadmium when both metals were present in equal concentrations (T1). This finding is consistent with the results of previous studies that have suggested a greater affinity of liver tissue for lead (Table 5). These studies also highlight the variability in metal accumulation depending on environmental conditions, analytical methods, and sampling sites.

Table 5. Accumulation of Cd and Pb in liver tissue of *Cyprinus carpio* reported in different studies.

Metal	Tissue	Liver Concentration $\mu g\ g^{-1}$ Dry wt.	Analytical Method	Source	Reference
Cd, Pb, Cu, Cr, Ni	Liver, gills, and muscle	Cd: 1.51 Pb: 9.45	Atomic absorption spectrophotometry	Seyhan River, Turkey	[14]
Cd, Pb, Cr, Ni	Liver, gills, kidney, and flesh	Cd: 1.69 Pb: 2.00	Atomic absorption spectrophotometry	Ponds of Tamilnadu, India	[19]
Cd, Pb, Zn, Cr, Ni, Cu	Liver, muscle, intestine, and gills	Cd: 58.0 Pb: 261	Atomic absorption spectrophotometry	Kabul River, Afghanistan	[44]
Cd, Pb, Cu, Cr, Zn	Liver, muscle, intestine, and gills	Cd: 39.8 Pb: 307	Atomic absorption spectrophotometry	Kabul River, Afghanistan	[45]
Cd, Pb, Fe, Cu, Zn, Hg	Liver, stomach, intestine, gills, skin, and muscle	Cd: 0.010 Pb: 0.024	ICP-MS	Kızılırmak River, Turkey	[46]
Cd, Pb	Liver	Cd: 10.0 Pb: 11.0	Proposed method	Colorado River, Argentina	This study

The concentrations of Cd and Pb reported in different studies show considerable variability, ranging from the lowest values observed in the Kızılırmak River, Turkey (Cd: 0.01 $\mu g\ g^{-1}$ dry wt.; Pb: 0.024 $\mu g\ g^{-1}$ dry wt., [46]), to the highest concentrations recorded in the Kabul River, Afghanistan (Cd: 39.8–58.0 $\mu g\ g^{-1}$; Pb: 261–307 $\mu g\ g^{-1}$, [44,45]). This wide range may be attributed to differences in environmental conditions, contamination levels, or even genetic and ecological variations among *Cyprinus carpio* populations. Factors such as water salinity, pH, and hardness are known to play a critical role in metal bioavailability and bioaccumulation.

The extreme concentrations reported for the Kabul River highlight the severity of environmental pollution in that region. As noted in previous studies, industrial and urban runoff significantly impact the Kabul River, leading to sublethal effects on *Cyprinus carpio* populations and potentially reducing their development. This underscores the influence of anthropogenic activities on metal accumulation in aquatic ecosystems.

Notably, our study is the first to employ an automated electrochemical system with a bismuth film electrode for the detection of cadmium and lead in aquatic bioindicators. The system's automation enables precise analyses with minimal operator intervention, offering a more accessible, cost-effective alternative to traditional techniques. This innovation eliminates the need for expensive instrumentation, while maintaining high sensitivity for trace metal detection, and it allows for continuous monitoring, making it a promising tool for routine environmental monitoring and toxic substance management.

In experimental T2, where the concentration of lead was four times greater, this trend of prioritizing lead accumulation over cadmium was maintained. These results further confirm the suitability of *Cyprinus carpio* as a bioindicator, particularly in ecosystems where species may be exposed to heavy metals and fluctuations in water conditions can favor toxin accumulation in critical tissues.

In contrast, in experimental T3, where the concentration of cadmium was four times greater than that of lead, the cadmium accumulation exceeded sublethal levels, demonstrating an accumulation proportional to the concentration present in the medium. This shift reinforces the capacity of *Cyprinus carpio* to respond predictably to variations in environmental metal concentrations, highlighting its utility in monitoring aquatic ecosystems [47].

Furthermore, while previous studies predominantly employed atomic absorption spectrophotometry (AAS) for metal analysis, our study utilized an automated flow-batch system with SWASV. This novel approach represents a significant advancement in analytical methodology, offering a comparable sensitivity while reducing operator intervention. Its automation and ease of use make it a promising tool for routine environmental monitoring.

3.4. Validation with Reference Method

The cadmium and lead content in the samples with 48 h of bioaccumulation in the three experimental tanks were validated by the inductively coupled atomic emission spectrometry technique (ICP-AES), and the results obtained were deemed satisfactory (Table 6). The results derived from the proposed method were compared with the ICP-AES values for both analytes. With regard to the element of cadmium, the identity line demonstrates a linear correlation, with an intercept of 0.376 $\mu g\ L^{-1}$, a slope of 0.979, and a correlation coefficient of r = 0.993. In the case of lead, the identity line yielded a linear correlation with an intercept of 0.023 $\mu g\ L^{-1}$, a slope of 0.994, and a correlation coefficient r = 0.988. These results confirm that no notable discrepancy exist between the proposed method and the reference technique, further validating the robustness of the automated system.

Table 6. Comparison of the proposed method with the ICP-AES reference method.

	Experimental Tank 1		**Experimental Tank 2**		**Experimental Tank 3**	
	Cd	**Pb**	**Cd**	**Pb**	**Cd**	**Pb**
			$\mu g\ g^{-1}$ Dry wt.			
Proposed method	10.4 ± 0.8	11.6 ± 0.6	2.0 ± 0.1	14.7 ± 0.5	13.2 ± 0.1	2.5 ± 0.1
ICP-AES	11.1 ± 0.5	10.8 ± 0.3	2.2 ± 0.4	15.2 ± 0.5	12.9 ± 0.2	2.7 ± 0.1
RSD %	6.3	7.4	9.1	3.3	2.3	7.4

4. Conclusions

This study demonstrates the bioaccumulation of cadmium and lead in *Cyprinus carpio*, thereby substantiating its value as a sensitive bioindicator for environmental contamination. The utilization of sophisticated electrochemical techniques, such as square-wave anodic stripping voltammetry (SWASV) with a bismuth film electrode integrated into an automated flow-batch system, offers a cost-effective and reliable methodology for the detection of trace metal concentrations. The results were validated against those obtained using traditional techniques, such as inductively coupled plasma–atomic emission spectrometry (ICP-AES), confirming the accuracy and potential of this innovative approach for monitoring heavy metals in aquatic ecosystems.

The findings revealed a significant degree of bioaccumulation of lead and cadmium in the liver of *Cyprinus carpio*, with lead demonstrating a stronger affinity and higher accumulation levels. These findings are particularly pertinent given the environmental stressors present during the study, including high salinity and altered flow conditions resulting from drought and the Casa de Piedra Dam. It is probable that elevated salinity exacerbated the toxic effects of metal exposure, thereby increasing bioaccumulation and the potential risks to aquatic organisms. These findings highlight the necessity for continuous monitoring of the Colorado River in order to mitigate the impacts of water crises and climate change on vulnerable ecosystems and to ensure the sustainable management of water resources.

Author Contributions: Conceptualization, S.S., M.S.D.N. and M.F.P.; Methodology, F.D.V. and S.S.; Software, F.D.V. and S.S.; Validation, F.D.V., M.T. and F.B.; Formal analysis, F.D.V., M.T. and F.B.; Investigation, F.D.V., S.S., M.T. and F.B.; Resources, A.J. and M.F.P.; Data curation, M.T. and F.B.; Writing—original draft, F.D.V. and S.S.; Writing—review & editing, F.D.V., M.T. and F.B.; Visualization, F.D.V., M.T., F.B. and M.S.D.N.; Supervision, M.S.D.N., A.J. and M.F.P.; Project administration, M.S.D.N., A.J. and M.F.P.; Funding acquisition, M.F.P. All authors have read and agreed to the published version of the manuscript.

Funding: This research received no external funding.

Data Availability Statement: All data are included in the manuscript.

Acknowledgments: The authors are grateful for the financial support of the Universidad Nacional del Sur (Dpto. de Química—INQUISUR). F. Vallese, S. Stupniki, M.Trillini, F. Belen, and A. Juan thank the Consejo Nacional de Investigaciones Científicas y Técnicas (CONICET, Argentina). M. F. Pistonesi is also grateful to the (CIC) Comisión de Investigaciones Científicas de la Provincia de Buenos Aires. Special thanks are extended to INTA (Instituto Nacional de Tecnología Agropecuaria – Hilario Ascasubi) and CORFO Río Colorado (Corporación de Fomento del Valle Bonaerense del Río Colorado) for providing important data for the development of this work.

Conflicts of Interest: The authors declare no conflicts of interest.

References

1. Häder, D.P.; Banaszak, A.T.; Villafañe, V.E.; Narvarte, M.A.; González, R.A.; Helbling, E.W. Anthropogenic Pollution of Aquatic Ecosystems: Emerging Problems with Global Implications. *Sci. Total Environ.* **2020**, *713*, 136586. [CrossRef] [PubMed]
2. Butnariu, M. Aquatic Pollution and Marine Ecosystems. In *Bacterial Fish Diseases*; Academic Press: Cambridge, MA, USA, 2022; pp. 1–29. [CrossRef]
3. Adewale, A.; Adegbola, P.I.; Owoade, A.O.; Aborisade, A.B.; Adewale, A.; Adegbola, P.I.; Owoade, A.O.; Aborisade, A.B. Fish as a Bioindicator of Polycyclic Aromatic Hydrocarbon Pollution in Aquatic Ecosystem of Ogun and Eleyele Rivers, Nigeria, and Risk Assessment for Consumer's Health. *J. Hazard. Mater. Adv.* **2022**, *7*, 100096. [CrossRef]
4. Barakat, A.; El Baghdadi, M.; Rais, J.; Aghezzaf, B.; Slassi, M. Assessment of Spatial and Seasonal Water Quality Variation of Oum Er Rbia River (Morocco) Using Multivariate Statistical Techniques. *Int. Soil Water Conserv. Res.* **2016**, *4*, 284–292. [CrossRef]

5. Vajargah, M.F.; Azar, H. Investigating the Effects of Accumulation of Lead and Cadmium Metals in Fish and Its Impact on Human Health. *J. Aquac. Mar. Biol.* **2023**, *12*, 209–213. [CrossRef]
6. Vutukuru, S.S. Chromium Induced Alterations in Some Biochemical Profiles of the Indian Major Carp, Labeo Rohita (Hamilton). *Bull. Environ. Contam. Toxicol.* **2003**, *70*, 118–123. [CrossRef] [PubMed]
7. Vutukuru, S.S. Acute Effects of Hexavalent Chromium on Survival, Oxygen Consumption, Hematological Parameters and Some Biochemical Profiles of the Indian Major Carp, Labeo Rohita. *Int. J. Environ. Res. Public Health* **2005**, *2*, 456–462. [CrossRef] [PubMed]
8. Tanna, K.; Bhavsar, D.; Mankad, A. Aquatic Toxicity Impacts on Behaviour and Survival of Fresh Water Fish: A Review. *Int. J. Fish. Aquat. Stud.* **2020**, *8*, 235–243.
9. Van Aggelen, G.; Ankley, G.T.; Baldwin, W.S.; Bearden, D.W.; Benson, W.H.; Chipman, J.K.; Collette, T.W.; Craft, J.A.; Denslow, N.D.; Embry, M.R.; et al. Integrating Omic Technologies into Aquatic Ecological Risk Assessment and Environmental Monitoring: Hurdles, Achievements, and Future Outlook. *Environ. Health Perspect.* **2010**, *118*, 1–5. [CrossRef]
10. Li, L.; Zheng, B.; Liu, L. Biomonitoring and Bioindicators Used for River Ecosystems: Definitions, Approaches and Trends. *Procedia Environ. Sci.* **2010**, *2*, 1510–1524. [CrossRef]
11. Altenburger, R.; Scholz, S.; Schmitt-Jansen, M.; Busch, W.; Escher, B.I. Mixture Toxicity Revisited from a Toxicogenomic Perspective. *Environ. Sci. Technol.* **2012**, *46*, 2508–2522. [CrossRef]
12. de Mello, K.; Taniwaki, R.H.; Macedo, D.R.; Leal, C.G.; Randhir, T.O. Biomonitoring for Watershed Protection from a Multiscale Land-Use Perspective. *Diversity* **2023**, *15*, 636. [CrossRef]
13. López, H.L.; Menni, R.C.; Protogini, L.C. Bibliografía de Los Peces de Agua Dulce de Argentina. 1994. Available online: https://www.produccion-animal.com.ar/produccion_peces/peces_argentinos/181-Biblio_peces.pdf (accessed on 27 November 2024).
14. Canli, M.; Ay, Ö.; Kalay, M. Levels of Heavy Metals (Cd, Pb, Cu, Cr and Ni) in Tissue of *Cyprinus carpio*, Barbus Capito and Chondrostoma Regium from the Seyhan River, Turkey. *Turk. J. Zool.* **1997**, *22*, 149–158.
15. Horváth, L.; Tamás, G.; Seagrave, C. *Carp and Pond Fish Culture*, 2nd ed.; Fishing New Books: Sydney, Australia, 2007.
16. Scarcia, P.; Calamante, G.; De La Torre, F. Biomarker Responses in Caged Carp (Cyprinuscarpio) and Native Collected Fish (Leporinus Obtusidens) in the Río de La Plata Estuary, Argentina. *Environ. Toxicol.* **2014**, *29*, 950–960. [CrossRef] [PubMed]
17. Murussi, C.R.; Costa, M.; Menezes, C.; Leitemperger, J.; Guerra, L.; Lópes, T.; Severo, E.; Zanella, R.; Loro, V.L. Integrated Assessment of Biomarker Response in Carp (*Cyprinus carpio*) and Silver Catfish (Rhamdia Quelen) Exposed to Clomazone. *Arch. Environ. Contam. Toxicol.* **2015**, *68*, 646–654. [CrossRef] [PubMed]
18. Nzeve, J.; Kitur, E.C.; Nzeve, J.K.; Kitur, E. Bioaccumulation of Heavy Metals in Common Carp (*Cyprinus carpio*) of Masinga Dam, Kenya. *IOSR J. Environ. Sci.* **2019**, *13*, 23–28. [CrossRef]
19. Vinodhini, R.; Narayanan, M. Bioaccumulation of Heavy Metals in Organs of Fresh Water Fish *Cyprinus carpio* (Common Carp). *Int. J. Environ. Sci. Technol.* **2008**, *5*, 179–182. [CrossRef]
20. Jamil Emon, F.; Rohani, M.F.; Sumaiya, N.; Tuj Jannat, M.F.; Akter, Y.; Shahjahan, M.; Abdul Kari, Z.; Tahiluddin, A.B.; Goh, K.W. Bioaccumulation and Bioremediation of Heavy Metals in Fishes—A Review. *Toxics* **2023**, *11*, 510. [CrossRef] [PubMed]
21. Chen, Y.; Li, M.; Fu, L.; Hou, X.; Jiang, X. Simultaneous Determination of Trace Cadmium and Lead in Single Human Hair by Tungsten Electrothermal Vaporization-Flame Atomic Fluorescence Spectrometry. *Microchem. J.* **2014**, *114*, 182–186. [CrossRef]
22. Cao, Y.; Deng, B.; Yan, L.; Huang, H. An Environmentally-Friendly, Highly Efficient, Gas Pressure-Assisted Sample Introduction System for ICP-MS and Its Application to Detection of Cadmium and Lead in Human Plasma. *Talanta* **2017**, *167*, 520–525. [CrossRef]
23. Zhang, N.; Shen, K.; Yang, X.; Li, Z.; Zhou, T.; Zhang, Y.; Sheng, Q.; Zheng, J. Simultaneous Determination of Arsenic, Cadmium and Lead in Plant Foods by ICP-MS Combined with Automated Focused Infrared Ashing and Cold Trap. *Food Chem.* **2018**, *264*, 462–470. [CrossRef] [PubMed]
24. Valasques, G.S.; dos Santos, A.M.; de Souza, V.S.; Teixeira, L.S.; Alves, J.P.; de Jesus Santos, M.; dos Santos, W.P.; Bezerra, M.A. Multivariate Optimization for the Determination of Cadmium and Lead in Crude Palm Oil by Graphite Furnace Atomic Absorption Spectrometry after Extraction Induced by Emulsion Breaking. *Microchem. J.* **2020**, *153*, 104401. [CrossRef]
25. Yang, M.; Zhang, Z.; Hu, Z.; Li, J. Differential Pulse Anodic Stripping Voltammetry Detection of Metallothionein at Bismuth Film Electrodes. *Talanta* **2006**, *69*, 1162–1165. [CrossRef] [PubMed]
26. Krepper, G.; Resende de Cerqueira, P.B.; Pistonesi, M.F.; Di Nezio, M.S.; Centurión, M.E. Determination of Cadmium Residues in Bee Products Using a 'Lab-Made' Bismuth Bulk Electrode. *Int. J. Environ. Anal. Chem.* **2016**, *96*, 1331–1340. [CrossRef]
27. Pierini, G.D.; Granero, A.M.; Di Nezio, M.S.; Centurión, M.E.; Zon, M.A.; Fernández, H. Development of an Electroanalytical Method for the Determination of Lead in Argentina Raw Propolis Based on Bismuth Electrodes. *Microchem. J.* **2013**, *106*, 102–106. [CrossRef]
28. Pierini, G.D.; Pistonesi, M.F.; Di Nezio, M.S.; Centurión, M.E. A Pencil-Lead Bismuth Film Electrode and Chemometric Tools for Simultaneous Determination of Heavy Metals in Propolis Samples. *Microchem. J.* **2016**, *125*, 266–272. [CrossRef]

29. Eggly, G.M.; Nabaes, M.; Di Nezio, M.S.; Centurión, M.E.; Santos, R.; Pistonesi, M.F. Embedded Flow-Batch System with Electrochemical Detection for the Determination of Lead in Propolis Samples. *Int. J. Environ. Anal. Chem.* **2017**, *97*, 922–934. [CrossRef]
30. APHA American Public Health Association; American Water Works Association. *Standard Methods for the Examination of Water and Waste Water*, 22nd ed.; American Public Health Association: Washington, DC, USA, 2012.
31. Directive 2010/63/EU of the European Parliament and of the Council of 22 September 2010 on the Protection of Animals Used for Scientific Purposes (Text with EEA Relevance). Available online: https://eur-lex.europa.eu/legal-content/EN/TXT/?uri=CELEX:32010L0063 (accessed on 27 November 2024).
32. Kawamura, K.; Kawamura, K. "Handbook of European Freshwater Fishes" by M. Kottelat and J. Freyhof (2007). *Ichthyol. Res.* **2008**, *55*, 99. [CrossRef]
33. Duarte, I.A.; Reis-Santos, P.; Fick, J.; Cabral, H.N.; Duarte, B.; Fonseca, V.F. Neuroactive Pharmaceuticals in Estuaries: Occurrence and Tissue-Specific Bioaccumulation in Multiple Fish Species. *Environ. Pollut.* **2023**, *316*, 120531. [CrossRef]
34. Fan, W.; Wang, X.; Li, X.; Xue, F. Determination of Metallothionein in Daphnia Magna by Modified Square Wave Cathodic Stripping Voltammetry. *Electrochem. Commun.* **2015**, *52*, 17–20. [CrossRef]
35. Trillini, M.; Pierini, J.O.; Vallese, F.D.; Dunel Guerra, L.; Pistonesi, M.F. Colorado River (Argentina) Water Crisis Scenarios and Influence on Irrigation Water Quality Conditions. *Sustainability* **2023**, *15*, 8457. [CrossRef]
36. Vallese, F.D.; Trillini, M.; Dunel Guerra, L.; Pistonesi, M.F.; Pierini, J.O. Data Analysis to Evaluate the Influence of Drought on Water Quality in the Colorado River Basin. *Water* **2024**, *16*, 2750. [CrossRef]
37. Programa Integral—COIRCO. Available online: https://www.coirco.gov.ar/calidad-de-aguas/programa-integral/ (accessed on 24 November 2024).
38. Cañedo-Argüelles, M.; Kefford, B.J.; Piscart, C.; Prat, N.; Schäfer, R.B.; Schulz, C.J. Salinisation of Rivers: An Urgent Ecological Issue. *Environ. Pollut.* **2013**, *173*, 157–167. [CrossRef]
39. Santos, B.; Andrade, T.; Domingues, I.; Ribeiro, R.; Soares, A.M.; Lopes, I. Influence of Salinity on the Toxicity of Copper and Cadmium to Zebrafish Embryos. *Aquat. Toxicol.* **2021**, *241*, 106003. [CrossRef]
40. Deane, E.E.; Woo, N.Y.S. Modulation of Fish Growth Hormone Levels by Salinity, Temperature, Pollutants and Aquaculture Related Stress: A Review. *Rev. Fish. Biol. Fish.* **2009**, *19*, 97–120. [CrossRef]
41. Kültz, D. Physiological Mechanisms Used by Fish to Cope with Salinity Stress. *J. Exp. Biol.* **2015**, *218*, 1907–1914. [CrossRef] [PubMed]
42. Schneider, L.; Belger, L.; Burger, J.; Vogt, R.C. Mercury Bioacumulation in Four Tissues of Podocnemis Erythrocephala (Podocnemididae: Testudines) as a Function of Water Parameters. *Sci. Total Environ.* **2009**, *407*, 1048–1054. [CrossRef]
43. Hao, Y.; Miao, X.; Song, M.; Zhang, H. The Bioaccumulation and Health Risk Assessment of Metals among Two Most Consumed Species of Angling Fish (*Cyprinus carpio* and *Pseudohemiculter dispar*) in Liuzhou (China): Winter Should Be Treated as a Suitable Season for Fish Angling. *Int. J. Environ. Res. Public Health* **2022**, *19*, 1519. [CrossRef] [PubMed]
44. Yousafzai, A.M.; Siraj, M.; Ahmad, H.; Chivers, D.P. Bioaccumulation of Heavy Metals in Common Carp: Implications for Human Health. *Pak. J. Zool.* **2012**, *44*, 489–494.
45. Ishaq MPhil, M.; Muhammad Yousafzai, A.; Ahmad, B.; Latif, M.; Ishaq, M. Investigation of Heavy Metals in Common Carp (*Cyprinus carpio*) Netted from River Kabul District, Nowshera. *J. Entomol. Zool. Stud.* **2018**, *6*, 2610–2614.
46. YARDIM, Ö.; BAT, L. Distribution of Metals in Organs and Tissues of *Cyprinus carpio* L., 1758 from Kızılırmak. *J. Anatol. Environ. Anim. Sci.* **2020**, *5*, 290–294. [CrossRef]
47. Gupta, S.; Singh, V.; Aggarwal, M.L. *Cyprinus carpio*: Bioindicator Of Heavy Metal Pollution In Yamuna River, Delhi Region. *J. Exp. Biol. Agric. Sci.* **2021**, *9*, 805–812. [CrossRef]

water

MDPI

Article

The Influence of Waters of Lake Baikal on the Spatiotemporal Dynamics of Phytoplankton in the Irkutsk Reservoir

Alena Firsova, Yuri Galachyants, Anna Bessudova *, Diana Hilkhanova, Lubov Titova, Maria Nalimova, Vasilisa Buzevich, Artyom Marchenkov, Maria Sakirko and Yelena Likhoshway

Limnological Institute, Siberian Branch of the Russian Academy of Sciences, 3 Ulan-Batorskaya, 664033 Irkutsk, Russia
* Correspondence: annabessudova@mail.ru

Abstract: On a model natural object, the Lake Baikal–Angara River–Irkutsk Reservoir (IR), we studied changes in the qualitative and quantitative characteristics of phytoplankton communities over three seasons in 2023 depending on seasonal changes in habitat parameters. Of the 151 identified taxa, Chrysophyta (57), Chlorophyta (41) and Bacillariophyta (24) predominated in diversity. Over the entire observation period, the highest values of total biomass and total abundance were detected in the IR in June (hydrological spring) at a water temperature of 10.0–12.7 °C, and the lowest in August, despite the fact that the water warmed up to 20 °C. No mass blooms of Cyanobacteria were observed. Statistical analysis of species abundance profiles revealed that phytoplankton community structure varied across time and space. The direct effect of cold lake waters on the structure of phytoplankton in the reservoir was observed only in early June. In summer and autumn, the structures of phytoplankton in the lake and in the reservoir differed, even at the same water temperature. Low concentrations of phosphates and nitrates, high species diversity, the presence of cold-water species and species with a wide range of temperature preferences formed a dynamic spatiotemporal structure of IR phytoplankton, distinct from other temperate reservoirs, including Lake Baikal. The results obtained are important for understanding the mechanisms of formation of the flora of artificial reservoirs of temperate latitudes and for their monitoring, taking into account seasonal dynamics and the context of global climate warming.

Keywords: freshwater; lake–reservoir transition; species composition; abundance; water temperature

1. Introduction

Phytoplankton are the most important part of the ecosystem, and their responses to environmental changes not only affect the grazing, feeding, growth, reproduction and survival of various aquatic organisms [1], along with biogeochemical cycles [2,3], but can also serve as an indicator of climate change and anthropogenic impacts [4–7]. Planktonic microalgae have features of seasonal growth, which depend on abiotic conditions in the annual cycle, affecting its composition and quantitative characteristics [8–10]. In recent years, climate warming has led to restructuring of the plankton community in large reservoirs [7,11–14], including a change in species and changes in intraspecific characteristics [15], including in Lake Baikal, where these changes are most pronounced in the southern part of the lake [16]. A significant feature of deep-water lakes of temperate latitudes is the spring bloom, in which the main part of the algae biomass is formed. The study of phytoplankton in different seasons showed that in large reservoirs, spring phytoplankton is significantly (1.5–6 times) greater than summer phytoplankton, and it most often consists of diatoms. For example, in Lake Huron and Erie, spring phytoplankton consisted of filamentous diatoms, primarily *Aulacoseira islandica* (O.Müller) Simonsen [17]. In coastal seas, such as the Baltic Sea, eutrophication can lead to an increase in the abundance and biomass of summer phytoplankton [11,18], as well as more frequent and intense blooms [19–21]. It has also been shown that the species composition of phytoplankton varies depending on the concentrations of nutrients [22]. Observations of seasonal changes help to identify changes in communities of planktonic microalgae and also make it possible to judge not only their biodiversity [8,18,23] but also the appearance of invasive species [24].

The increase in the number of reservoirs in the world over the last century is associated with their significant economic and geographical role, and assessment of their environmental condition is very important. Differences in the geographic distributions of lakes and reservoirs determine the geographic features of their catchments. In reservoirs, the structure of the water balance, the period of water exchange, fluctuations in the water level, water inflow and runoff and water temperature change affect the structure and functioning of aquatic organisms [25]. First of all, phytoplankton react to these changes, in which structural and quantitative rearrangements occur. For example, the flow regime of the Wimmera River (Australia) was significantly changed after the construction of a reservoir, and this led to a change in the species structure of algae along the entire length of the river throughout the year. The diversity of Chlorophyta, Cyanobacteria and Chrysophyta taxa gradually augmented from the upstream to the downstream regions under normal flow conditions prior to water release. Conversely, diatom abundance was observed to be higher in the upstream area and to increase downstream following water release [26]. Depending on the type of reservoir, river or lake, which have differences in hydrodynamics and water retention times, the structure of the phytoplankton community also differs. Using the example of the Jiuqiuwan (river-type) and Taihu (lake-type) reservoirs, in the area of the source of the Dongjiang River in the Pearl River basin (China), it has been shown that over time, the number of phytoplankton species in the river-type reservoir has decreased, while in the lake-type reservoir, the number has increased versus that before the flooding of the reservoir [27]. The Irkutsk Reservoir was formed in 1956–1958 as a result of the creation of a dam on the Angara River. The hydroelectric power station (HPP) is located 56 km from Lake Baikal in the city of Irkutsk. The hydrochemical composition of the Irkutsk Reservoir is influenced by the waters of the southern part of Lake Baikal [28]. The waters of the lake belong to the hydrocarbonate class of the calcium group, characterized by very low mineralization (no higher than 100 mg m^{-3}), a low content of organic matter and high concentrations of dissolved oxygen [29]. In terms of the composition of planktonic microalgae, the Irkutsk Reservoir is poorer than the underlying sections of the river after the dam [30–32]. The temperature regime of the reservoir is heterogeneous. During the open water period, the heated part of the reservoir near the dam is the warmest, with maximum temperatures in August–September of up to 16.2 °C [31]. The bays of the reservoir are also characterized by heterogeneity of the temperature regime and slow warming of deep waters; this is especially noticeable in July–August, when the difference in temperature with depth is 5–8 °C [31,32]. Thus, in Kurminsky Bay in July 1958, the surface temperature reached 19–21 °C, while at a depth of 5 m, it was only 12–13 °C [30]. Further studies have shown that, when including in the dam section of the reservoir, the temperature difference between the surface and bottom layers during the period of most intense heating (July–August) can exceed 14 °C. Observations of changes in the state of phytoplankton in the Irkutsk Reservoir have been carried out sporadically since its creation [30–32]. The total number of phytoplankton over the entire observation period has varied from 0.30×10^5 cells L^{-1} to 2.51×10^5 cells L^{-1}, and the biomass has varied from 0.12×10^5 mg L^{-1} to 0.85×10^5 mg L^{-1} [30–32]. The maximum growth of phytoplankton during the year can occur in different months; for example, in 1973, it occurred in August, in 1980, in June, in 1982, in August, in 1986, in July and in 1987, in May. The water temperature during these periods was 8.3–16.2 °C. Moreover, the dominant complex of species in the reservoir in the spring was similar to that in Lake Baikal, while in the summer, it was different [31]. The level of Cyanobacteria growth in the reservoir has always remained low [30–32].

The dominant phytoplankton complex of the pelagic zone and bays of Lake Baikal includes cold-water species that develop at low temperatures—*Aulacoseira baicalensis* (K.

Meyer) Simonsen at 1–8 °C, *Aulacoseira islandica* (=*A. skvortzowii* Edlund, Stoermer & Taylor) at 1–5 °C and *Stephanodiscus meyeri* Genkal et Popovskaya at up to 10 °C—and species with a wider range of temperature tolerance, for example, *Nitzschia graciliformis* Lange-Bertalot et Simonsen emend. Genkal et Popovskaya at 1–16 °C, *Asterionella formosa* Hass. At 8–19 °C, *Fragilaria radians* (Kützing) Williams et Round at up to 15 °C and *Ulnaria acus* (Kützing) Aboal at up to 16 °C [33]. Since the Irkutsk Reservoir is connected to Southern Baikal, but is shallow and warms up better, it was interesting to determine the degree of influence of the lake waters on the phytoplankton of the reservoir. In our previous works, we studied the phytoplankton of Southern Baikal and the Irkutsk Reservoir in the hydrological spring (June) [34,35] and summer (August) of 2023 [35,36]. The aim of this work was, based on a comparison of the species structure and abundance of phytoplankton in Southern Baikal and the Irkutsk Reservoir in the seasonal dynamics of changing habitat parameters, to assess the degree of influence of lake waters on the spatiotemporal distribution of phytoplankton in the reservoir.

2. Materials and Methods

2.1. Sampling and Microscopy

Sampling was carried out during expeditions within the framework of the Russian Science Foundation project in Southern Baikal (SB) and the Irkutsk Reservoir (IR) on October 15–17. For comparison, we used the results of the analysis of samples taken on June 22–26 (hydrological spring) [34] and August 17–20 [36] 2023. Additionally, samples were taken at a station 3 km from Listvyanka Village (51°49.033 N 104°54.616 E) after the ice melted, on 1 June, 16 July and 14 September 2023 (Figure 1).

During sampling, temperature and pH were measured, and water transparency was determined using a Secchi disk (S, m). In the laboratory, frozen water samples were thawed at room temperature. The mineral forms of the biogenic elements were determined after filtration using membrane cellulose acetate filters with 0.45 μm pores (Vladisart, Vladimir, Russia) to remove suspended matter, and we carried out hydrochemical analyses of the water for its Si content, PO_4^{3-}, NO_2^-, NO_3^- and NH_4^+ (Table S1). The methods for the sampling and analysis of hydrochemical parameters were given previously [34].

2.2. Statistical Analysis

Environmental factors and numerical community variables were analyzed for collinearity. Pearson correlation coefficients and their *p*-values were computed for each pair of explanatory variables using R packages corrplot [37] and Hmisc [38]. The correlation matrix was visualized with the corrplot function using hierarchical clustering to group variables. Exploratory analyses of community composition were performed using R package vegan v.2.5-6 [39]. Environmental variables were centered and scaled to have zero means and standard deviations of one. This standardized environmental matrix was used for ordination. The species abundance data were transformed using the Hellinger procedure and subjected to transformation-based principal component (tbPCA) and redundancy (tbRDA) analyses. For unconstrained ordination, linear regression of explanatory variables was performed by the envfit function of the package vegan followed by the adjustment of permutation-based regression *p*-values by the FDR procedure. Environmental factors with an adjusted *p*-value below 0.05 were drawn on the ordination plane. For constrained ordination, eJ9, Jl9 and S9 profiles were excluded as these samples lacked data on environmental variables to use in RDA (Table S1). Both forward selection and backward elimination approaches were tested to produce an RDA model.

To generate the species abundance heatmap, the R package pheatmap [40] was used. The 50 most abundant species were selected, and an abundance table was transformed with $\log_2(x + 1)$ and passed to the pheatmap function. UPGMA-assisted clustering of the Bray–Curtis pairwise distance matrices was used to produce sample-wise and OTU-wise clustering trees. The package apcluster [41] was used for cluster analysis of β-diversity. The pairwise distance matrix computed with the Bray–Curtis similarity index was used to generate clusters by affinity propagation, followed by exemplar-based agglomerative clustering.

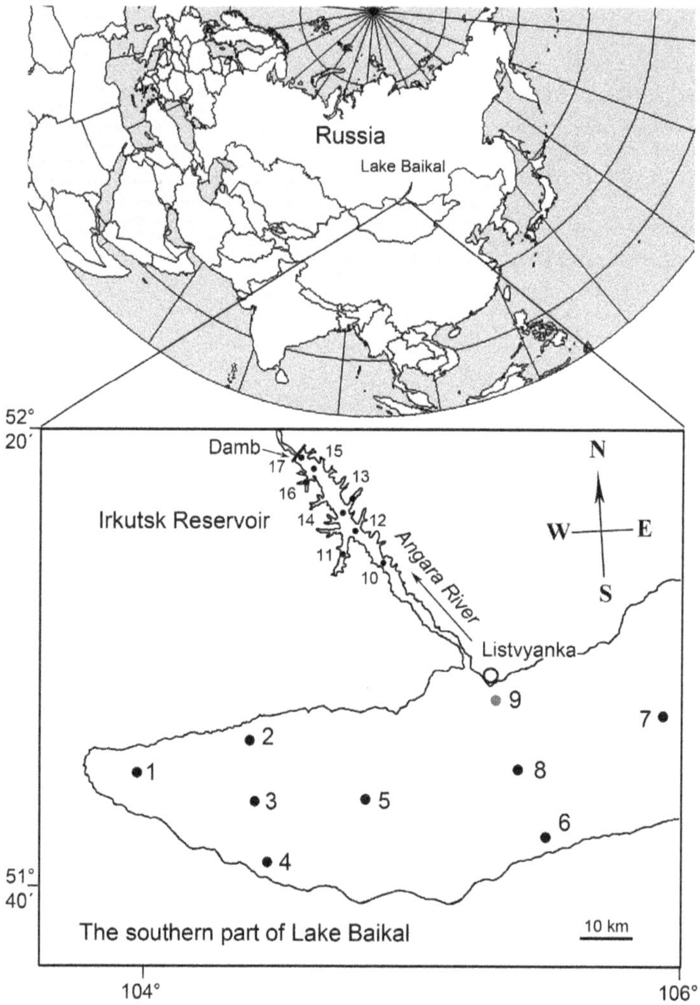

Figure 1. Map of sampling in 2023. Red dot—monthly sampling during the open water period at a station 3 km from Listvyanka Village (St. 9); black dots indicate sampling in October 2023 at stations that were sampled earlier in June [33] and August [35]. Bays of Irkutsk Reservoir: 11—Kurminsky; 13—Elovy; 16—Ershovsky.

3. Results and Discussion

3.1. Temporal Monthly Changes in the Phytoplankton Community at the SB Station near the Source of the Angara River During the Open Water Period in 2023

To determine which phytoplankton are transported from the SB to the IR, we examined their composition at St. 9 near the source of the Angara River monthly during the open water period (from June to October) (Figures 1 and 2).

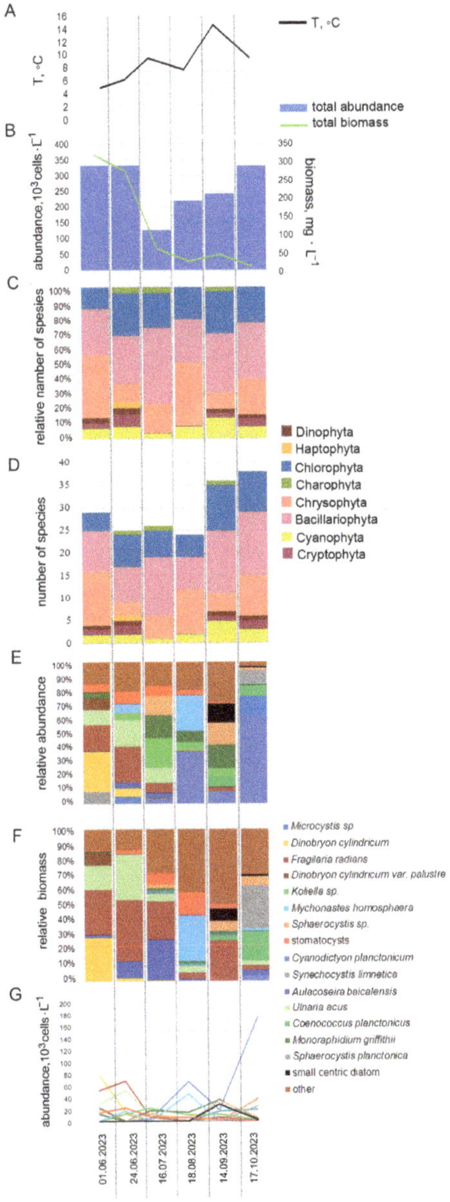

Figure 2. Quantitative and qualitative characteristics of phytoplankton at St. 9 near the source of the Angara River during the open water period in 2023: water temperature (**A**), total abundance and biomass of phytoplankton (**B**), relative share of large taxonomic groups (divisions) of microalgae (**C**), number of species in divisions (**D**), contribution of dominant species, the number of which exceeds 20×10^3 cells L^{-1}, to the total abundance (**E**) and the total biomass (**F**) of phytoplankton, the absolute abundance of the dominant species (**G**).

During the research period, the water temperature changed from 4.3 °C to 7 °C (Figure 2A), and increases in the quantitative indicators of phytoplankton and their structure were observed (Figure 2B–G). The total abundance and biomass of phytoplankton varied from 111 × 10^3 to 292 × 10^3 cells L^{-1} and from 13 × 10^3 to 366 × 10^3 mg L^{-1}, respectively (Figure 2B). The abundance levels at the beginning and end of June had similar values, but the biomass at the beginning of June was higher (366 × 10^3 mg L^{-1}) due to the growth of large species of Bacillariophyta *F. radians* and Chrysophyta of the genus *Dinobryon* (*D. cylindricum* and *D. cylindricum* var. *palustre*). By the end of June, the biomass of small-celled Chlorophyta species such as *Mychonastes homosphaera* and *Koliella* cf. *variabilis* increased, and *Microcystis* sp. During the entire study period, the number of species at the station was dominated by Bacillariophyta, with the exception of June 1 and August 18, when Chrysophyta of the genus *Dinobryon* actively developed (Figure 2G). In June at St. 9, the species richness of Bacillariophyta was formed of the large-celled species *A. baicalensis, A. islandica, F. radians* and *U. acus*; in July, the number of species increased, and small centric diatom *Stephanodiscus minutulus* appeared; and in August, large-celled diatoms disappeared, such as spring species of the genus *Aulacoseira*. The representative of Chlorophyta, *M. homosphaera,* was found sporadically during open water periods, developing noticeably in August; however, its number did not exceed 50 × 10^3 cells L^{-1} and its biomass 8 × 10^3 mg L^{-1} (Table S1). In July, when the total abundance of phytoplankton was minimal (111 × 10^3 cells L^{-1}), the main place in the community composition was occupied by Chlorophyta, dominated by *Koliella* sp. (23 × 10^3 cells L^{-1}) and *Monoraphidium griffithii* (19 × 10^3 cells L^{-1}). The largest contribution to the total biomass, amounting to 68 mg L^{-1}, was made by large-celled Bacillariophyta *F. radians* (15 × 10^3 mg L^{-1}) and *A. baicalensis* (16 × 10^3 mg L^{-1}), despite their low abundance. Species richness increased significantly in the autumn months due to small centric diatoms such as *Cyclostephanos makarovae, Cyclostephanos dubius, Discostella pseudostelligera, Stephanocyclus meneghinianus* and *Thalassiosira pseudonana*, which were absent in June and July (Figure 2, Table S1).

Thus, at a station near the source of the Angara River during 2023, temporary changes in the structure of phytoplankton communities were observed. At the next stage of research, in order to determine the degree of influence of the waters of Southern Baikal on the Irkutsk Reservoir, a spatial comparison of quantitative and qualitative indicators of phytoplankton in the lake and reservoir was carried out in three seasons (June—which corresponds to the hydrological spring, July, August and September—to the hydrological summer, and October—to the hydrological autumn).

3.2. Seasonal Dynamics of Quantitative Characteristics and Species Structure of Phytoplankton in Southern Baikal and the Irkutsk Reservoir in 2023

Figure 3 shows that during the open water period of 2023, the quantitative characteristics and structure of phytoplankton were changing.

The total abundance and biomass of microalgae in spring (June) at SB stations varied within the ranges of (255–1333) × 10^3 cells L^{-1} and (61–473) × 10^3 mg L^{-1}, respectively. It should be noted that at St. 1 (12 km from Kultuk), we found 2779 × 10^3 cells L^{-1} due to the growth of three Cyanobacteria: *Cyanodictyon planctonicum, Cyanodictyon* sp. and *Microcystis* sp., while their total biomass had a low value of 5.2 × 10^3 mg L^{-1} (Figure 3B after [34]). The abundance and biomass of phytoplankton in the IR had the highest values for all seasons: (960–2350) × 10^3 cells L^{-1} and (544–1679) × 10^3 mg L^{-1}, respectively (Figure 3B).

In summer, the total abundance and biomass of microalgae in SB varied widely, at (190–2779) × 10^3 cells L^{-1} and (26–427) × 10^3 mg L^{-1}, respectively. In the IR, including bays, abundance and biomass decreased significantly and had the lowest values over the period of our observations—(186–310) × 10^3 cells L^{-1} and (41–140) × 10^3 mg L^{-1}, respectively (Figure 3B according to [36]).

The autumn period was characterized by the smallest differences in water temperature between SB and IR (Table S1). In SB, the abundance varied within the range of (147–427) × 10^3 cells L^{-1}, and the biomass of microalgae, due to the predominance of small-celled forms, was significantly lower than in IR and its bays—(12–45) × 10^3 mg L^{-1}. The number of microalgae in the IR in autumn varied within the range of (160–494) × 10^3 cells L^{-1}, and the biomass at (123–239) $\times$ 10^3 mg L^{-1}. The highest values of abundance and biomass were noted in Kurminsky Bay—494 $\times$ 10^3 cells L^{-1} and 333 $\times$ 10^3 mg L^{-1}, respectively (Figure 3B).

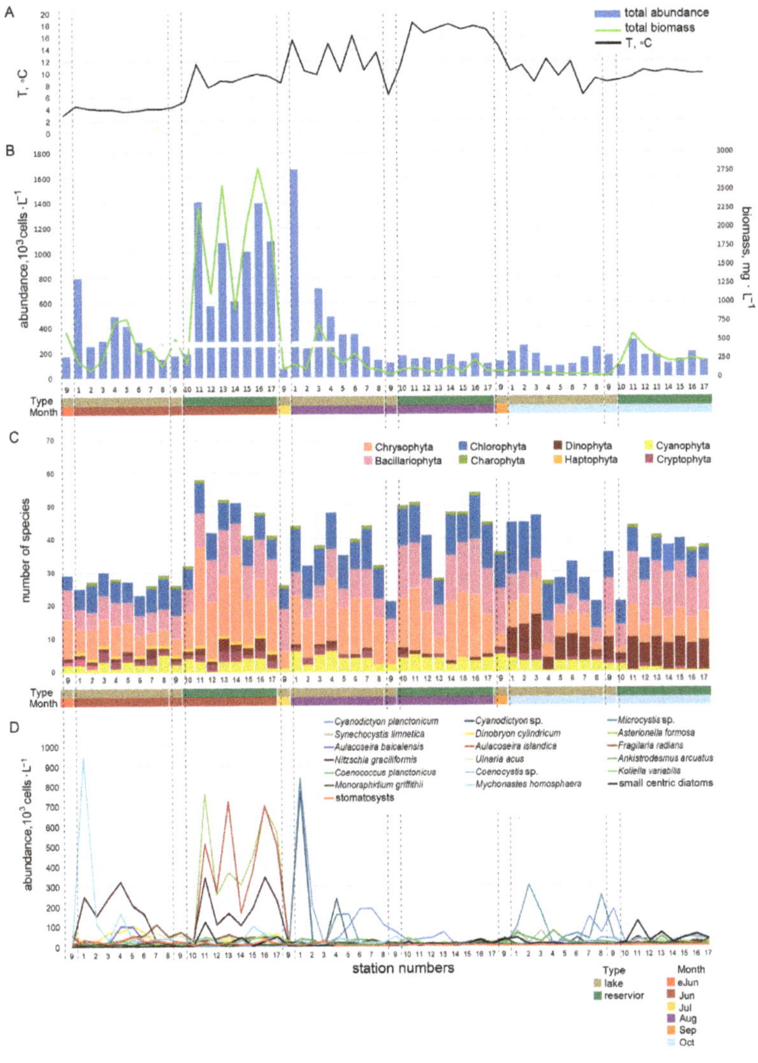

Figure 3. Seasonal dynamics of water temperature (**A**) and phytoplankton structure (**B**–**D**) in Southern Baikal and the Irkutsk Reservoir in 2023 during the period of open water: (**B**)—total abundance and biomass, (**C**)—number of species of different taxonomic groups, (**D**)—abundance of dominant species.

A total of 151 taxa of microalgae were discovered during the observation period (Table S2), 109 in SB and 129 in IR. We expanded the species list of phytoplankton in IR compared to previous data (118 species [32]) due to greater coverage of stations in the water area reservoirs and a thorough study of silica-scaled chrysophytes within the class Chrysophyceae using scanning QUANTA 200 SEM (FEI Company; Hillsboro, OR, USA) and transmission Leo 906 E TEM (Zeiss, Germany) microscopy [36].

Figure 3C shows the change in departmental structure. The proportion of Chrysophyta in SB and IR varied from 43% in spring to 27% in autumn of the total number of species. The Chlorophyta group was represented by 41 taxa, their number increasing from spring (15%) to autumn (34%). The same trend was noted for Bacillariophyta, the number of which increased due to small-celled centric diatoms from 19% in spring and summer to 22% in autumn.

The species richness of Cyanobacteria, both in SB and IR, was not high (9–11%), increasing slightly by summer in well-warmed bays of the reservoir (St. 11, St. 13 and St. 16).

The predominance of species diversity of microalgae Chrysophyta in the IR distinguishes it from other reservoirs, consisting of Bacillariophyta, Chlorophyta and Cyanobacteria. The predominance of these groups in reservoirs of both temperate [42–44] and tropical latitudes [45–47] is typical. An analysis of studies in temperate Russian reservoirs of the Volga River—Rybinsk, Sheksna and Kuibyshev—previously showed a high diversity of Chrysophyta [48], but they were not among the three dominant microalgae in those reservoirs.

Figure 3D shows that SB and IR phytoplankton differed in the range of dominant species across all seasons.

Cluster analysis of the abundance of phytoplankton species (Figure 4A) showed that in all seasons, the communities of IR stations form clusters, separate from the clusters of SB stations.

At the same time, St. 9 (eJ9, J9, Jl9, A9, S9, O9), located near the source of the Angara River (3 km from Listvyanka Village), does not fall into the cluster of reservoir stations at any time. But the profile of St. 10, located in the reservoir at a distance of 26 km from the source, is grouped with the spring profiles of Southern Baikal. These observations indicate that the direct influence of lake waters and the transfer of phytoplankton species from the lake to the reservoir occurred only in the spring.

Figure 4B shows that divisions occurred in community groups due to changes in species, changes in the abundance of the same species and changes in dominant species. The divisions occurred into four large clusters. The first cluster combined spring samples from SB and IR; the second, summer–autumn samples from IR. Summer and autumn SB samples formed two more separate clusters. This may indicate that the phytoplankton of the reservoir develop in relative independence from the phytoplankton of the lake in the summer and autumn periods.

3.3. Correlation of Parameters of the Aquatic Environment with Quantitative Indicators of Phytoplankton and Its Structure

We have shown differences in the degrees of correlation between environmental parameters and quantitative characteristics of phytoplankton in different seasons of the year (Figure 5A). For example, in spring (June), biomass is positively correlated with water temperature, with a coefficient of 0.85, in summer (July), it does not depend on temperature, and in autumn (October), it is again positively correlated, but with a lower correlation coefficient (0.53) than in spring.

During the open water period, environmental parameters had different relationships (Figure 5A). In June, there was a high positive correlation between water transparency and the concentration of phosphates and nitrates, as well as between water temperature and general indicators of community productivity (abundance, biomass). There was a negative correlation between these two groups of parameters. In August and October, the interdependence of these two groups of factors became less pronounced, but was partially preserved.

If we consider the codependency of environmental factors throughout the entire period of open water (Figure 5B), then the concentrations of phosphates and nitrates had a positive correlation. In addition, PO_4^{3-} and NO_3^- concentrations were positively correlated with water clarity. These three measures were negatively correlated with temperature. The silicon concentration, on the contrary, had a strong positive correlation with water temperature. No strong or moderate correlation of phytoplankton quantitative parameters with parameters of the aquatic environment was observed (Figure 5B). Concentrations of nitrites and ammonium in all periods considered were at the detection limits of the methods used, and, accordingly, their connection with the growth of phytoplankton was not identified (Table S1).

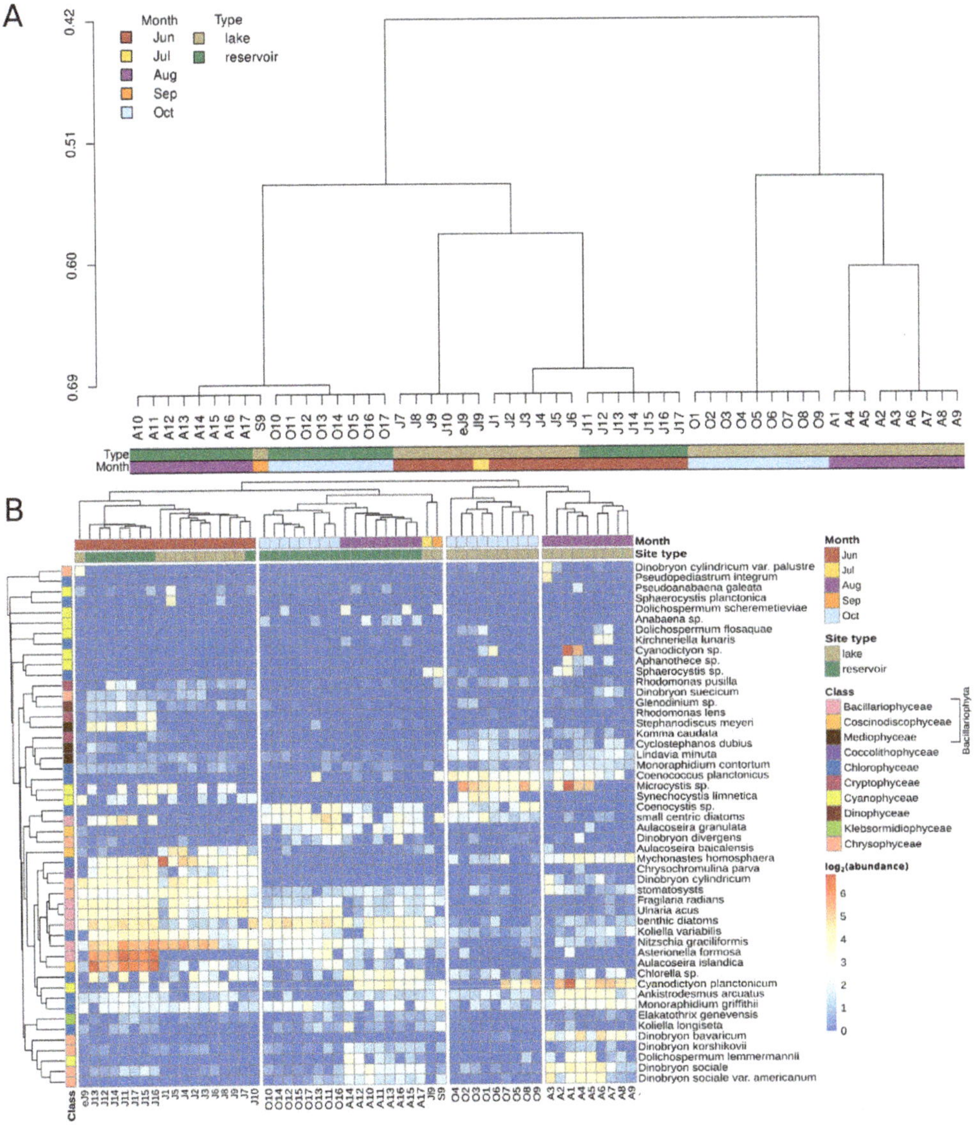

Figure 4. Cluster analysis of species abundance community profiles using affinity propagation (**A**) and heatmap (**B**) of the species abundance profiles generated with a set of the 50 most abundant species. Color annotations below the cluster dendrogram and above the heatmap describe the spatial (**Type**) and temporal (**Month**) categories of communities. Color annotation on the left of heatmap denotes the species taxonomic affiliation at the "Class" level.

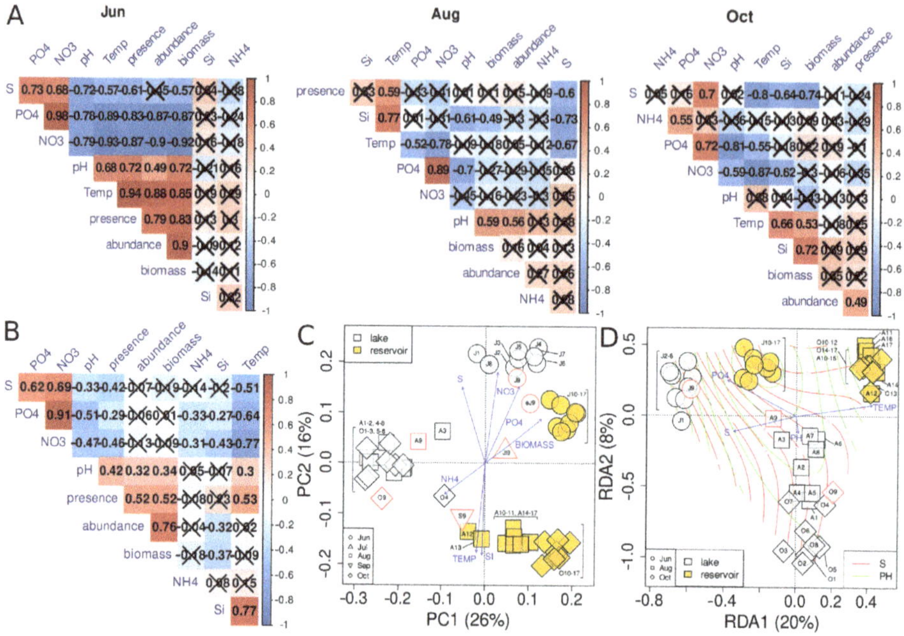

Figure 5. Correlation of environmental parameters (**A**,**B**) and exploratory analysis of environmental parameters and species abundance data (**C**,**D**): (**A**) Correlation of environmental parameters and summary numerical variables by month of sampling. Numerical values are Pearson correlation coefficients with the color legend on the right. Strikeout cells are non-significant correlations ($p > 0.05$). Profiles eJ9, Jl9 and S9 were not analyzed. (**B**) Correlation of environmental parameters and summary numerical variables of all community profiles sampled during 2023. (**C**) Unconstrained ordination of species abundance data using tbPCA. Shape of the point designates the month of sampling, and color denotes the sampling site type: Lake Baikal or Irkutsk Reservoir (Table S1). (**D**) Constrained ordination of species abundance profiles, excluding eJ9, Jl9 and S9, using tbRDA. Color and shape of the points as in Figure 4B. Red and green isolines show the gradient of S and pH, respectively.

The results of the unconstrained ordination of all community profiles (Figure 5C) support the results of the cluster analysis (Figure 4A) and show that in all seasons, IR stations cluster separately from SB stations. The sample taken near the source of the Angara River (St. 9) on 1 June (eJ9) is grouped with samples from IR stations (J10–J17) on June 24–25 (Figure 5C), which may indicate a direct influence of lake waters on the reservoir. In the results of the cluster analysis of the full species profiles (Figure 5D), eJ9 falls into the group J7–10 + Jl9, and this group is somewhat distant from J1–6 (SB)/J11–17. That is, it turns out that in June, the IR profiles are similar to the profiles of Baikal communities with St. 1–6. However, in the remaining months of observation, no such connection was detected.

We used the following continuous environmental variables to build a model of factors influencing the community structure: phosphate, nitrite anion concentrations, concentration of ammonium cation, water transparency, water temperature and pH. The adjusted total explained variance of a community composition matrix was only 28%. By using the "forward selection" approach, a model was generated that included the phosphate anion concentration, water temperature, transparency and pH as significant quantitative model variables (Figure 5D). The adjusted total explained variance of the model was 25.5%. The ordination pattern obtained with the constrained approach was similar to that of the unconstrained technique (Figure 5C).

3.4. Changes in the Concentration of Nutrients and the Growth of Different Groups of Microalgae

If the species composition and abundance of phytoplankton depend on temperature, and temperature does not depend on the presence of microalgae, then interactions with other environmental parameters are more complex for microalgae. Water transparency depends not only on suspended matter but also on the presence of algae themselves in the environment. On the one hand, nutrients are necessary for the growth of algae, and on the other hand, they are susceptible to grazing, and in different ways by different groups of microalgae. Therefore, when interpreting the correlation between the concentration of nutrients and the growth of phytoplankton, it is necessary to consider the species composition of the dominant species.

High concentrations of phosphates (0.015–0.023 mg L^{-1}) and nitrates (0.29–0.41 mg L^{-1}) were observed in the cold (3.6–4.5 °C) SB waters in spring (Table S1). Silicon concentrations during this period varied over a wide range, as a rule, depending on the dominant microalgae in abundance and biomass. For example, at St. 1 and St. 2, higher silicon concentrations were noted (0.52 and 0.53 mg L^{-1}, respectively), with the dominance of Chlorophyta *M. homosphaera*, which does not consume silicon. At St. 4 and St. 5, the drop in silicon concentration (to 0.17 and 0.20 mg L^{-1}, respectively) was associated not only with the growth of Bacillariophyta *A. baicalensis*, *N. gracilliformis*, *F. radians* and *U. acus* but also with the formation of siliceous stomatocysts of the chrysophyte *D. cylindricum*. During the same period, in warmer (7.6–11.5 °C) IR waters, the concentrations of phosphates (0.007–0.016 mg L^{-1}) and nitrates (0.04–0.24 mg L^{-1}) were lower than in SB due to more active growth of microalgae. The silicon concentration also dropped to 0.40 mg L^{-1} (Table S1) as a result of its consumption by the dominant Bacillariophyta *A. formosa*, *A. islandica*, *N. gracilliformis*, *F. radians* and *U. acus* (Figure 3D from [34]).

At SB stations in summer, the temperature in the surface layers of water mainly varied between 10 and 16 °C. The abundance and biomass were dominated by Chrysophyta genus *Dinobryon* and Chlorophyta *M. homosphaera*, which is not involved in silicon consumption, and the growth of which led to a decrease in the concentrations of phosphates (to 0.007 mg L^{-1}) and nitrates (to 0.06 mg L^{-1}). St. 9 (3 km from Listvyanka) differed from other SB stations in the summer, where at the lowest water temperature (6.3 °C), the highest concentrations of nutrients (phosphates—0.023 mg L^{-1} and nitrates—0.35 mg L^{-1}, respectively) and the lowest values of the abundance and biomass of microalgae were observed (according to [36]). Summer water temperatures in the IR reached the highest values for the period of our observations (16.6–18.3 °C), large species of spring Bacillariophyta had low numbers (Figure 3A,D according to [36]) and the silicon concentration had the highest values, up to 0.91 mg L^{-1} for the entire period of our study. The drop in concentrations to the lowest values for the entire period of our study for phosphates (up to 0.008 mg L^{-1}) and nitrates (up to 0.06 mg L^{-1}) (Table S1) was associated with their consumption by small-celled Cyanobacteria (dominated by *C. planctonicum*, *Microcystis* sp., *Dolichospermum lemmermannii*) and Chlorophyta (*Koliella longiseta*, *M. griffithii* and *Chlorella* sp. dominated). The abundance of large-celled species was low, with only *Dinobryon sociale* and *D. sociale* var. *americanum*.

In autumn, water temperature (Figure 3A) and nutrient concentrations (Table S1) were generally comparable between SB and IR. At St. 3 and St. 7, the highest concentrations of phosphates (0.016 and 0.015 mg L^{-1}, respectively) and nitrates (0.30 and 0.29 mg L^{-1}, respectively) were observed; the productivity of microalgae was not high, but the maximum abundance of the small-celled cyanobacterium *C. planctonicum* was noted (at St. 7—144 × 10^3 cells L^{-1}) (Figure 3B). In addition, these two stations had the highest transparency (9.5 and 9 m, respectively). The abundance of phytoplankton in SB, as in the summer, was formed by small-celled Cyanobacteria—*Microcystis* sp., *C. planctonicum*, *Synechocystis limnetica*—and Chlorophyta—*Coenococcus planctonicus*, *Coenocystis* sp., *Lindavia minuta*, which were dominant among Bacillariophyta.

Although silicon concentrations were comparable between SB and IR in autumn, in contrast to SB, large species of Chrysophyta, Bacillariophyta and Dinophyta predominated in IR. Also in IR, a high content of benthic diatoms was observed, the number of which exceeded 50% of the total number of species and amounted to up to 147 × 10^3 cells L^{-1}. At the same time, at all IR stations, including bays, transparency was low, at 1.5–2 m, with the exception of the very first station of the reservoir, where transparency was 7 m. Cyanobacteria were practically absent. The abundance was formed by Chlorophyta—*Coenocystis* sp., *M. griffithii*, *Koliella variabilis*—as well as small centric diatoms *C. dubius*, *C. makarovae*, *Discostella pseudostelligera*, *S. minutulus* and *T. pseudonana*. The abundance of Chrysophyta in IR compared to SB was higher due to *D. divergens*, and Bacillariophyta due to *N. graciliformis*, *A. formosa* and *U. acus*. The IR bays differed in composition, both among themselves and from the central part of the reservoir. In Kurminsky Bay, the dominant species were Bacillariophyta *A. formosa*, *A. granulata*, *N. graciliformis*, *U. acus* and *A. islandica*. In Elovy Bay, Chlorophyta *C. planctonicus* and Bacillariophyta *A. formosa* were dominant. In Ershovsky Bay, *Coenocystis* sp., *N. graciliformis*, *U. acus*, *A. granulata* and Chrysophyta *Dinobryon divergens* were dominant. All bays were dominated by small centric diatoms, the maximum abundance of which was noted in Kurminsky Bay—120 × 10^3 cells L^{-1}.

In autumn, the concentrations of phosphates (0.008–0.012 mg L^{-1}) and nitrates (0.06–0.10 mg L^{-1}) in the bays were not high, as in other parts of the IR, due to their consumption by actively developing microalgae. Silicon concentrations in the bays were comparable to other IR stations and varied from 0.61 to 0.76 mg L^{-1}.

Thus, in autumn, despite similar habitat parameters, the species compositions in SB and IR differed. Small-celled Cyanobacteria and Chlorophyta continued to develop in SB, while large-celled Chrysophyta, Bacillariophyta and Dinophyta species predominated in IR, which was reflected in higher phytoplankton biomass values in IR compared to SB (Figure 3D).

3.5. Comparison of the Obtained Data with Previous Studies

As our studies showed, the species compositions of phytoplankton in SB and IR throughout all three seasons had both similarities and differences (Figure 6). Studies conducted in the Irkutsk Reservoir previously [30–32] also showed significant differences in both quantitative indicators and species structure of phytoplankton throughout the year. Comparison of the species structure of phytoplankton in SB and IR in the same seasons revealed differences in the growth of the dominant Baikal species. As shown above, their distribution is influenced by a complex of factors, the most significant of which is water temperature. A comparison of the phytoplankton communities of SB and IR (not including bays) in different seasons showed that the largest number of common species is observed in summer (August), when the difference in water temperature evens out. The number of common species is lowest in spring and autumn (Figure 6).

In spring, these differences are most pronounced in SB, with a spring peak in the growth of cold-loving Bacillariophyta species, some of which do not develop in IR. At the same time, a peak in the growth of heat-loving Bacillariophyta species not found in SB was observed in IR. The spring growth of Bacillariophyta in IR was also shown previously [32]. However, analysis of seasonal dynamics in IR in different years indicates significant variability may occur in spring [32], summer [30,31] and autumn [30]. As a rule, the growth of large species of Bacillariophyta in spring and early summer is typical for many reservoirs in both temperate [42,43,49] and tropical zones [50].

Previous studies [30–32] showed that in spring, *A. baicalensis* is capable of developing in the waters of the reservoir, and its abundance depends on its abundance in Lake Baikal. In our study, *A. baicalensis* did not reach great growth in Baikal, and it did not spread far with the water flow in IR. If we evaluate the level of phytoplankton growth in 2023 on the productivity scale [51], then in SB, the year was unproductive, since the biomass did not exceed 500 × 10^3 mg L^{-1}. At the same time, IR should be classified as highly productive, since the biomass exceeded 1600 × 10^3 mg L^{-1}. This high productivity was ensured by the absolute dominance of three diatom species—*A. formosa*, *N. graciliformis* and *A. islandica*. At St. 11, these three species accounted for 80% of the total abundance. *A. islandica* is an oligotrophic cold-water species that replaced *A. baicalensis*, developing maximally at temperatures of 7.66–11.55 °C (Sts 11–17). Also, *S. meyeri* developed at these stations, reaching its maximum values during the observation period at the same temperatures (Figure 3D), which corresponds to its autecology.

Figure 6. Venn diagram. Species composition of phytoplankton in SB and IR during different seasons in 2023.

As was shown earlier, the cosmopolitan species *A. formosa* has a temperature optimum of 10–15 °C; temperatures above this inhibit the growth of this species [30], making it a typical representative of summer phytoplankton for Baikal [52]. In our study, the peak of its development occurred at the end of June, falling within the range of these temperatures in the warmer IR waters of Sts 11–17 at temperatures of 7.66–11.55 °C.

It was previously noted that in the spring of 1980, *N. graciliformis* was actively developing in SB, reaching 2×10^5 cells L^{-1}, and remained dominant in IR. In summer, the growth of *N. graciliformis* began to decline. It was replaced by *A. formosa* and *A. islandica* [31]. In our study, *N. graciliformis* reaches its peak growth in spring in the IR. At the same time, *A. formosa* and *A. islandica* were its subdominants, remaining in the summer phytoplankton in insignificant quantities (less than 10×10^3 cells L^{-1}).

In summer, species diversity increases significantly in both SB and IR and the number of common species increases (Figure 6). The role of small-celled Cyanobacteria and Chlorophyta is increasing. At the same time, Cyanobacteria *C. planctonicum*, colonial cyanobacteria typical of surface waters of mesotrophic lakes in the summer, dominate in abundance. This species of Cyanobacteria was previously shown to be part of the dominant group in the coastal areas of SB [53], together with two representatives of summer phytoplankton—*Microcystis* sp. and *D. lemmermannii*. These species, while actively developing in SB, significantly reduced their quantitative indicators in IR. Thus, *C. planctonicum* had the maximum abundance in SB at St. 1 (840×10^3 cells L^{-1}), at other stations its abundance was significantly lower (Figure 3D) and at IR stations it did not exceed 68×10^3 cells L^{-1}. Importantly, the active growth in summer in SB of small-celled Cyanobacteria and Chlorophyta, which have the characteristics of R-strategists, was noted previously [52,53]. The total biomass of this ultrananoplanktonic group of algae, which includes Chlorophyta, Chlorococcales and Cyanobacteria with a size of no more than 4 microns, first described by O.M. Kozhova (1964) [30] as "green bacteria", during the year varied from 1×10^3 mg L^{-1} to 238×10^3 mg L^{-1} with a maximum in July–September. Changes in the community structure towards small cells under conditions of summer warming of waters, as a rule, indicate ecological instability [43,49,54]. However, since in our studies, the quantitative indicators of this group of planktonic algae were low and did not exceed 60×10^3 mg L^{-1} (among which *M. homosphaera* predominated), we cannot consider them indicators of deterioration in water quality. Among Chlorophyta, as before, *Koliella longiseta* and *M. griffithii* predominated [32]. In the present study, the summer growth of *Chlorella* sp. and the number of large-celled species has decreased significantly, leaving only *Dinobryon sociale* and *D. sociale* var. *americanum*. The total abundance and biomass of microalgae in IR, including bays, decreased significantly in summer from 186×10^3 cells L^{-1} to 310×10^3 cells L^{-1} and 41×10^3 mg L^{-1} to 140×10^3 mg L^{-1}, respectively [36].

In autumn in IR, we observed the growth of the phytoplankton community, previously characteristic of the summer period [32]. In addition, we recorded a second peak in the growth of large-celled microalgae, mainly Bacillariophyta. The quantitative indicators of IR did not exceed those previously discovered. As before, Cyanobacteria species did not reach high values, unlike other reservoirs of the Angara cascade [31,55,56].

4. Conclusions

For the first time, we carried out simultaneous studies of the southern part of Lake Baikal and the Irkutsk Reservoir during three seasons of the year, which made it possible to identify the dynamics of phytoplankton growth, assess the influence of environmental parameters on the distribution and growth of qualitative and quantitative indicators of phytoplankton and trace the relationship between the drop in nutrient concentrations and the levels of growth of various microalgae. During the most productive period in the growth of phytoplankton in the reservoir, in spring (June), the main abundance and biomass are created by diatoms. During this season, water temperature has a major influence on the spatial distribution of algae. The most sensitive to changes in the temperature regime in the phytoplankton community were silica-scaled chrysophytes and diatoms. In the summer and autumn periods, the flow of water with a low content of phosphates and nitrates from the oligotrophic lake into the reservoir limited the growth of Cyanobacteria and small-celled Chlorophyta; they were present both in the lake and in the reservoir, but their numbers and biomass were not high. The influence of Southern Baikal on the Irkutsk Reservoir primarily lies in the direct transfer of water from the lake to the Angara River, on which a dam was built and a reservoir was formed. The high diversity of species, both cold-water and those with broader temperature preferences, forms a spatiotemporal structure of the reservoir's phytoplankton that differs from those of other temperate reservoirs.

Supplementary Materials: The following supporting information can be downloaded at https://www.mdpi.com/article/10.3390/w16223284/s1, Table S1. Water parameters at stations in Southern Baikal and the Irkutsk Reservoir in different seasons in 2023. For localization of stations, see Figure 1. Table S2. The species composition of phytoplankton in the Southern Baikal (SB) and the Irkutsk Reservoir (IR) during the open water period in 2023 (station numbering corresponds to Figure 1 and Table S1). Data for June [34] and August [36] were published earlier.

Author Contributions: A.F., Y.L. and A.B., literature search, interpretation of the results, writing of the first version of the manuscript; Y.G., statistical analysis and interpretation of the results; L.T., A.F. and V.B., light microscopy, counting of the phytoplankton, determination of diatoms' proportion in the abundance and biomass of phytoplankton; A.F. and A.B., electron microscopy, identification of scaled chrysophytes; M.S., hydrochemical analysis; A.M., D.H., V.B. and M.N., sampling, measurements during fieldwork, preparation of samples for laboratory research; Y.L., writing—review and editing. All authors have read and agreed to the published version of the manuscript.

Funding: This work was performed with financial support from the Russian Science Foundation, project No. 23-14-00028 "Communities of microeukaryotes in Angara Cascade Reservoirs" https://rscf.ru/en/project/23-14-00028/ (accessed on 24 September 2024).

Data Availability Statement: The original contributions presented in the study are included in the article, further inquiries can be directed to the corresponding author.

Acknowledgments: The authors express their gratitude to Daria Petrova for her valuable comments during the preparation of the manuscript. The microscopy studies were performed at the Electron Microscopy Center of the Shared Research Facilities "Ultramicroanalysis" of the Limnological Institute, http://www.lin.irk.ru/copp/ (accessed on 4 September 2024).

Conflicts of Interest: The authors declare no conflicts of interest.

References

1. Hällfors, H.; Uusitalo, L. Early Warning Indicators: Phytoplankton. Indicators of the Good Environmental Status of Food Webs in the Baltic Sea. Status of Food Webs in the Baltic Sea GES-REG Project Final Report on Food Web Indicators, September 2013. 2013, pp. 52–64. Available online: http://gesreg.msi.ttu.ee/fi/tulokset (accessed on 24 September 2024).
2. Tamelander, T.; Heiskanen, A.-S. Effects of spring bloom phytoplankton dynamics and hydrography on the composition of settling material in the coastal northern Baltic Sea. *J. Mar. Syst.* **2004**, *52*, 217–234. [CrossRef]
3. Spilling, K.; Lindström, M. Phytoplankton life cycle transformations lead to species-specific effects on sediment processes in the Baltic Sea. *Cont. Shelf Res.* **2008**, *28*, 2488–2495. [CrossRef]
4. Cloern, J.E.; Foster, S.Q.; Kleckner, A.E. Phytoplankton primary production in the world's estuarine-coastal ecosystems. *Biogeosciences* **2014**, *11*, 2477–2501. [CrossRef]
5. Hällfors, H.; Backer, H.; Leppänen, J.-M.; Hällfors, S.; Hällfors, G.; Kuosa, H. The northern Baltic Sea phytoplankton communities in 1903–1911 and 1993–2005: A comparison of historical and modern species data. *Hydrobiologia* **2013**, *707*, 109–133. [CrossRef]
6. Salmaso, N.; Tolotti, M. Phytoplankton and anthropogenic changes in pelagic environments. *Hydrobiologia* **2021**, *848*, 251–284. [CrossRef]
7. Viitasalo, M.; Bonsdorf, E. Global climate change and the Baltic Sea ecosystem: Direct and indirect effects on species, communities and ecosystem functioning. *Earth Syst. Dyn.* **2022**, *13*, 711–747. [CrossRef]
8. Vidal, T.; Calado, A.J.; Moita, M.T.; Cunha, M.R. Phytoplankton dynamics in relation to seasonal variability and upwelling and relaxation patterns at the mouth of Ria de Aveiro (West Iberian Margin) over a four-year period. *PLoS ONE* **2017**, *12*, e0177237. [CrossRef]
9. Assmy, P.; Kvernvik, A.C.; Hop, H.; Hoppe, C.J.M.; Chierici, M.D.; David, T.; Duarte, P.; Fransson, A.; García, L.M.; Patuła, W.; et al. Seasonal plankton dynamics in Kongsfjorden during two years of contrasting environmental conditions. *Prog. Oceanogr.* **2023**, *213*, 102996. [CrossRef]
10. Xu, S.; Xiao, Y.; Youwei, X.; Su, L.; Cai, Y.; Zhanhui, Q.; Liu, Y.; Chen, Z.; Lakshmikandan, M. Effects of seasonal variations and environmental factors on phytoplankton community structure and abundance in Beibu Gulf, China. *Ocean Coast. Manag.* **2024**, *248*, 106982. [CrossRef]
11. Suikkanen, S.; Pulina, S.; Engström-Öst, J.; Lehtiniemi, M.; Lehtinen, S.; Brutemark, A. Climate change and eutrophication induced shifts in northern summer plankton communities. *PLoS ONE* **2013**, *8*, e66475. [CrossRef]
12. Jeppesen, E.; Meerhoff, M.; Davidson, T.A.; Trolle, D.; Sondergaard, M.; Lauridsen, T.L.; Beklioğlu, M.; Brucet, S.; Volta, P.; Gonzalez-Bergonzoni, I.; et al. Climate change impacts on lakes: An integrated ecological perspective based on a multi-faceted approach, with special focus on shallow lakes. *Int. J. Limnol.* **2014**, *73*, 88–111. [CrossRef]
13. Henson, S.A.; Beaulieu, C.; Ilyina, T.; John, J.G.; Long, M.; Seferian, R.; Tjiputra, J.; Sarmiento, J.L. Rapid emergence of climate change in environmental drivers of marine ecosystems. *Nat. Commun.* **2017**, *8*, 14682. [CrossRef] [PubMed]
14. Saraiva, S.; Meier, H.E.M.; Andersson, H.; Höglund, A.; Dieterich, C.; Gröger, M.; Hordoir, R.; Eilola, K. Baltic Sea ecosystem response to various nutrient load scenarios in present and future climates. *Clim. Dyn.* **2018**, *52*, 3369–3387. [CrossRef]
15. Salo, T.; Mattila, J.; Eklöf, J. Long-term warming affects ecosystem functioning through species turnover and intraspecific trait variation. *Oikos* **2020**, *129*, 283–295. [CrossRef]
16. Izmest'eva, L.R.; Moore, M.V.; Hampton, S.E.; Ferwerda, C.J.; Gray, D.K.; Woo, K.H.; Pislegina, H.V.; Krashchuk, L.S.; Shimaraeva, S.V.; Silow, E.A. Lake-wide physical and biological trends associated with warming in Lake Baikal. *J. Great Lakes Res.* **2016**, *42*, 6–17. [CrossRef]
17. Reavie, E.D.; Cai, M.; Twiss, M.R.; Carrick, H.J.; Davis, T.W.; Johengen, T.H.; Gossiaux, D.; Smith, D.E.; Palladino, D.; Burtner, A.; et al. Winter–spring diatom production in Lake Erie is an important driver of summer hypoxia. *J. Great Lakes Res.* **2016**, *42*, 608–618. [CrossRef]
18. Uusitalo, L.; Fleming-Lehtinen, V.; Hällfors, H.; Jaanus, A.; Hällfors, S.; London, L. A novel approach for estimating phytoplankton biodiversity. *ICES J. Mar. Sci.* **2013**, *70*, 408–417. [CrossRef]
19. Finni, T.; Kononen, K.; Olsonen, R.; Wallström, K. The history of cyanobacterial blooms in the Baltic Sea. *Ambio* **2001**, *30*, 172–178. [CrossRef] [PubMed]

20. Carstensen, J.; Henriksen, P.; Heiskanen, A.S. Summer algal blooms in shallow estuaries: Definition, mechanisms, and link to eutrophication. *Limnol. Oceanogr.* **2007**, *52*, 370–384. [CrossRef]
21. Kahru, M.; Elmgren, R. Multidecadal time series of satellite-detected accumulations of cyanobacteria in the Baltic Sea. *Biogeosciences* **2014**, *11*, 3619–3633. [CrossRef]
22. Jurgensone, I.; Carstensen, J.; Ikauniece, A.; Kalveka, B. Long-term changes and controlling factors of phytoplankton community in the Gulf of Riga (Baltic Sea). *Estuaries Coasts* **2011**, *34*, 1205–1219. [CrossRef]
23. Forsström, L.; Sorvari, S.; Korhola, A.; Rautio, M. Seasonality of phytoplankton in subarctic Lake Saanajärvi in NW Finnish Lapland. *Polar Biol.* **2005**, *28*, 846–861. [CrossRef]
24. Olenina, I.; Wasmund, N.; Hajdu, S.; Jurgensone, I.; Gromisz, S.; Kownacka, J.; Toming, K.; Vaici, D.; Olenin, S. Assessing impacts of invasive phytoplankton: The Baltic Sea case. *Mar. Pollut. Bull.* **2010**, *60*, 1691–1700. [CrossRef]
25. Datsenko, Y.S. Features and differences of abiotic components in lakes and reservoirs ecosystems (review). *Geoecology Russ. J. Appl. Ecol.* **2022**, *1*, 39–47. (In Russian) [CrossRef]
26. Atazadeh, E.; Gell, P.; Mill, K.; Barton, A.; Newall, P. Community structure and ecological responses to hydrological changes in benthic algal assemblages in a regulated river: Application of algal metrics and multivariate techniques in river management. *Environ. Sci. Pollut. Res.* **2021**, *28*, 39805–39825. [CrossRef]
27. Wu, Q.; Qiuhua, L.; Luo, H.; Chen, Q.; Chen, H.; Dong, Y.; Li, S. Comparison in phytoplankton diversity-productivity-community stability between river-type reservoir and lake-type reservoir. *J. Oceanol. Limnol.* **2022**, *40*, 1485–1507. [CrossRef]
28. Karnaukhova, G.A. Hydrochemistry of the Angara and reservoirs of the Angara cascade. *Water Resour.* **2008**, *35*, 71–79. (In Russian) [CrossRef]
29. Grachev, M.A. *On the Current State of Lake Baikal Ecosystem*; Sibirskoe Otdelenie Rossiyskoy Akademii Nauk: Novosibirsk, Russia, 2002; pp. 1–153. (In Russian)
30. Kozhova, O.M. *Fitoplankton Irkutskogo Vodokhranilishcha (Phytoplankton of the Irkutsk Reservoir)*; Nauka: Moscow, Russia, 1964; pp. 41–114. (In Russian)
31. Vorobyova, S.S. *Fitoplankton Vodoemov Angary (Phytoplankton of Water Bodies Formed on the Angara River)*; Nauka: Novosibirsk, Russia, 1995; pp. 1–126. (In Russian)
32. Popovskaya, G.I.; Firsova, A.D.; Bessudova, A.Y.; Sakirko, M.V.; Suturin, A.N.; Likhoshway, Y.V. Phytoplankton of the Irkutsk Reservoir as an indicator of water quality. *Oceanol. Hydrobiol. Stud.* **2012**, *2*, 29–38. [CrossRef]
33. Popovskaya, G.I.; Genkal, S.I.; Likhoshway, Y.V. *Diatoms of the Plankton of Lake Baikal: Atlas and Key*; Nauka: Novosibirsk, Russia, 2016; pp. 1–180.
34. Firsova, A.; Galachyants, Y.; Bessudova, A.; Titova, L.; Sakirko, M.; Marchenkov, A.; Hilkhanova, D.; Nalimova, M.; Buzevich, V.; Mikhailov, I.; et al. Environmental factors affecting distribution and diversity of phytoplankton in the Irkutsk Reservoir ecosystem in June 2023. *Diversity* **2023**, *15*, 1070. [CrossRef]
35. Bessudova, A.; Galachyants, Y.; Firsova, A.; Hilkhanova, D.; Marchenkov, A.; Nalimova, M.; Sakirko, M.; Likhoshway, Y. Seasonal dynamics of the silica-scaled chrysophytes as potential markers of climate change in natural model: Deep cold lake–shallow warmer reservoir. *Sustainability* **2024**, *16*, 7299. [CrossRef]
36. Firsova, A.; Galachyants, Y.; Bessudova, A.; Mikhailov, I.; Titova, L.; Marchenkov, A.; Hilkhanova, D.; Nalimova, M.; Buzevich, V.; Likhoshway, Y. Summer phytoplankton species composition and abundance in the southern basin of Lake Baikal and Irkutsk Reservoir. *Limnol. Freshw. Biol.* **2023**, *6*, 204–228. [CrossRef]
37. Taiyun, W.; Simko, V.R. Package 'Corrplot': Visualization of a Correlation Matrix (Version 0.92). 2021. Available online: https://github.com/taiyun/corrplot (accessed on 29 September 2023).
38. Harrell, F.E., Jr. Hmisc: Harrell Miscellaneous. R Package Version 4.7-1. 2022. Available online: https://CRAN.R-project.org/package=Hmisc (accessed on 24 September 2024).
39. Oksanen, J.; Simpson, G.L.; Blanchet, G.F.; Kindt, R.; Legendre, P.; Minchin, P.R.; O'Hara, R.B.; Solymos, P.; Stevens, M.H.H.; Szoecs, E.; et al. Vegan: Community Ecology Package. R Package Version 2.5-6. 2022. Available online: https://CRAN.R-project.org/package=vegan (accessed on 5 August 2024).
40. Kolde, R. pheatmap: Pretty Heatmaps_. R package Version 1.0.12. 2019. Available online: https://CRAN.R-project.org/package=pheatmap (accessed on 24 September 2024).
41. Bodenhofer, U.; Kothmeier, A.; Hochreiter, S. APCluster: An R package for affinity propagation clustering. *Bioinformatics* **2011**, *27*, 2463–2464. [CrossRef] [PubMed]
42. Belyaeva, P.G. Spatial-temporal changes of phytoplankton of Kama Reservoir. *Proc. Samara Sci. Cent. Russ. Acad. Sci.* **2015**, *17*, 733–738. (In Russian)
43. *Struktura i Funktsionirovaniem Ekosistemy Rybinskogo Vodokhranilishcha v Nachale XXI veka (Structure and Functioning of the Ecosystem of the Rybinsk Reservoir at the Beginning of the 21st Century)*; Rossiyskaya Akademia Nauk: Moscow, Russia, 2018; pp. 1–456. (In Russian)
44. Gvozdareva, M.; Lubina, O.; Melnikova, A. The development of plankton communities in the Kuibyshev Reservoir in the zone of influence of the Cheboksary hydroelectric power plant. *Hydroecology* **2021**, *3*, 23–29. (In Russian) [CrossRef]
45. Liu, C.; Sun, X.; Su, L.; Cai, J.; Zhang, L.; Guo, L. Assessment of phytoplankton community structure and water quality in the Hongmen Reservoir. *Water Qual. Res. J.* **2021**, *56*, 19–30. [CrossRef]

46. Tran, T.-H.-Y.; Trang, L.T.; Dang, P.D.; Tran, T.T.; Nguyen, T.V.; Nguyen-Ngoc, L.; Pham, T.L. Seasonal variation of phytoplankton functional groups in Tuyen Lam Reservoir, Central Highlands, Vietnam. *Dalat Univ. J. Sci.* **2022**, *13*, 25–35. [CrossRef]
47. Nwonumara, G.N.; Elebe, F.A.; Nwibo, O.D. The physico-chemical variables and phytoplankton of Ufiobodo and Ebonyi Reservoirs, Ebonyi State, Nigeria. *Zoologist* **2023**, *21*, 41–48. [CrossRef]
48. Korneva, L.G.; Solovyeva, V.V. Golden algae (Chrysophyta) in plankton of the Volga River reservoirs: Taxonomic structure, dynamics of diversity, and abundance. *Inland Water Biol.* **2017**, *10*, 168–175. [CrossRef]
49. Korneva, L.G. *Phytoplankton of Reservoirs of the Volga Basin*; Kostroma Printing House: Kostroma, Russia, 2015; p. 284.
50. Ochieng, B.; Mbao, E.O.; Zhang, Z.; Shi, L.; Liu, Q. Phytoplankton community structure of Tang-Pu Reservoir: Status and ecological assessment in relation to physicochemical variability. *Environ. Monit. Assess* **2022**, *194*, 382. [CrossRef]
51. Popovskaya, G.I. Phytoplankton dynamics of pelagial of Baikal (1964–1974). In *Biologicheskaya Produktivnost' Pelagiali Baykala i yeye Izmenchivost' [Biological Productivity of the Pelagial of Baikal and Its Variability]*; Nauka: Novosibirsk, Russia, 1977; pp. 5–39. (In Russian)
52. Izmest'eva, L.R.; Kozhova, O.M. *Dolgosrochnoye Prognozirovaniye Sostoyaniya Ekosistem (Long-Term Forecasting of Ecosystem Conditions)*; Nauka: Novosibirsk, Russia, 1988; pp. 1–235. (In Russian)
53. Malashenkov, D.V.; Mosharova, I.V.; Ilinskiy, V.V.; Mosharov, S.A. Use of phytoplankton functional classification and microbiological parameters for environmental assessment of coastal waters of Southern Baikal. *Inland Water Biol.* **2022**, *15*, 3–13. [CrossRef]
54. Sakharova, E.G.; Korneva, L.G. Phytoplankton in the mouth area of the Rybinsk Reservoir tributary. *Inland Water Biol.* **2023**, *16*, 61–66. [CrossRef]
55. Vorobyova, S.S. Phytoplankton of the Ust'-Ilimsk Reservoir. In *Biologiya Ust'Ilimskogo Vodokhranilishcha (Biology of the Ust'-Ilimsk Reservoir)*; Nauka: Novosibirsk, Russia, 1987; pp. 8–75. (In Russian)
56. Kozhova, O.M.; Shirobokova, N.P. Phytoplankton of the Bratsk Reservoir and prediction of its state. In *Dolgosrochnoe Prognozirovanie Sostoyaniya Ekosistem (Long-Term Prognosis of Ecosystem State)*; Nauka: Novosibirsk, Russia, 1988; pp. 69–92. (In Russian)

water

MDPI

Article

Assessment of the Health Status of Whitefish (*Coregonus lavaretus* Linnaeus, 1758) and the Quality of Its Habitat in Lake Sevan (Armenia) Using a Multi-Biomarker Approach

Hranush Melkonyan [1,2], Grigorii Chuiko [3], Nelli Barseghyan [1], Tigran Vardanyan [1], Evelina Ghukasyan [1], Hripsime Kobelyan [1] and Bardukh Gabrielyan [1,*]

[1] Scientific Center of Zoology and Hydroecology NAS RA, 7 P. Sevak Str., Yerevan 0014, Armenia; hranush.melkonyan@rau.am (H.M.); nelli.barseghyan@yahoo.com (N.B.); vardtigran@mail.ru (T.V.); e_ghukasyan@yahoo.com (E.G.); hripsimekobelyan@mail.ru (H.K.)

[2] Institute of Biomedicine and Pharmacy, Russian-Armenian University, 123 H. Emin Str., Yerevan 0051, Armenia

[3] Papanin Institute for Biology of Inland Waters, Russian Academy of Sciences, Yaroslavl 152742, Russia; gchuiko@ibiw.ru

* Correspondence: gabrielb@sci.am

Abstract: Lake Sevan is a freshwater reservoir in the Caucasus region. Since the first half of the 20th century, the lake has undergone significant changes caused by human activity and anthropogenic pressure. To identify the current ecological state of two bays—Lchashen and Lichk—located in different parts of the lake, a study was conducted in 2022–2023 using a multi-biomarker approach. For this purpose, biomarkers for assessing the health status of fish and the quality of their living conditions were used such as the activity of acetylcholinesterase (AChE) in the brain, glutathione-S-transferase (GST), superoxide dismutase (SOD), catalase (CAT), and the concentration of reduced glutathione (RGS) and malondialdehyde (MDA) in the liver and gills of the whitefish (*Coregonus lavaretus* Linnaeus, 1758). In addition, hydrochemical and ichthyological analyses were conducted. This study demonstrated seasonal dynamics for all biomarkers. Comparative analysis of biomarkers and hydrochemical and ichthyological data showed that the whitefish in Lchashen Bay is in worse health, and its living conditions there are less favorable than in Lichk Bay.

Keywords: fish health; habitat conditions; assessment; multi-biomarker approach; whitefish; condition factor

1. Introduction

Anthropogenic pollution of aquatic ecosystems globally is one of the pressing environmental problems of modern society. Pollution of water bodies disrupts the functioning of their ecosystems and related ecosystem services. In particular, wastewater from the food processing industry and municipal and agricultural wastewater cause an additional flow of easily oxidized organic matter and biogenic elements into water bodies, which leads to an increase in their trophicity and saprobity. This, in turn, can cause a decrease in the concentration of dissolved oxygen leading to the creation of hypoxia or anoxia and deterioration of the living conditions of aquatic organisms. The discharge of industrial wastewater containing heavy metals and inorganic and organic toxic compounds cause pathological disorders in the body of aquatic animals and their death. All this leads to a change in the structure of aquatic communities and ecosystems [1–6].

Lake Sevan, one of the largest freshwater high-mountain lakes in the world, is the largest freshwater reservoir in the Caucasus region. It consists of two parts, Small and Big Sevan, which were formed in the late Miocene and then significantly transformed in the

Pleistocene and Holocene [7]. The ecosystem of Lake Sevan is a complex object of great natural and socio-economic significance representing a strategic source of drinking water, providing water for recreation, irrigation, hydropower generation, and fish resources [8,9].

Since the first half of the 20th century, the lake has experienced significant changes caused by human activities: declining water levels, loss of biodiversity, habitat degradation, intensive nutrient input from the catchment area, and overfishing. As a result, since the mid-20th century, the oligotrophic lake has turned into a mesotrophic one [8].

Bays of Lake Sevan are one of the main habitats of different fishes. Young whitefish (*Coregonus lavareus*) mainly feed in the bays of the lake, the Lichk and Lchashen Bays also serve as spawning grounds for whitefish. The whitefish was introduced into the lake in the late 1920s and 1930s and became the main commercial species of the lake by the 1960 [10,11]. While they have been the main pelagic consumer of zooplankton [12], the whitefish have large feeding plasticity and can feed on any available prey [13].

However, the bays are subject to anthropogenic input associated with industrial activity and municipal and agricultural wastewater [14], and the whitefish (*Coregonus lavaretus*) belonging to the Salmonidae family is sensitive to changes in the water quality. For example, the feeding of Atlantic salmon may be reduced to as low as 60% at an oxygen level around 40% saturation [15]. Changes in the biochemical parameters of fish may be early signs of deteriorated feeding conditions long before the irreversible changes are shown at the population level [16]. For example, the activity of acetylcholinesterase (AChE) is a widely used method for monitoring pollution, mainly due to its high sensitivity to anticholinergic chemicals such as organophosphate pesticides and carbamates [17]. AChE has the ability to stoichiometrically bind organophosphorus (OP) and carbamate compounds (Cs), which leads to irreversible inhibition of ChE activity, thereby causing disruption of the normal functioning of the organism, ultimately leading to its death [17,18]. Moreover, there are data of AChE activity inhibition via cyanotoxins [19]. In this connection, a decrease in cholinesterase activity in animal tissues is a specific and long-term biomarker of fish being poisoned with these compounds [20]. On the other hand, it has been shown that the activity of AChE in the fish brain can increase during acute and chronic stress, induced by different factors [21].

At the molecular biological level, the damaging effect of xenobiotics is caused by excessive production of reactive oxygen species (ROS) in living organisms. Under normal conditions, ROS are produced in small quantities as a by-product of metabolism. ROS are strong oxidizers and extremely active compounds that destroy submolecular cellular structures and functional molecules. At the same time, the intensity of lipid peroxidation (LPO) and oxidative damage to DNA and proteins increases and the activity of the antioxidant system (AOS) is changed [22]. Increased accumulation of lipid peroxidation products, in particular malondialdehyde (MDA), is one of the signs indicating oxidative stress and the presence of active oxygen species in the cell. The cellular antioxidant defense system neutralizes the damaging effects of ROS. Important elements of AOS are the enzymes catalase (CAT), superoxide dismutase (SOD), glutathione reductase (GSH), glutathione peroxidase (GPO), and glutathione transferase (GST).

Hence, the aim of this work is to evaluate the health of whitefish and its habitat conditions in the two bays of Lake Sevan (Lichk and Lchashen) via biochemical methods (biomarkers of oxidative stress, AChE, and LPO) using a multi-biomarker approach. In parallel, ichthyological investigations were conducted to evaluate the changes in the Condition Factor (CF) of fish which is indicative of the fish feeding state in the bays.

2. Materials and Methods

2.1. Study Area and Fish Sampling

The studies were conducted on Lake Sevan (Armenia), located at an altitude of about 1900 m above sea level and consisting of two morphometrically different parts: Small Sevan (SS) in the northwest and Big Sevan (BS) in the southeast, connected by a narrow strait (Figure 1). Samples were collected in May, August, and October of 2022 and 2023 years at two stations located along the eastern coast of the lake at a distance of 30 m from the shore: one in the northern part of Small Sevan in Lchashen Bay (40°30′45.0″ N 44°57′25.3″ E) and the other in the southern part of Big Sevan in Lichk Bay (40°11′12.4″ N 45°14′39.3″ E). Both bays are ecosystems with rich biodiversity due to their shallow depth and unique hydrological regime. Due to the relatively shallow depth compared to other parts of the lake and the direct influence of runoff from the drainage areas, the most intensive processing of nutrients and energy transfer occurs in these bays. The Lichk Bay receives the input of two rivers, while in Lchashen Bay, the water is mostly stagnant due to the absence of inflowing rivers. Nevertheless, both bays are under the influence of anthropogenic loads associated with industrial activity and domestic and agricultural wastewater [14].

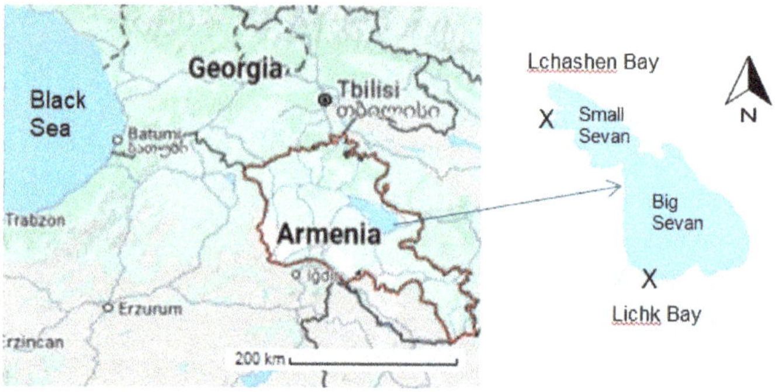

Figure 1. Location of sampling sites in Lchashen and Lichk Bays of Lake Sevan (Armenia).

Whitefish (*Coregonus lavaretus* Linnaeus, 1758) were caught using gill nets (length 100 m, width 3 and 5 m, mesh size 45 × 45 mm), which were set in the previous evening at 19:00–20:00 and removed around 06:00–07:00 the following morning. The total number of whitefish studied was 120 individuals—10 specimens in each season in each bay.

2.2. Ichthyological Analysis

The fish were taken out of the nets, and the length to the tips of the caudal fins and the weight of whole fish were measured as quickly as possible. The fish were killed via cervical transection, dissected, and sexed. Biometric and morphological analyses were carried out using standard ichthyological methods [23–26]. Age structure was determined through counting annual rings on a scale [24,26]. Size structure was identified through measuring the total length. The total body weight (W) was recorded. The sex structure—sex and the level of development of gonads (according to a 6-grade system) were identified [24]. Food components were investigated according to a guide for the investigation of fish feeding [27,28].

CF was calculated according to the following formula [27,28]:

$$CF = (\text{whole body weight, g}/\text{total length}^3\text{, cm}) \times 100$$

2.3. Biochemical Analysis

Immediately after biometric analyses, fish were placed on ice and dissected: the brain, liver, and gills were removed, weighed, washed in a chilled 0.1 M phosphate buffer with pH 7.5, and dried on filter paper. The samples were placed in a sealed bag and stored at a temperature of no more than −86 °C until biochemical analysis was performed. Each tissue sample was homogenized in ice-cold phosphate buffer pH 7.5 at a 1:5 ratio (w/v) for enzyme activity assays (1 g tissue in 5 mL buffer) and a 1:1 ratio for reduced glutathione (GSH) assay (1 g tissue in 1 mL buffer) using an Ultra-Turrax T10 Basic homogenizer (IKA). Homogenates were centrifuged at 12,000× g for 10 min at 4 °C in a Hettich Mikro 22 R (Germany) centrifuge. After centrifugation, the lipid phase was removed, and supernatants were collected for further analysis. The brain was used for the determination of acetylcholinesterase activity (AChE), and the liver and gill were used for the determination of glutathione S-transferase (GST) activity, superoxide dismutase (SOD) activity, catalase (CAT) activity, reduced glutathione (GSH) concentrations, and lipoperoxidation (LOP). Biochemical assays were carried out using an MRC Spectro UV-18 (Israel) spectrophotometer. Each sample was measured in duplicate. Protein contents were measured using the method of Bradford (1976) [29] with bovine serum albumin as a standard.

The AChE activity was determined following the procedure of Ellman (1961) [30]. The enzyme activities were assayed via the addition of a substrate of acetylthiocholine iodide (Sigma-Aldrich®, Darmstadt, Germany) (AThCh; 4.3×10^{-4} M final concentration) and dithiobisnitrobenzoic acid (Sigma-Aldrich®, Germany) (DTNB; 7.1×10^{-5} M final concentration) mixture. After 20 min of incubation, the reaction was stopped with the addition of prostigmine (Sigma-Aldrich®, Germany) to each of the tubes except the blank control, and the enzyme activities were determined. Samples were measured at 412 nm. The activity of AChE was calculated as nmol/μg protein/min.

The contents of MDA as a measure of LOP intensity were assayed via color reaction with 2-thiobarbituric acid [31]. Trichloroacetic acid (Component-Reaktiv, Moscow, Russia) solution at 30%, 0.5 M HCl (Component-Reaktiv, Russia) solution, and 0.75% 2-thiobarbituric acid (Sigma-Aldrich®, Germany) solution were successively added to tissue homogenate. The mixture was stirred vigorously and heated for 20 min in a bath of boiling water. After cooling, samples were separated via centrifuging. An ethanol solution of ionol (Sigma-Aldrich®, Germany) was added to the supernatant (final concentration 5 mM) in order to prevent lipid peroxidation in the sample. Samples were measured at 532 nm. A molar extinction coefficient of 1.56×10^5 M^{-1} cm^{-1} was used for calculations. The content of MDA was calculated as pmol/μg protein.

The CAT activity was measured by following the rate of 0.3% H_2O_2 decomposition, which can be determined directly by the decrease in absorbance at 410 nm, using a molar extinction coefficient of 22.2×10^3 M^{-1} cm^{-1}. Change in absorbance per minute was measured and calculated as pmol/μg protein/min [32].

The SOD activity was measured by the inhibition of nitroblue tetrazolium reduction in reaction with phenazine methosulfate and NADH under basic conditions [33]. Measurement was performed in 0.15 M phosphate buffer pH 7.8 at 20 °C. The reaction mixture contained 0.00033 M EDTA (Sigma-Aldrich®, Germany), 0.0006 M phenazine methosulfate (Sigma-Aldrich®, Germany), 0.00135 M nitroblue tetrazolium (Sigma-Aldrich®, Germany), 0.078 M NADH (Sigma-Aldrich®, Germany), and 0.3 mL sample aliquot. SOD activity determination was performed at 540 nm and calculated as the increase in concentration ($\Delta E \times 10^{-6}$) of nitroformazan produced per μg protein in the sample per 1 min of reaction time.

The GSH concentrations were detected in reaction with 5,5′-dithiobis (2-nitrobenzoic acid) (DTNB). Trichloroacetic acid (30%) was added to homogenate aliquot, mixed thoroughly, and placed on ice for 1 h. The resulting precipitate was separated via centrifugation at 10,000× *g* for 10 min at 4 °C. Each sample cuvette contained 0.1 M phosphate buffer pH 7.5, 0.001 M DTNB (Sigma-Aldrich®, Germany), and 0.5 mL supernatant fraction. The colored complex was registered at 412 nm. GSH content was calculated using a molar extinction coefficient of 13.6×10^3 M^{-1} cm^{-1} and calculated as pmol/μg protein [34].

The GST activity was determined by monitoring the conjugation of GSH with 1-chloro-2,4-dinitrobenzene used as a substrate [35]. The reaction mixture was composed of 0.1 M phosphate buffer pH 7.5, 0.1 M GSH, 0.05 M 1-chloro-2,4-dinitrobenzene (Sigma-Aldrich®, Germany), and 0.1 mL of the sample aliquot. Absorbance was measured at 340 nm, 20 °C. The increase in absorbance was recorded for 3 min. Enzyme activity was measured using a molar extinction coefficient of 9.6×10^3 M^{-1} cm^{-1} and calculated as the concentration of 1-cloro-2,4-dinitrobenzene produced per mg protein in the sample per minute of reaction time (nmol/μg protein/min).

All reagent work solutions for analysis were prepared in the laboratory and commercial kits were not used.

2.4. Hydrochemical and Hydrophysical Analysis

Simultaneously with the catch of fish, water samples were taken in the same places using a Molchanov bathometer (Zawod Gidrometpribor, Russia) from 5 to 10 m depth of the lake. Water samples were collected for physical–chemical analysis using 1 L clean polyethylene bottles (ten bottles from each bay) after being rinsed three times with a few milliliters of lake water sample.

Dissolved oxygen (DO) concentration, biochemical oxygen demand (BOD_5), permanganate oxidation, nutrients (nitrates, nitrites, ammonium ions, and phosphates), metals (K, Na, Ca, Mg, total Fe, Cu, Zn, Pb, and Cd) concentrations, and pH were determined. Nitrate ion was analyzed via an ion chromatograph with a photoelectrocolorimetric detector Metrohm 940 Professional IC Vario (Switzerland); nitrites, ammonium, and phosphate ions were measured via the photoelectrocolorimetric method using a Specord 210 PLUS Analytic Jena spectrophotometer (Germany). BOD_5 was measured via the electrochemical method with a portable dissolved oxygen meter H19146 (Germany). The concentrations of metals (K, Na, Ca, Mg, total Fe, Cu, Zn, Pb, and Cd) in water were measured on a 7900 ICP-MS Agilent inductively coupled plasma mass spectrometer (US).

2.5. Statistical Analysis

The data are presented as the means with the standard errors (mean ± SEM). The data were statistically processed using the Statistica 8.0 software package. The differences in the levels of AChE, antioxidants, and MDA and the length and weight of the whitefish among the season sampling points were examined using one-way ANOVA analysis with Tukey post hoc tests or Student's t-test between bays at a $p = 0.05$ significance level.

3. Results

3.1. Hydrochemical and Hydrophysical Analyses

Hydrochemical and hydrophysical parameters of the water taken from the Lichk and Lchashen Bays are presented in Tables 1 and 2.

Table 1. Hydrochemical and hydrophysical parameters of the water in Lichk and Lchashen Bays.

Bay	Temperature °C	pH	Dissolved Oxygen (Mg L^{-1})	BOD_5 (mg L^{-1})	Permanganate Oxidation (mg O L^{-1})	Ammonium Ion (mg N L^{-1})	Nitrate Ion (mg N L^{-1})	Nitrite Ion (mg N L^{-1})	Phosphate Ion (mg P L^{-1})
			May, 2022						
Lchashen	22	8.4	6.8	3.9	6	0.1	0.24	0.007	0.06
Lichk	18	8.4	8.3	3.5	5.7	0.07	1.1	0.012	0.09
			August 2022						
Lchashen	20	8.5	2.4	4.5	6.8	0.27	0.18	0.005	0.09
Lichk	23	8.4	6.1	3.9	6.2	0.1	0.16	0.004	0.08
			October, 2022						
Lchashen	16.7	8.4	5.1	4.2	6.2	0.21	0.23	0.008	0.07
Lichk	19.7	8.4	6.9	3.6	5.9	0.08	0.19	0.005	0.09
			May, 2023						
Lchashen	21	8.3	6.5	3.8	5.9	0.13	0.2	0.006	0.08
Lichk	19	8.4	8.3	3.2	5.3	0.08	1.14	0.01	0.06
			August, 2023						
Lchashen	23	8.5	3.7	4.6	6.4	0.18	0.17	0.004	0.1
Lichk	22	8.3	6.4	3.7	5.8	0.11	0.16	0.007	0.07
			October, 2023						
Lchashen	17	8.5	5.8	4	6	0.17	0.19	0.007	0.09
Lichk	19	8.4	6.9	3.3	5.4	0.1	0.17	0.006	0.06
National surface water quality criteria *	-	6.5–8.5	≥6	3.0	5	0.39	9.0	0.02	3.5

Notes: * National surface water quality criteria adopted by the Ministry of Environment of Armenia (http://env.am/en/environment/environmental-monitoring, accessed 10 July 2024).

Table 2. Concentrations (mg L^{-1}) of select elements present in the water samples of Lichk and Lchashen Bays.

Bay	K^+	Na^+	Ca^{2+}	Mg^{2+}	Fe Total	Cu^{2+}	Zn^{2+}	Pb^{2+}	Cd^{2+}
				May, 2022					
Lchashen	13	62	26	45	0.02	<dL	<dL	<dL	<dL
Lichk	7.6	40	29	18	0.21	<dL	<dL	<dL	<dL
				August, 2022					
Lchashen	31.1	61	26.1	44	0.02	<dL	<dL	<dL	<dL
Lichk	19.4	56	27	34	0.09	<dL	<dL	<dL	<dL
				October, 2022					
Lchashen	12.8	58	27	44.3	0.01	<dL	<dL	<dL	<dL
Lichk	14	47	27.2	39	0.02	<dL	<dL	<dL	<dL
				May, 2023					
Lchashen	14	59	27	46	0.04	<dL	<dL	<dL	<dL
Lichk	7.3	41	28	23	0.19	<dL	<dL	<dL	<dL
				August, 2023					
Lchashen	26	60	27.1	46	0.03	<dL	<dL	<dL	<dL
Lichk	17	54	27	38	0.07	<dL	<dL	<dL	<dL
				October, 2023					
Lchashen	13.1	56	26.7	48	0.02	<dL	<dL	<dL	<dL
Lichk	12.6	49	27	41	0.02	<dL	<dL	<dL	<dL
"average" quality *	4 × B	4 × B	200	100	0.5	0.05	0.2	0.025	B + 0.002
"good" quality *	2 × B	2 × B	100	50	2 × B	B + 0.02			
"excellent" quality *	B	B	B	B	B	B			

Notes: * Armenia surface water quality guidelines. B—background value of a given element in a local environment; dL—detection limit: 0.002 mg L^{-1}.

The water temperature in both bays varied in the range of 16.7–23 °C, i.e., during the entire observation period the seasonal temperature range was no more than 6.3 °C, which corresponds to the usual seasonal temperature values and their variability in these parts of Lake Sevan (Table 1). In May, the water temperature in the Lchaschen Bay was always slightly higher, and in October, it was lower than in the Lichk Bay. In August, the temperature in both bays was almost the same and the highest of the season, and in October, the water temperature was minimal. Water pH values varied little during the entire observation period, equaling 8.3–8.5, and were within the norms for the previous period of 2017–2022 (http://env.am/en/environment/environmental-monitoring, accessed 10 July 2024). No pronounced seasonal variability was revealed.

The DO content varied over the observation period in a wide range from 2.4 to 8.3 mg L^{-1} (Table 1). In both bays, it was highest in May, then decreased to a minimum in August, and increased again in October, but did not reach the May values. It should be noted that in Lchashen Bay, the values of this indicator were always lower than in Lichk Bay. At the same time, in Lchashen Bay in May, the dissolved oxygen content corresponded to the national standard level, while in August and October, it was 1.62–2.5- and 1.03–1.18-times lower, respectively, depending on the year. Note that in Lichk Bay, the content of DO for all months was above the national standard level. The reduced concentrations of DO in Lchashen Bay are most likely associated with the increased content of easily oxidized organic matter, the oxidation of which consumes oxygen.

The organic matter content measured as BOD_5 and permanganate oxidation in both bays was high during the observation period, varying in the range from 3.2 to 6.8 mg L^{-1} (Table 1). It was highest in August compared to May and October. It should be noted that the organic content in Lchashen Bay was always higher than in Lichk Bay. At the same time, the BOD_5 and permanganate oxidation values in both bays exceeded the national water quality standards (by 1.13–1.5 and 1.2–1.4 times, respectively).

The content of nitrates and phosphates in both bays varied in the range of 0.004–1.14 mg L^{-1} (Table 1). The concentration of nitrates and nitrites was highest in May, then decreased in August and increased again in October but did not reach the May values. In contrast, the ammonium and phosphate content were higher in August than in May and October. According to the data, the nitrate and phosphate values in Lchashen Bay were always higher than those in Lichk Bay, which was due to the increase in the concentration of ammonium ions and additionally indicates some obvious deterioration in water quality. The nitrate, nitrite, and phosphate content in both bays corresponded to the national standard level (Table 1).

The concentrations of all studied metal ions (K, Na, Ca, Mg, total Fe, Cu, Zn, Pb, and Cd) varied throughout the observation period in the range of 0.02–62 and were much lower than the "average" water quality indicator, mostly corresponding to the "good" and even "excellent" water quality criterion (Table 2) according to the national surface water quality criteria adopted by the Ministry of Environment of Armenia (http://env.am/en/environment/environmental-monitoring, accessed 10 July 2024).

3.2. Ichthyological Data

Biometric parameters of the lake's whitefish in the study seasons are presented in Table 3.

Table 3. Biometric indicators of whitefish from Lichk and Lchaschen Bays of Lake Sevan.

	Age	N	Length, cm	Weight, g	Condition Factor
			Lichk Bay		
May, 2022	1 + 2+	10 0	28.61 ± 0.346 ^{a}	224.6 ± 7.45 ^{a}	0.96 ± 0.01 ^{a}
August, 2022	1 + 2+	4 6	31.52 ± 0.59 ^{b}	318.3 ± 20.7 ^{b}	1.01 ± 0.05 ^{a,b}
October, 2022	1 + 2+	7 3	30.2 ± 0.65 ^{b,c} *	291.7 ± 21.1 ^{b} *	1.05 ± 0.02 ^{a,b,c}
May, 2023	1 + 2+	9 1	29.05 ± 0.42 ^{a}	255.3 ± 12.3 ^{a} *	1.04 ± 0.03 ^{a,b,c} *
August, 2023	1 + 2+	9 1	31.71 ± 0.43 ^{b,c,d}	358.3 ± 14.3 ^{a,b,c} *	1.12 ± 0.04 ^{b,c} *
October, 2023	1 + 2+	0 10	33.18 ± 0.55 ^{b,c,d} *	385.8 ± 11.1 ^{b,c} *	1.06 ± 0.02 ^{a,b,c}
			Lchashen Bay		
May, 2022	1 + 2+	10 0	28.4 ± 0.247 ^{a}	224.9 ± 3.59 ^{a}	0.98 ± 0.01 ^{a}
August, 2022	1 + 2+	2 8	32.11 ± 0.367 ^{b}	344.6 ± 11.7 ^{b}	1.04 ± 0.03 ^{a,b}
October, 2022	1 + 2+	10 0	28.4 ± 0.296 ^{a} *	223.3 ± 5.37 ^{a} *	0.98 ± 0.02 ^{a,b}
May, 2023	1 + 2+	10 0	28.65 ± 0.224 ^{a}	224.6 ± 6.32 ^{a} *	0.95 ± 0.02 ^{a,b,c} *
August, 2023	1 + 2+	9 1	30.88 ± 0.262 ^{c}	290.5 ± 8.9 ^{c} *	0.99 ± 0.02 ^{a,b,c} *
October, 2023	1 + 2+	6 4	31.99 ± 0.38 ^{b,c} *	329.4 ± 20.2 ^{b,c} *	1.00 ± 0.04 ^{a,b,c}

Note: a,b,c,d values marked with the same letters in each column are not significantly different (ANOVA and LSD test, $p = 0.05$). * Means for the same date differ significantly between stations (Student's t-test, $p = 0.05$).

3.3. Biochemical Studies

Comparative analysis of the data showed that the values of all studied biochemical parameters of fish during the observation period of 2022–2023 were consistently and statistically significantly higher in fish from Lchashen Bay than in fish from Lichk Bay (Student's t-test, $p = 0.05$). In both bays, fish showed similar seasonal dynamics for each indicator, and seasonal differences within each year, with some exceptions, were statistically significant, while interannual differences for each month were mostly absent or minimal (ANOVA Tukey test, $p = 0.05$). In the liver, the values of AOS indicators were slightly higher than in the gills.

AChE activity. The value of this indicator in the brain of whitefish from Lchashen Bay was approximately 2-times higher than that of fish from Lichk Bay and varied within the range of 0.3543–0.5215 and 0.1692–0.2546 nmol/µg protein/min, respectively (Table 4). The maximum seasonal activity was observed in May; in August, it decreased by 1.4–1.5 times, and in October, it increased again but did not reach the maximum value, remaining 1.3-times lower than the May values. The differences between the May (maximum) and August (minimum) values were statistically significant, and the October values, which occupied an intermediate position, did not statistically differ depending on the year from either the May or August values.

Table 4. Values of acetylcholinesterase (AChE) in the brain of whitefish specimens.

Month, Year	Mean Values of Parameters ± S.E.	
	AChE nmol per 1 µg Protein per 1 min	
	Lichk Bay	Lchashen Bay
May, 2022	0.2546 ± 0.010^{a}	0.5117 ± 0.004^{a} *
August, 2022	0.1692 ± 0.008^{b}	$0.3623 \pm 0.012^{b,d}$ *
October, 2022	$0.1934 \pm 0.009^{b,c}$	0.3973 ± 0.007^{c} *
May, 2023	$0.2389 \pm 0.010^{a,d}$	0.5215 ± 0.004^{a} *
August, 2023	$0.1807 \pm 0.004^{b,c}$	0.3543 ± 0.009^{b} *
October, 2023	$0.2081 \pm 0.005^{c,d}$	$0.3903 \pm 0.0025^{c,d}$ *

Note: a,b,c,d values marked with the same letters in each column are not significantly different (ANOVA and LSD test, $p = 0.05$). * Means for the same date differ significantly between stations (Student's t-test, $p = 0.05$). The same note for Tables 5 and 6.

Table 5. Values of lipid peroxidation and antioxidant system in the liver of whitefish specimens.

Month, Year	Mean Values of Parameters ±S.E.				
	MDA	GSH	GST	Catalase	SOD
	pmol per 1 µg Protein		nmol per 1 µg Protein per 1 min		ΔE × 10^{-6} per 1 µg Protein per 1 min
Lichk Bay					
May, 2022	$0.5213 \pm 0.026^{a,c}$	8.21 ± 0.27^{a}	2.27 ± 0.18^{a}	24.1 ± 0.35^{a}	7.6 ± 0.18^{a}
August, 2022	0.6140 ± 0.016^{b}	12.9 ± 0.15^{b}	$1.15 \pm 0.03^{b,c}$	29.7 ± 0.17^{b}	16.2 ± 0.13^{b}
October, 2022	$0.5835 \pm 0.017^{a,b}$	11.1 ± 0.16^{c}	1.54 ± 0.03^{c}	27.4 ± 0.13^{c}	13.8 ± 0.08^{c}
May, 2023	0.5102 ± 0.013^{c}	8.27 ± 0.16^{a}	2.21 ± 0.15^{a}	23.7 ± 0.17^{a}	8.2 ± 0.13^{d}
August, 2023	0.6320 ± 0.009^{b}	13.3 ± 0.13^{b}	1.11 ± 0.02^{b}	29.3 ± 0.127^{b}	15.7 ± 0.13^{b}
October, 2023	$0.5792 \pm 0.003^{a,b}$	12.8 ± 0.095^{b}	$1.27 \pm 0.05^{b,c}$	28.4 ± 0.14^{d}	14.9 ± 0.02^{e}
Lchashen Bay					
May, 2022	0.6147 ± 0.020^{a} *	9.72 ± 0.21^{a} *	3.67 ± 0.26^{a} *	36.3 ± 0.27^{a}	14.9 ± 0.24^{a}
August, 2022	0.7420 ± 0.012^{b} *	14.8 ± 0.1^{b} *	2.73 ± 0.12^{b} *	41.3 ± 0.14^{b}	28.3 ± 0.12^{b}
October, 2022	0.6912 ± 0.007^{c} *	13.3 ± 0.18^{c} *	2.97 ± 0.103^{b} *	39.7 ± 0.053^{c}	$26.9 \pm 0.12^{c,e}$
May, 2023	$0.6412 \pm 0.005^{a,d}$ *	9.77 ± 0.045^{a} *	3.73 ± 0.05^{a} *	34.9 ± 0.02^{d}	14.2 ± 0.046^{d}
August, 2023	0.7390 ± 0.008^{b} *	14.1 ± 0.04^{d} *	2.67 ± 0.096^{b} *	40.7 ± 0.072^{e}	27.1 ± 0.11^{e}
October, 2023	$0.6827 \pm 0.007^{c,d}$ *	13.5 ± 0.09^{c} *	2.91 ± 0.13^{b} *	39.1 ± 0.02^{f}	26.3 ± 0.2^{c}

Note: a,b,c,d values marked with the same letters in each column are not significantly different (ANOVA and LSD test, $p = 0.05$).

Table 6. Values of lipid peroxidation and antioxidant system in the gill of whitefish specimens.

Month, Year	MDA	GSH	GST	Catalase	SOD
	Mean Values of Parameters ±S.E.				
	pmol per 1 µg Protein		nmol per 1 µg Protein per 1 min		$\Delta E \times 10^{-6}$ per 1 µg Protein per 1 min
Lichk Bay					
May, 2022	0.3217 ± 0.026 ^{a}	5.26 ± 0.24 ^{a}	1.97 ± 0.13 ^{a}	17.3 ± 0.35 ^{a}	6.9 ± 0.16 ^{a}
August, 2022	0.3671 ± 0.014 ^{a}	8.90 ± 0.15 ^{b}	1.09 ± 0.04 ^{b}	21.8 ± 0.19 ^{b}	14.8 ± 0.11 ^{b}
October, 2022	0.3583 ± 0.008 ^{a}	7.40 ± 0.16 ^{c}	1.37 ± 0.07 ^{b}	19.5 ± 0.08 ^{c}	12.9 ± 0.11 ^{c}
May, 2023	0.3172 ± 0.016 ^{a}	5.71 ± 0.10 ^{a}	1.81 ± 0.14 ^{a}	16.7 ± 0.13 ^{a}	7.6 ± 0.12 ^{d}
August, 2023	0.3431 ± 0.010 ^{a}	8.31 ± 0.13 ^{b,d}	1.16 ± 0.014 ^{b}	23.1 ± 0.18 ^{d}	13.9 ± 0.13 ^{e}
October, 2023	0.3479 ± 0.011 ^{a}	7.80 ± 0.095 ^{c,d}	1.31 ± 0.04 ^{b}	21.2 ± 0.14 ^{b}	12.3 ± 0.12 ^{f}
Lchashen Bay					
May, 2022	0.5134 ± 0.017 ^{a}	7.12 ± 0.12 ^{a}	2.73 ± 0.08 ^{a}	24.8 ± 0.16 ^{a}	11.9 ± 0.24 ^{a}
August, 2022	0.6052 ± 0.006 ^{b}	13.1 ± 0.11 ^{b}	1.89 ± 0.10 ^{b,c}	32.3 ± 0.15 ^{b}	26.1 ± 0.11 ^{b}
October, 2022	0.5943 ± 0.006 ^{b}	11.9 ± 0.07 ^{c}	2.18 ± 0.10 ^{c}	29.4 ± 0.13 ^{c}	23.2 ± 0.11 ^{c}
May, 2023	0.5371 ± 0.004 ^{a}	7.79 ± 0.04 ^{d}	2.87 ± 0.07 ^{a}	25.9 ± 0.02 ^{d}	12.8 ± 0.13 ^{d}
August, 2023	0.6149 ± 0.007 ^{b}	12.67 ± 0.16 ^{e}	1.71 ± 0.11 ^{b}	34.2 ± 0.33 ^{e}	27.4 ± 0.09 ^{e}
October, 2023	0.5818 ± 0.007 ^{b}	11.43 ± 0.08 ^{f}	2.03 ± 0.10 ^{b,c}	30.7 ± 0.16 ^{f}	24.9 ± 0.12 ^{f}

Note: a,b,c,d values marked with the same letters in each column are not significantly different (ANOVA and LSD test, $p = 0.05$).

MDA content. The values of the indicator in the liver and gills of fish from Lchashen Bay were statistically significantly higher than that of the fish from Lichk Bay by 17–26% and 60–79%, respectively (Tables 5 and 6).

In the liver, they varied in the range of 0.6147–0.7420 and 0.5102–0.6320 pmol/µg protein, respectively (Table 5). In May, the minimum seasonal values were observed; in August, they increased by 15–24%, and in October, they decreased again but did not reach the minimum, remaining 6–14% higher than the May values. The differences between the minimum (May) and maximum (August) values of the indicator were statistically significant, and the October values, which occupied an intermediate position, either statistically significantly differed from the other two months or did not.

In the gills, the indicator varied in the range of 0.5134–0.6149 and 0.3172–0.3671 pmol/µg protein, respectively (Table 6). For fish from Lchashen Bay, minimum seasonal values were observed in May; in August, they increased by 8–18%, reaching maximum values and remained at this level in October. The differences between the May (minimum) and August (maximum) values were statistically significant. In fish from Lichk Bay, the trend of seasonal dynamics was similar to that in fish from Lchashen Bay, but there were no statistically significant seasonal differences in the indicator within each year. Interannual differences in the same months in fish from both bays were mostly absent or statistically significant, but minimal.

GSH content. The value of the indicator in the liver and gills of fish from Lchashen Bay were statistically significantly higher than in the fish from Lichk Bay by 5–20% and 60–79% (Tables 5 and 6).

In the liver, it varied in the range of 9.72–14.8 and 8.21–13.3 pmol/µg protein, respectively (Table 5). Seasonal minimum values were observed in May; in August, they increased by 44–61%, and in October, they decreased again but did not reach the minimum, remaining 35–55% higher than May values. Interannual differences in the same months were mostly absent or minimal.

In the gills they varied in the range of 7.12–12.67 and 5.26–8.90 pmol/µg protein (Table 6). The minimum values were noted in May; in August, they increased by 46–84%, and in October, they decreased again but did not reach the minimum, remaining 37–67% higher than the May values. There were no interannual differences in the same months in fishes from Lichk Bay, but in fishes from Lchashen Bay, they were minimal but statistically significant.

GST activity. The values of the indicator in the liver and gills of fish from Lchashen Bay were statistically significantly higher than in fishes from Lichk Bay by 62–193% and by 39–73% (Tables 5 and 6).

In the liver, they varied in the range of 2.73–3.73 and 1.11–2.27 nmol/μg protein/min, respectively (Table 5). The maximum values were observed in May; in August, they decreased by 28–50%, while in October, they increased again but did not reach the maximum, remaining 19–43% lower than the May values. The differences between the maximum (May) and August minimum (August)) values were statistically significant, but the October values, which occupied an intermediate position, did not differ from the August ones. Interannual differences in the same months were either absent or minimal.

In the gills, they varied in the range of 1.71–2.87 and 1.09–1.97 nmol/μg protein/min, respectively (Table 6). In May, the maximum values were noted; in August, they decreased by 31–45%, and in October, they increased again but did not reach the maximum, remaining 13–27% higher than the May values.

CAT activity. The values of the indicator in the liver and gills of fish from Lchashen Bay were statistically significantly higher than in fishes from Lichk Bay by 38–51% and 45–55% (Tables 5 and 6).

In the liver, they varied in the range of 34.9–41.3 and 23.7–29.7 nmol/μg protein/min, respectively (Table 5). The minimum values were observed in May; in August, they were maximum, increasing by 14–24%, and in October, they decreased again but did not reach the May minimum, remaining 9–20% higher. Interannual differences in the same months were either absent or minimal.

In the gills, they varied in the range of 24.8–34.2 and 16.7–23.1 nmol/μg protein/min (Table 6). In May, the minimum values were noted; in August, they were maximum, increasing by 26–38%, and in October, they decreased again but did not reach the May minimum, remaining 9–20% higher. Interannual differences in the same months in fish from Lichk Bay were absent, and for the fishes from Lchashen Bay, they were minimal but statistically significant.

SOD activity. The values of the indicator in the liver and gills of fish from Lchashen Bay were statistically significantly higher than in the liver of fishes from Lichk Bay by 73–96% and 68–124% (Tables 5 and 6).

In the liver, they varied in the range of 14.2–28.3 and 7.6–16.2 $\Delta E \times 10^{-6}$/μg protein/min, respectively (Table 5). The minimum values were observed in May; in August, they were maximum, increasing by 90–113%, and in October, they decreased again but did not reach the May minimum, remaining 81–85% higher. Interannual differences in the same months were mostly absent or minimal.

In the gills, this indicator varied in the range of 11.9–27.4 and 6.9–14.8 $\Delta E \times 10^{-6}$/μg protein/min (Table 6). The minimum values were observed in May; in August, they were maximum, increasing by 83–119%, and in October, they decreased again but did not reach the May minimum, remaining 62–95% higher. The interannual differences in the same months were minimal but statistically significant.

4. Discussion

In this study, the brain AChE activity of the whitefish taken out from both bays of Lake Sevan varied within the range of 0.1692–0.5215 nmol/μg protein/min. This is in good agreement with previously published data, showing that the enzyme activity in the same whitefish, crucian carp (*Carassius auratus gibelio (Bloch, 1782)*, and khramicarp (*Capoeta sevangi* De Filippi, 1865) from Lake Sevan averages 0.548, 5.01, and 4.25 nmol/μg protein/min, respectively [36]. Similar values were obtained for roach (*Rutilus rutilus* Linnaeus, 1758) and bream (*Abramis brama* Linnaeus, 1758) of the Rybinsk Reservoir, equal to 0.280–0.603 [37] and 0.15–0.20 nmol/μg protein/min [38], respectively, depending on the season. In bream (*A. brama*) from the Danube River in Serbia [39] and carp (*Cyprinus carpio* Linnaeus, 1758) from three reservoirs in Tunisia [40], the brain AChE activity was equal to 0.05 and 0.125–0.195, respectively. In capoeta (*Capoeta umbla* Heckel, 1843) from the Pülümür River (Pülümür Stream, Turkey), it varies within the range of 0.04–0.54 nmol/µg protein/min [41]. In the ten-spotted live-bearer fish (*Cnesterodon decemmaculatus*, Jenyns 1842) from the Lujan River basin (Argentina), the enzyme activity varied within the range of 0.113–0.285 nmol/µg protein/min throughout the year [42]. In the Mediterranean rainbow wrasse (*Coris julis* Linnaeus, 1758), the brain AChE activity depending on habitat varied between 0.048 and 0.95 nmol/µg protein/min [43].

The data on the AOS values in the liver and gills of whitefish are generally consistent with the results obtained earlier by different authors on other fish species. Thus, in the liver of bream (*A. brama* L.) of the Rybinsk Reservoir, the values of MDA, GSH, GST, CAT, and SOD activity depending on the fishing location varied within the range of 0.311–0.601 and 7.99–11.95 pmol/µg protein, 1.03–3.57, 25.4–37.6 nmol/µg protein/min, and 11.0–34.5 $\Delta E \times 10^{-6}$/µg protein/min, respectively [44]. In another study, GST and CAT activities in the liver of bream (*A. brama*) from the Danube River, depending on the catch location, averaged 0.2–0.3 and 12–14 nmol/µg protein/min, respectively [39]. In carp (*C. carpio*) from three reservoirs in Tunisia, GST activities ranged from 0.3 to 0.8 nmol/µg protein/min [40].

Our study revealed two main trends in the status of the studied biochemical parameters in whitefish in Lake Sevan. First, all of the parameters demonstrated stable seasonal dynamics, repeating without noticeable changes during two years of observations. Secondly, the values of all parameters were higher in whitefish from Lchashen Bay than in fish from Lichk Bay.

It is well known that seasonal changes in the metabolism of living organisms are adaptive in nature and are aimed at ensuring their normal functioning under periodically changing external natural conditions during the annual cycle. The main drivers of the rhythm of seasonal biological cycles in mid-latitudes are photoperiod and temperature. Moreover, the first serves as a trigger that launches the sequence and direction of biochemical, physiological, structural–morphological, and behavioral reactions of the body, and the temperature allows for the operational regulation of these processes, accelerating or slowing them down [37,44,45]. In general, when there is a longer night phase of the day and lower temperatures in the autumn–winter season, the intensity of morphological and functional processes is reduced, and in the spring–summer, when the duration of the light phase of the day and average daily temperatures are increased, it is increased. An additional factor influencing the metabolism and seasonal dynamics of the internal processes of the body is spawning.

The seasonal dynamics of brain AChE activity in the whitefish from Lake Sevan generally corresponds to the identified patterns. The highest enzyme activity over a two-year period (2022–2023) was observed in May; in August, it decreased and then increased slightly again but did not reach the May values. Such natural dynamics of AChE activity, in our opinion, may be due to the action of two factors. In May and August, it reflects the seasonal cyclicity of the photoperiod, and the observed increase in activity in October is caused by the beginning of fish preparation for spawning, which for whitefish in Lake Sevan occurs in late November-early December. Note that the similar seasonal dynamics of AChE activity were previously found in the brain of the European sardine (*Sardina pilchardus* Walbaum, 1792) [46].

Correspondence of the fish brain AChE activity to natural seasonal rhythms has also been shown for perch (*P. fluviatilis* Linnaeus, 1758), roach (*R. rutilus* Linnaeus, 1758), and bream (*A. brama* Linnaeus, 1758) in the Rybinsk Reservoir (Russia) [37,38], capoeta (*C. umbla* Heckel, 1843) from the Pülümür River (Pülümür Stream, Turkey) [41], and ten-spotted live-bearer fish (*C. decemmaculatus*) from the Lujan River basin (Argentina) [42]. In winter, the AChE activity in the brain is the lowest. In spring, when the water temperature is still close to 0 °C, but the light phase of the day is already increased, it also begins to increase. Subsequently, the activity increases with an increase in the average water temperature, reaching a maximum in the summer months. In fish, with a one-time spawning (whitefish, bream, roach, perch, capoeta, etc.), an additional increase in enzyme activity is observed in the prespawning–spawning period.

Changes in enzyme activity during the prespawning and spawning period compared to other seasonal stages of the life of fish and other aquatic organisms are associated with hormonal changes and stress states that they experience at this time. An increase in AChE activity has been experimentally demonstrated during the stress response induced by hormones (by adrenaline and dexamethasone) and by "handling" in the fish brain [21], as well as in sea urchin coelomocytes upon exposure to cold [47].

Seasonal dynamics of AOS indicators both in the liver and in the gills of whitefish from Lake Sevan were generally similar and also cyclical. Moreover, changes in different indicators over a two-year period (2022–2023) had a multidirectional nature. Thus, in May, the content of MDA and GSH, SOD, and CAT activity in both organs of whitefish showed the lowest seasonal values, while GST activity was the highest. In August, the values of the first four indicators have reached a seasonal maximum, and the last one decreased to a seasonal minimum. In October, the changes in indicators were in the opposite direction, but they did not reach the extreme values observed in May. Previously, the seasonal dynamics of AOS indicators was studied in bream (*A. brama* Linneaus 1758.) from the Rybinsk Reservoir [44]. It was found that all these AOS indicators demonstrate minimum values in the winter months and maximum values in spring or summer. It should be noted that the MDA content in both whitefish and bream had the lowest interseasonal variability, which amounted to 15 and 29%, respectively, and was not always statistically significant. That is, during the seasonal cycle, the fish organism managed to maintain the content of lipid peroxidation products (LPOs) at a relatively constant level, and, consequently, the content of ROS also remained constant due to the active work of the antioxidant defense (AOD) system. The intensification of the work of the AOD system is indicated by a wider range of seasonal changes in the values of its components: from 30% and higher for GSH and up to more than 2 times for SOD.

In our study, throughout the entire observation period, the values of all the studied biochemical markers of whitefish from Lchashen Bay were higher than those from Lichk Bay. These differences may be due to both internal biological and external environmental factors. Comparative analysis of biometric indicators of whitefish in the bays across the seasons showed consistently lower weight, length, or CF in the fish (1+ and 2+ years old) of Lchashen Bay starting from the October 2022 (Table 3). These values may primarily indicate worsened feeding conditions in Lchashen Bay, compared to the Lichk, caused by lesser food availability due to different anthropogenic pressures, e.g., organic pollution. The DO content was lower, while biological (BOD) and chemical (COD) oxygen demand, as well as the concentration of ammonium in the water throughout the year, were significantly higher in Lchashen Bay than in Lichk Bay. At the same time, the nitrate content was higher in Lichk bay. The remaining indicators, as well as the concentrations of the main elements, including metals, differ to a lesser extent and are significantly below the level of national water quality standards (http://env.am/en/environment/environmental-monitoring, accessed 10 July 2024). The most pronounced differences were observed for the DO indicator, the values of which in Lchashen Bay in August and October were below the national standard level ($\geq$6 mg L^{-1}), decreasing especially sharply in August (2.7–3.4 mg L^{-1}), which is even below the critical level (4 mg L^{-1}) necessary for the normal existence of the fish. The reason for the decrease in the DO level in Lchashen Bay in August and partly in October may be the higher level of organic matter, which is oxidized using dissolved oxygen. This assumption is supported by the increased values of BOD, COD, and ammonium concentration in the water in Lchashen Bay compared to Lichk Bay. As Lchashen Bay is characterized by water stagnation due to the absence of inflowing rivers, pollution from neighboring lands may apparently have greater impact on the water chemistry and the biota inhabiting the bay. It can be assumed that Lchashen Bay is more susceptible to anthropogenic impact than Lichk Bay.

Hence, the elevated levels of brain AChE activity and AOS indices in the liver and gills of the whitefish from Lchashen Bay found in our case are due to hypoxia. In Lichk Bay, the oxygen regime of fish has corresponded to normal conditions and the values of all biochemical indicators were lower than those of fish from Lchashen Bay. The effect of hypoxia on the activity of brain AChE was previously demonstrated in rats. It has been established that after exposure to hypoxia in the prenatal period, an increased level of activity of the membrane-bound form of AChE is observed in the brains of animals throughout their life [48].

It is also known that the decrease in the DO level in water and hypoxia results in the development of AOS in fish [49–52]. In our study, the whitefish from Lchashen Bay had an increased MDA level in both organs, indicating an increase in the intensity of ROS formation. At the same time, the values of the AOD system indicators were also increased, indicating its activation. Normally, the intensity of ROS formation and neutralization in the cell, recorded via the MDA content, is balanced, and their content is at a stationary level [49,53]. With short-term exposure to a stress factor, acute oxidative stress is developed when the balance is shifted towards the formation of ROS. However, activation of the AOD system quickly returns their level to the normal values. With long-term action of the stress factor, chronic oxidative stress develops, and a return to the initial level of ROS content requires more intensive work of the AOD system for a long time. However, in some cases, the ROS content does not return to the initial level, despite the high activity of the AOD system, but is established at a new, higher level, called quasi-stationary. Based on the data we received, we can say that this is exactly the condition observed in whitefish from Lchashen Bay. It should be noted that such a state is less stable than a stationary one, and under the influence of additional stress factors, for example, elevated temperatures or increased anthropogenic load, a depletion of the AOD system and an uncontrolled increase in the content of ROS in the cell may occur. This can result in the development of numerous pathological processes in the body [50,52].

5. Conclusions

The AChE activity in the brain and AOS indicators in the liver and gills of whitefish from Lchashen and Lichk Bays of Lake Sevan demonstrate a stable seasonal cyclicity during the studied period 2022–2023. Comparative analysis of biochemical parameters showed that the whitefish in Lchashen Bay are in a worse condition, and their habitat conditions are less favorable than in Lichk Bay. The most probable reason for this is a low DO value and the increased level of organic pollution, which is confirmed by the data from hydrochemical analysis. The concentration of main elements and metals in waters of bays do not play a significant role in the fish biochemistry responses. The ichthyological data also demonstrated worse feeding condition for whitefish in Lchashen Bay compared to Lichk Bay. It can be assumed that in the future, additional anthropogenic load on Lchashen Bay can lead to an even greater deterioration in the health of fish and pronounced pathological disorders in their organisms.

Author Contributions: Conceptualization, B.G. and G.C.; methodology, H.M., N.B. and T.V.; formal analysis, H.M. and G.C.; investigation, H.M., N.B., T.V., E.G. and H.K.; resources, B.G.; writing-original draft preparation, H.M., N.B. and G.C.; writing-review and editing, B.G.; supervision, B.G.; project administration, B.G.; funding acquisition, B.G. All authors have read and agreed to the published version of the manuscript.

Funding: This research was funded by the Higher Education and Science Committee of Armenia, grant No. 21T-1F183, and with the financial support from the Papanin Institute for Biology of Inland Waters, Russian Academy of Sciences (theme registration number 124032500015-7).

Data Availability Statement: Data are available from the corresponding author on request.

Conflicts of Interest: The authors declare no conflicts of interest.

References

1. Li, X.; Gong, S.; Shi, Q.; Fang, Y. A review of ecosystem services based on bibliometric analysis: Progress, challenges, and future directions. *Sustainability* **2023**, *15*, 16277. [CrossRef]
2. Ferreira, V.; Bini, L.M.; Gonzales Sagrario, M.A.; Kovalenko, K.E.; Naselli-Flores, L.; Padial, A.A.; Padisak, J. Aquatic ecosystem services: An overview of the Special Issue. *Hydrobiologia* **2023**, *850*, 2473–2483. [CrossRef]
3. Zandebasiri, M.; Jahanbazi Goujani, H.; Iranmanesh, Y.; Azadi, H.; Vira, A.H.; Habibi, M. Ecosystem services valuation: A review of concepts, systems, new issues, and considerations about pollution in ecosystem services. *Environ. Sci Pollut. Res* **2023**, *30*, 83051–83070. [CrossRef] [PubMed]
4. Rizhinashvili, A.L. An Outline of the Theory of the Functioning of Aquatic Ecosystems: Nutrient Limitation. *Biol. Bull. Rev.* **2022**, *12*, 596–608. [CrossRef]
5. Grizzetti, B.; Lanzanova, D.; Liquete, C.; Reynaud, A.; Cardoso, A.C. Assessing water ecosystem services for water resource management. *Environ. Sci. Policy* **2016**, *61*, 194–203. [CrossRef]
6. Vári, Á.; Podschun, S.A.; Erős, T.; Hein, T.; Pataki, B.; Ioja, I.C.; Adamescu, C.V.; Gerhardt, A.; Gruber, T.; Dedic, A.; et al. Freshwater systems and ecosystem services: Challenges and chances for cross-fertilization of disciplines. *Ambio* **2022**, *51*, 135–151. [CrossRef] [PubMed]
7. Wilkinson, I.P. Lake Sevan: Evolution, Biotic Variability and Ecological Degradation. In *Large Asian Lakes in the Changing World*; Mischke, S., Ed.; Springer International Publishing: Cham, Switzerland, 2020; pp. 35–63. [CrossRef]
8. Gabrielyan, B.; Khosrovyan, A.; Schulze, M. A review of anthropogenic stressors on Lake Sevan, Armenia. *J. Limnol.* **2022**, *81*, 2061. [CrossRef]
9. Gevorgyan, G.; von Tuempling, W.; Shahnazaryan, G.; Friese, K.; Schultze, M. Lake-wide assessment of trace elements in surface sediments and water of Lake Sevan. *J. Limnol.* **2022**, *81*, 2096. [CrossRef]
10. Pavlov, P.I. Results of the introduction of whitefish into Lake Sevan. *Proc. Sevan Hydrobiol. Stn.* **1947**, *8*, 113–141. (In Russian)
11. Mailyan, R.A. Whitefish of Lake Sevan. *Proc. Sevan Hydrobiol. Stn.* **1957**, *15*, 137–196. (In Russian)
12. Simonyan, A.A. *Zooplankton of Lake Sevan*; National Academy of Sciences of Armenia: Yerevan, Armenia, 1991.
13. Gabrielyan, B.; Vardanyan, T.; Barseghyan, N.; Khosrovyan, A. Estimation of the Potential Wild Fish Stock Biomass to be Supported by Available Food Base in a Lake. *Inland Water Biol.* **2022**, *15*, 331–340. [CrossRef]
14. Khosrovyan, A.; Avalyan, R.; Atoyants, A.; Aghajanyan, E.; Hambaryan, L.; Aroutiounian, R.; Gabrielyan, B. *Tradescantia*-Based Test Systems Can Be Used for the Evaluation of the Toxic Potential of Harmful Algal Blooms. *Water* **2023**, *15*, 2500. [CrossRef]
15. Remen, M.; Oppedal, F.; Torgersen, T.; Imsland, A.; Olsen, R. Effects of cyclic environmental hypoxia on physiology and feed intake of post-smolt Atlantic salmon: Initial responses and acclimation. *Aquaculture* **2012**, *326–329*, 148–155. [CrossRef]
16. Oven, L.S.; Rudneva, I.I.; Shevchenko, N.F. Responses of *Scorpaenaporcus* (Scorpaenidae) to Anthropogenic Impact. *Vopr. Ikhtiol.* **2000**, *40*, 70–73.
17. Fulton, M.H.; Key, P.B. Acetylcholinesterase inhibition in estuarine fish and invertebrates as an indicator of organophosphorus insecticide exposure and effects. *Environ. Toxicol. Chem.* **2009**, *20*, 37–45. [CrossRef]
18. Durieux, E.; Farver, T.; Fitzgerald, P.; Eder, K.; Ostrach, D. Natural factors to consider when using acetylcholinesterase activity as neurotoxicity biomarker in Young-Of-Year striped bass (*Morone saxatilis*). *Fish Physiol. Biochem.* **2011**, *37*, 21–29. [CrossRef] [PubMed]
19. Falfushynska, H.; Kasianchuk, N.; Siemens, E.; Henao, E.; Rzymski, P. A Review of Common Cyanotoxins and Their Effects on Fish. *Toxics* **2023**, *11*, 118. [CrossRef]
20. Marsillach, J.; Costa, L.G.; Furlong, C.E. Protein adducts as biomarkers of exposure to organophosphorus compounds. *Toxicology* **2013**, *307*, 46–54. [CrossRef]
21. Pavlov, D.F.; Chuiko, G.M.; Shabrova, A.G. Adrenaline induced changes of acetylcholinesterase activity in the brain of perch (*Perca fluviatilis* L.). *Comp. Biochem. Physiol.* **1994**, *108C*, 113–115. [CrossRef]
22. Møller, P.; Wallin, H.; Knudsen, L.E. Oxidative Stress Associated with Exercise, Psychological Stress and Life Style Factors. *Chem. Biol. Interact.* **1996**, *102*, 17–36. [CrossRef]
23. Nikolsky, G.V. *Selected Research Works, Vol. 1: Theory of Fish Herd Dynamics*; VNIRO: Moscow, Russia, 1974.
24. Moiseev, P.A.; Azizova, N.A.; Kuranova, I.I. *Ichthology*; Legkaya I Pishchevaya Promishlennost: Moscow, Russia, 1981; 384p. (In Russian)
25. Vinberg, G.G. Length and weight of animals. *J. Gen. Biol.* **1971**, *32*, 714–722.
26. Morphology and Physiology of Fish. *Laboratory Practice: Learning Material*; Usov, M.M., Ed.; Morphology and Physiology of Fish: Gorki, Russia, 2017; 114p.
27. *Guide for Investigation of Fish Feeding in Natural Conditions (1961)*, NAS USSR ed.; USSR: Moscow, Russia, 1961; 263p.
28. *Guide for Investigation of Fish Feeding*; USSR: Vladivostok, Russia, 1986; 32p. (In Russian)
29. Bradford, M.M. A rapid and sensitive method for the quantitation of microgram quantities of protein utilizing the principal of protein-dye binding. *Anal. Biochem.* **1976**, *72*, 248–254. [CrossRef] [PubMed]
30. Ellman, G.L.; Courtney, K.D.; Andres, V.; Featherstone, I.; Featherstone, R.M. A new and rapid colorimetric determination of acetylcholinesterase activity. *Biochem. Pharmacol.* **1961**, *70*, 88–95. [CrossRef] [PubMed]
31. Buege, J.A.; Aust, S.D. Microsomal lipid peroxidation. *Methods Enzym.* **1978**, *52*, 302–310. [CrossRef]
32. Aebi, H. Catalase in vitro. *Methods Enzym.* **1984**, *105*, 121–126. [CrossRef]

33. Chevari, S.; Chaba, I.; Sekei, I. Role of superoxide dismutase in cellular oxidative processes and the method of its determination in biological materials. *Lab. Delo* **1985**, *11*, 678–681. (In Russian)
34. Moron, M.S.; Depierre, J.W.; Mannervik, B. Levels of glutathione, glutathione reductase and glutathione S-transferase activities in rat lung and liver. *Biochim. Biophys. Acta* **1979**, *582*, 67–78. [CrossRef] [PubMed]
35. Habig, W.H.; Pabst, M.J.; Jacoby, W.B. Glutathione S-transferases. The first enzymatic step in mercapturic acid formation. *J. Biol Chem.* **1974**, *249*, 7130–7139. [CrossRef]
36. Chuiko, G.M.; Podgornaya, V.A. Biochemical properties of cholinesterases in blood serum and some organs of fish from Lake Sevan. In *Ecology of Lake Sevan during the Period of Water Level Rise*; Nauka DNC: Makhachkala, Russia, 2010; Chapter 3.4; pp. 279–284. (In Russian)
37. Chuiko, G.M.; Zhelnin, Y.Y.; Podgornaya, V.A. Seasonal fluctuations in brain acetylcholinesterase activity and soluble protein content in roach (*Rutilus rutilus* L.): A freshwater fish from Northwest Russia. *Comp. Biochem. Physiol.* **1997**, *117*, 251–257. [CrossRef]
38. Chuiko, G.M.; Golovkina, E.I.; Podgornaya, V.A. Seasonal dynamics of acetylcholinesterase activity and protein content in different parts of the brain of freshwater teleost fish *Abramis brama* L. from the Rybinsk Reservoir. *Proc. Inst. Biol. Inland Waters Russ. Acad. Sci.* **2015**, *72*, 117–125. (In Russian) [CrossRef]
39. Tenji, D.; Micic, B.; Sipos, S.; Miljanovic, B.; Teodorovic, I.; Kaisarevic, S. Fish biomarkers from a different perspective: Evidence of adaptive strategy of *Abramis brama* (L.) to chemical stress. *Environ. Sci. Eur.* **2020**, *32*, 47. [CrossRef]
40. Tlili, S.; Jebali, J.; Banni, M.; Haouas, Z.; Mlayah, A.; Helal, A.N.; Boussetta, H. Multimarker approach analysis in common carp *Cyprinus carpio* sampled from three freshwater sites. *Environ. Monit. Assess.* **2009**, *168*, 285–298. [CrossRef] [PubMed]
41. Pala, A.; Serdar, O. Seasonal Variation of Acetylcholinesterase Activity as a Biomarker in Brain Tissue of *Capoeta umbla* in Pülümür Stream. *LimnoFish* **2018**, *4*, 98–102. [CrossRef]
42. Menéndez-Helman, R.J.; Ferreyroa, G.V.; dos Santos Afonso, M.; Salibián, A. Circannual rhythms of acetylcholinesterase (AChE) activity in the freshwater fish *Cnesterodon decemmaculatus*. *Ecotoxicol. Environ. Saf.* **2015**, *111*, 236–241. [CrossRef] [PubMed]
43. Fasulo, S.; Marino, S.; Mauceri, A.; Maisano, M.; Giannetto, A.; D'Agata, A.; De Domenico, E. A multibiomarker approach in *Coris julis* living in a natural environment. *Ecotoxicol. Environ. Saf.* **2010**, *73*, 1565–1573. [CrossRef]
44. Morozov, A.A.; Chuiko, G.M.; Brodskii, E.S. Functional state of the liver antioxidant system of the bream *Abramis brama* (L.) from Rybinsk Reservoir regions with different anthropogenic loads. *Inland Water Biol.* **2012**, *5*, 147–152. [CrossRef]
45. Bayir, A.; Sirkecioğlu, A.N.; Polat, H.; Mevlüt Aras, N. Biochemical profile of blood serum of siraz *Capoeta capoeta umbla*. *Comp. Clin. Pathol.* **2007**, *16*, 119–126. [CrossRef]
46. Nunes, B.S.; Travasso, R.; Gonçalves, F.; Castro, B.B. Biochemical and physiological modifications in tissues of Sardina pilchardus: Spatial and temporal patterns as a baseline for biomonitoring studies. *Front. Environ. Sci.* **2015**, *3*, 7. [CrossRef]
47. Angelini, C.; Amaroli, A.; Falugi, C.; Di Bella, G.; Matranga, V. Acetylcholinesterase activity is affected by stress conditions in *Paracentrotus lividus* coelomocytes. *Mar. Biol.* **2003**, *143*, 623–628. [CrossRef]
48. Kochkina, E.G.; Plesneva, S.A.; Zhuravin, I.A.; Turner, A.J.; Nalivaeva, N.N. Effect of hypoxia on cholinesterase activity in rat sensorimotor cortex. *J. Evol. Biochem. Physiol.* **2015**, *51*, 107–116. [CrossRef]
49. Lushchak, V.I. Environmentally induced oxidative stress in aquatic animals. *Aquat. Toxicol.* **2011**, *101*, 13–30. [CrossRef] [PubMed]
50. Chowdhury, S.; Saikia, S.K. Oxidative Stress in Fish: A Review. *J. Sci. Res.* **2020**, *12*, 145–160. [CrossRef]
51. Menon, S.V.; Kumar, A.; Middha, S.K.; Paital, B.; Mathur, S.; Johnson, R.; Kademan, A.; Usha, T.; Hemavathi, K.N.; Dayal, S.; et al. Water physicochemical factors and oxidative stress physiology in fish, a review. *Front. Environ. Sci.* **2023**, *11*, 1240813. [CrossRef]
52. Gu, Y.; Sun, J.L.; Yao, F.C.; Jiang, T.; Jin, C.X.; Shi, L.P.; Sun, S.K.; Song, F.B.; Luo, J. Long-term hypoxia and reoxygenation induced oxidative stress lead to immunosuppression and apoptosis in golden pompano (*Trachinotus blochii*). *Front. Mar. Sci.* **2023**, *10*, 1212571. [CrossRef]
53. Lushchak, V.I. Adaptive response to oxidative stress: Bacteria, fungi, plants and animals. *Comp. Biochem. Physiol. Part C Toxicol. Pharmacol.* **2011**, *153*, 175–190. [CrossRef]

water

MDPI

Article

The Distribution and Succession of Filamentous Algae in the Southern Taihang Catchment under Different Nutrient Regimes

Bo Yang [1,2,†], Yiguang Zhang [1,2,†], Man Zhang [1,2,*], Xucong Lv [1,2], Yuhua Li [1,2], Jingxiao Zhang [1,2], Xianfeng Wang [1,2], Xiaofei Gao [1,2], Xueqin Zhao [1] and Xiufen Wang [1,2]

1 College of Fisheries, Henan Normal University, Xinxiang 453007, China; yangbo19981027@163.com (B.Y.); zhangjingxiao@htu.edu.cn (J.Z.); gaoxiaofei@htu.edu.cn (X.G.); zxueqin0708@163.com (X.Z.)

2 Engineering Technology Research Center of Henan Province for Aquatic Animal Cultivation, Henan Normal University, Xinxiang 453007, China

* Correspondence: zm0378@163.com

† These authors contributed equally to this work.

Abstract: Human activities have resulted in the eutrophication of rivers, leading to heightened concerns regarding the occurrence of filamentous algal blooms. With the increasing utilization of rivers by humans, the occurrence of these nuisance filamentous algae is expected to increase in frequency in the future. Blooms primarily occur due to energy congestion at the trophic level of primary producers, resulting from inefficient energy flow in both the bottom-up and top-down pathways. To investigate the mechanism underlying the outbreak of filamentous algae, two streams in the southern Taihang catchment with different nutrient conditions were selected for this study. The objective of this study was to understand the effects of nutrient levels and other potential factors on the distribution and succession of filamentous algae. Our findings revealed a positive correlation between nutrient conditions and the biomass of filamentous algae. *Cladophora* and *Spirogyra* were identified as the dominant species among filamentous algae, each exhibiting unique distribution patterns in the two streams. *Spirogyra* thrived predominantly in the Baligou stream, where lower nutrient levels and warmer temperatures prevailed. In contrast, *Cladophora* flourished in the nutrient-rich Nanping stream at colder temperatures. Results from the generalized linear model indicated that the biomass of *Cladophora* was influenced by nutrient concentration, water depth, water temperature, and macrobenthic biomass. The biomass of *Spirogyra*, on the other hand, was primarily determined by water temperature, nutrient concentrations, water depth, and velocity. The positive correlation between *Cladophora* and macrobenthos revealed a possible mutually beneficial relationship, suggesting that macrobenthos may promote the growth of *Cladophora* by inhibiting periphytic diatoms. In return, the macrobenthos benefit from a secure refuge and an environment conducive to foraging and reproduction. This study suggested that to alleviate energy flow congestion in the benthic food chain, it is advisable to address this issue by either reducing nutrient loadings in rivers or enhancing the presence of benthivorous fishes in streams.

Keywords: eutrophication; filamentous algae; periphytic diatoms; macrobenthos; interpretation model

1. Introduction

The muddy substrates commonly observed in rivers and shallow lakes often reveal a prevalent dominance of either macrophytes or phytoplankton, attracting the interest of researchers who are particularly focused on investigating the regime shift dynamics between these two ecological components [1,2]. However, it is worth emphasizing that a distinct contrast arises when considering the rocky and artificial cement substrates within river environments, as these substrates exhibit a higher vulnerability to being overwhelmed by filamentous green algae, which are a diverse group of photosynthetic organisms characterized by long, thread-like structures [3]. The proliferation of filamentous green algae not only significantly diminishes the aesthetic appeal of water bodies but also gives rise to blockage issues in water supply channels, ultimately leading to disruptions in the water distribution network.

Eutrophication refers to the excessive input of nutrients (such as nitrogen and phosphorus) into water bodies due to human activities, leading to the overgrowth of algae [4], which is a significant issue that impacts rivers globally. This phenomenon triggers the overgrowth of nuisance filamentous algae, leading to a shift from well-maintained water bodies to nuisance ones with visible coverage on the water surface [5]. The formation of algal mats not only creates aesthetic and odor problems but also threatens water conservation and recreational value [6]. For example, the explosive growth of *Spirogyra*, a filamentous green alga, in the littoral zone of Lake Baikal is chiefly caused by groundwater contamination stemming from untreated sewage. The excessive growth of *Spirogyra* not only deteriorates water quality but also presents risks to the endemic species inhabiting the lake and endangers the lake's unique ecosystem [7]. Localized blooms of *Cladophora* in the western basin of Lake Erie are mainly driven by nutrient enrichment from agricultural runoff and the presence of invasive dreissenid mussels, resulting in significant ecological degradation and posing risks to recreational activities due to beach fouling and water quality deterioration [8].

Filamentous algae play a vital role in river and stream ecosystems, providing food and habitats for other organisms [9]. Species from the genera *Spirogyra*, *Cladophora*, and *Oedogonium* are common bloom-forming algae. In recent years, the frequency of filamentous algae outbreaks has significantly increased, raising concerns about the negative effects of these blooms on water quality and ecosystem health [10–12]. These algae can grow rapidly and form dense mats, depleting oxygen and blocking sunlight, which adversely affects aquatic organisms and habitats [13]. Researchers have observed that the biomass of *Spirogyra* and *Cladophora* species in filamentous algae tends to increase as nutrient levels increase [12]. However, there is currently a lack of comprehensive understanding of the distribution and succession patterns of filamentous algae communities under different nutrient conditions.

The southern Taihang catchment is a crucial region with abundant streams, multiple conservations, and biodiversity. The impact of human activities on each stream ecosystem varies significantly, resulting in notable differences in water trophic levels across different regions. This discrepancy can be attributed to the distinct management strategies employed in each conservation area. In recent years, frequent outbreaks of filamentous algae have occurred in these regions. Notably, each species of filamentous algae exhibits specific nutrient preferences during the vegetative season [14]. This emphasizes the importance of understanding the influence of trophic levels, such as nitrogen and phosphorus levels, on the growth and proliferation of filamentous algae. These factors can lead to notable changes in the distribution and succession patterns of filamentous algae communities.

Although the trophic level has been recognized as a crucial factor influencing filamentous algae biomass, our study adopted a comprehensive approach by investigating additional ecological variables, such as temperature, water velocity, light availability, and biotic interactions. These variables also significantly influence the distribution of filamentous algae communities [15,16]. Through a comprehensive examination of dense mat composition within aquatic ecosystems and analysis of environmental gradients, our research enhances the fundamental comprehension of filamentous algal blooms and provides practical applications for controlling.

The southern Taihang catchment encompasses several conservation zones, each of which primarily depends on rainfall for its water supply [17]. Nonetheless, the variation in management strategies among these zones renders the area particularly conducive to scientific research. The objective of this study was to explore the distribution and succession patterns of filamentous algae communities under two nutrient regimes within the southern Taihang catchment. By assessing filamentous algae species composition and biomass, we aim to deepen the understanding of their adaptive strategies to different nutrient conditions, as well as their occupation of diverse ecological niches and establishment of succession patterns. Specifically, this study aimed to (i) compare the distribution patterns of filamentous algae in the two regions, (ii) identify the factors influencing the presence of *Cladophora* and *Spirogyra* in the catchment, and (iii) develop effective interpretive models of filamentous algae for water management. In summary, this study significantly contributes to the understanding of the ecological dynamics of filamentous algae while providing valuable insights for the effective management and conservation of water resources.

2. Materials and Methods

2.1. Study Sites

The study sites for this research are two independent conservation areas, Nanping and Baligou, located in the southern Taihang catchment of central China (34°34′–40°43′ N, 110°14′–114°33′ E) (Figure 1). Both streams within these conservation zones primarily rely on rainfall as their water source [17], but their tributary networks are not interconnected. The management strategy in Nanping conservation is relatively lenient, with a local population of approximately 15,000. In contrast, Baligou conservation has a stricter management strategy, with a resident population of less than 500, resulting in distinct and independent water trophic levels. The lengths of the Nanping and Baligou streams are 70 km and 60 km, respectively, and both are punctuated by several waterfalls and deep pools. The dry season spans from October to May, while the wet season spans from June to September and accounts for approximately 70% of the annual precipitation. The catchment area receives an annual precipitation range of 390–750 mm [18]. The average annual temperature is 13 °C, with the lowest (−1.6 °C) and highest (26.3 °C) monthly temperatures occurring in January and July, respectively.

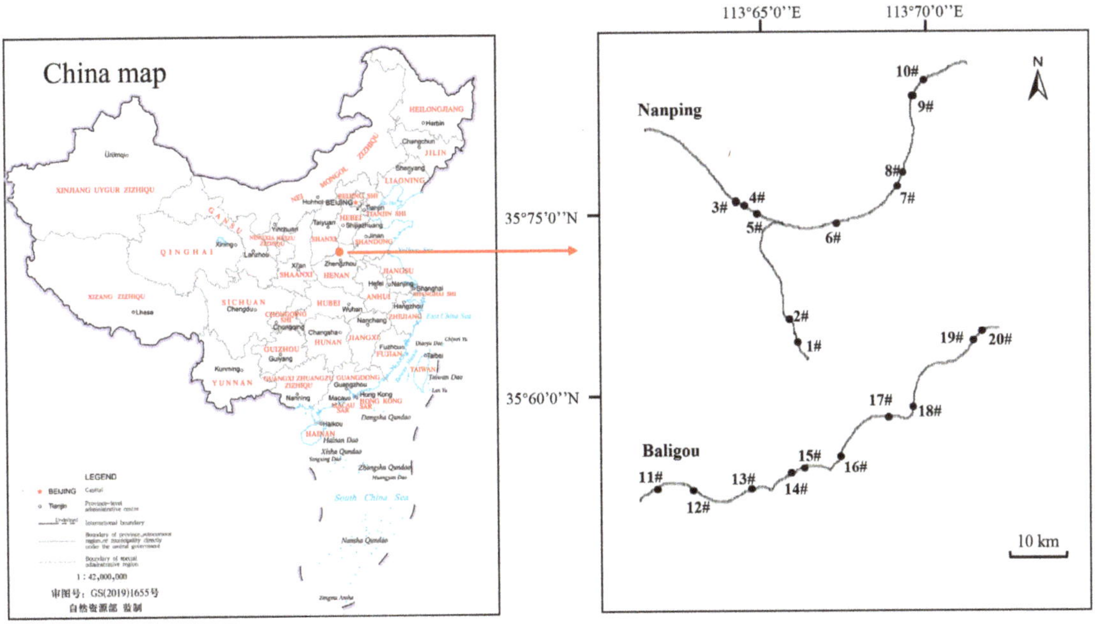

Figure 1. Locations of the Nanping and Baligou streams in the southern Taihang black dots show the sampling sites in the two streams. 1# to 10# are the sampling sites of Nanping; 11# to 20# are the sampling sites of Baligou.

2.2. Filamentous Algae Collection, Identification, and Measurement

Sampling was carried out during the peak growth season from mid-July to mid-September in 2023. Ten designated sites within each stream were sampled once every two weeks

(Figure 1), with a total of five sampling sessions. Filamentous algae samples were collected from the tops of rocks or boulders within the streams using a grab bucket within a defined area of 0.2 m^2. To ensure accuracy, replicate samples (n = 5) were collected from the top layer of the rocks, typically within the range of 20 to 50 cm below the water-surfaced depth to assess the filamentous algal biomass. Then, the defined area was scraped with brushes to collect all the filamentous algae on the rocks. The combined samples were then carefully washed using a series of sieves to remove any traces of mud or silt. To remove excess water, the samples were placed between layers of paper towels before their biomass was weighed using an electronic scale [19]. The collected samples were then preserved in polythene bags treated with a 4% formalin solution and transported to the laboratory for further analysis. All the collected algal samples were stored in a refrigerator for future research. In the laboratory, a light microscope (Nikon Ci-L, Tokyo, Japan) with a magnification of $\times 100$ was used to examine the samples, allowing for algae identification and calculation of the relative abundance of each species. The identification of filamentous algae followed the guidelines provided by the Flora Algarum Sinicarum Aquae Dulcis [20]. The spatial and temporal distribution profiles of the filamentous algae samples were analyzed.

2.3. Periphytic Algae Collection, Identification, and Measurement

Periphytic algae were collected simultaneously with filamentous algae samples by using natural substrates such as stones and boulders. The quantitative samples were brushed from the substrate surfaces under 10–30 cm of water using toothbrushes. At each site, samples from three locations were mixed to create a composite sample, which was then rinsed several times with distilled water. The composite sample was placed in a 100 mL plastic bottle and preserved with a 4% formalin solution. The examination of diatom taxa was conducted using an inverted microscope (Nikon ECLIPSE TS2; Tokyo, Japan) and the identification references used were Krammer and Lange-Bertalot and Diatoms of North America (http://diatoms.org/ (accessed on 24 July 2024)) [21,22]. To evaluate the competitive impact of periphytic algae on filamentous algae, the relative abundance of periphytic algae was assessed by counting the number of cells per square centimeter under a microscope [23]. To estimate the periphytic algae biomass, at least 10 individuals for each species were measured and then approximations to geometric solids were applied to calculate individual biovolume [24]. Periphytic algae biomass was estimated from the biovolume, assuming that 10^6 μm^3 corresponds to 1 μg of biomass.

2.4. Macrobenthos Collection, Identification, and Measurement

During the sampling of filamentous algae, macrobenthic samples were simultaneously collected using a Ponar grab. The collection area covered an area of 1 m^2 and focused on substrates predominantly composed of sand, gravel, or cobbles. To ensure effective sampling, riverbank areas containing rocks and boulders were avoided. The grab samples were washed and sieved on-site using a 0.5 mm mesh. The macrobenthos were then extracted using forceps and placed in sampling bottles. The macrobenthic samples were preserved in 70% alcohol, and additional sorting was conducted in the laboratory. The samples were examined under a stereomicroscope (Nikon SMZ645, Tokyo, Japan) at $\times 100$ magnification for identification in accordance with the methods described by Campos et al. [25]. The wet biomass of the macrobenthos was determined using an electronic scale with 10^{-5} g of precision [26].

2.5. Fish Collection, Identification, and Measurement

Fish samples were collected from the two streams using fish traps constructed with 16 steel frames measuring 25 cm $\times$ 12 cm $\times$ 15 cm and spaced 6 cm apart. The fish entrance had a vertical depth of 15 cm and a mesh size of 4 mm, with a total trap length of 3 m. The traps were set in the evening at 18:00 and retrieved the following morning at 6:00, allowing for a full night of sampling. In the laboratory, the collected fish were carefully identified and classified following the methods described by Joseph S. Nelson [27]. Subsequently, the samples were accurately counted, weighed (using an electronic scale with a precision of 0.1 g), and measured for length (with an accuracy of 0.1 cm). The catch per unit effort (CPUE) of fish was calculated using the formula provided by Harley et al. [28] and Gupta et al. [29]:

$$CPUE = \frac{Total\ catch}{Sum\ of\ efforts}$$

The *sum of effort* was measured in terms of time and the number of traps.

2.6. Abiotic Conditions

Various abiotic conditions were measured to analyze their impacts on filamentous algae. These conditions included water temperature (T), pH, dissolved oxygen (DO), electrical conductivity (EC), water velocity (V), water depth (Dep), light intensity, elevation, hardness, soluble reactive phosphorus (SRP), nitrate (NO_3-N), nitrite (NO_2-N), ammonium (NH_4-N), ferrite (Fe^{2+}), silicate ($SiO_3{}^{2-}$), total carbon (TC), and inorganic carbon (IC) concentrations. We utilized several instruments to collect data for our water quality analysis. A multiparameter water quality monitor (YSI Pro20; Ohio, USA) was used to measure T, pH, DO, EC, and Dep at the water surface. A flowmeter (Xiangruide LS300-A; Weifang, China) was used to measure water velocity. An illuminometer (Spectrum 3415FX LightScout; San Antonio, TX, USA) was used to measure light intensity, and a mobile positioning device (Huawei Pocket; Shenzhen, China) was used to measure elevation. For water sample collection, we employed a 5 L iron sampler. The water was collected at a depth ranging from 0.2 m to 0.5 m.These samples were subsequently analyzed for SRP, NO_3-N, NO_2-N, NH_4-N, hardness, Fe^{2+}, and $SiO_3{}^{2-}$ concentrations in accordance with the Chinese National Standards for Water Quality and the Environmental Protection Agency of the USA [30]. Subsequently, 20 mL of the water samples were filtered through a 0.45-µm membrane filter and subjected to a carbon analyzer (Multi N/C 3100, Jena, Germany) for TC and IC measurements.

2.7. Data Analysis

2.7.1. Patterns in the Filamentous Algae Community and Environmental Contributions

Principal coordinate analysis (PCoA) was performed using the 'VEGAN' package (v2.6-6.1) in R based on Bray-Curtis distance metrics calculated for the species compositions of the filamentous communities [31]. The data points were color-coded and shaped based on two independent streams (Nanping and Baligou). Ellipses with 90% confidence intervals were constructed to indicate differences between the two streams.

A detailed correspondence analysis using the Decorana function was conducted with R's 'VEGAN' package. The results indicated that the first axis had a gradient length of 1.98, suggesting that a linear model was appropriate for further analysis. To identify the key factors influencing filamentous algal species composition, a forward-selection procedure (ordistep function) with Monte Carlo permutation tests (999 permutations) was performed using the ordistep function to select a parsimonious set of factors explaining a statistically significant ($p < 0.05$) amount of variation in the filamentous algae data in each sample [32]. The explanatory variables were sequentially tested using the variance inflation factor (VIF) to eliminate collinearity among the selected factors, employing the VIF function until all the VIF values were less than 20. Finally, a linear model redundancy analysis (RDA) was conducted to examine patterns of interaction between filamentous algae and environmental determinants [33].

To determine the contributions of potential explanatory variables to specific filamentous algae, a variation partitioning analysis (VPA) based on the RDA algorithm was utilized. Prior to conducting the VPA, Hellinger standardization was applied to the filamentous algae biomass data. Furthermore, feature scaling transformation was employed to normalize the environmental explanatory conditions, thereby avoiding differential weighting. VPA was performed using the 'RDACCA.HP' package (v1.1-0) [34].

2.7.2. Generalized Linear Models

To assess the effects of the explanatory variables on filamentous algae biomass, a generalized linear model (GLM) was used to determine the relative contributions of these variables. The 'best' model from the GLMs was selected using a forward stepwise procedure based on the Akaike information criterion (AIC) and Bayesian information criterion (BIC). The model incorporating the set of explanatory factors with the lowest AIC and BIC values was selected [35]. All GLM analyses were performed using the 'GLMULTI' package (v1.0.8) in R.

2.7.3. Association Network between the Filamentous Algae and Macrobenthos

All macrobenthic species and two filamentous algae (*Cladophora glomerata* (Linnaeus) Kützing, 1843 and *Spirogyra communis* (Hassall) Kützing, 1849), were utilized for network analyses aiming to determine the correlations between the filamentous algae and macrobenthos. To simplify the data sets, only species that were present in more than ten samples were included in the construction of networks. Subsequently, all possible pairwise Spearman's rank correlations (r) between these species were calculated using the 'PICANTE' R package. Only correlations that were both effective ($r > 0.4$ or $r < -0.4$) and statistically significant ($p < 0.05$) were used for the network analyses [36]. To visualize the network and perform modular analysis, Gephi 0.9.2 was used. Additionally, node-level topological properties such as degree, betweenness, closeness, and eigenvector were calculated using the 'IGRAPH' R package. Statistical differences in node-level attributes across different taxa were determined using the nonparametric Mann-Whitney U test. Macrobenthos that had a high degree (>100) and low betweenness centrality values (<5000) were considered keystone species for the *C. glomerata* and *S. communis* networks [37].

2.7.4. Other Statistical Analysis

Independent *t*-tests were used to assess differences in abiotic and biotic variables between the two streams.

3. Results

3.1. Characteristics of Nutrients, Other Biotic, and Abiotic Conditions in the Two Streams

During the two-month observation period, there was a peak in the growth of filamentous algae followed by a gradual decrease, marking this sampling period as a significant stage in the succession of filamentous algal communities. Table 1 presents the mean values of nutrients and biotic and abiotic conditions in the Baligou and Nanping streams. All 15 abiotic factors and 9 biotic parameters exhibited wide ranges, demonstrating diverse spatial and temporal disparities.

Independent T-tests revealed significant increases in the SRP and dissolved inorganic nitrogen (DIN, NO_3^--N + NO_2^--N + NH_4^+-N) in the Nanping stream compared to those in the Baligou stream (SRP: $p = 0.005$, DIN: $p < 0.001$) (Figure 2a,b). The Baligou stream has a significantly higher water temperature than Nanping ($p = 0.021$) (Figure 2c). No significant differences were found between the two streams in terms of pH, DO, EC, V, Dep, light intensity, elevation, hardness, ferrite, silicate, TC, or IC ($p > 0.05$).

Macrobenthos are grazers of benthic algae [38]. It was found that the abundance and biomass of macrobenthos were significantly higher in Nanping than in the Baligou stream (abundance: $p = 0.003$; biomass: $p = 0.003$) (Table 1, Figure 2d). Fish are the top predators in benthic food chains [39]. The CPUE of total fish was significantly greater in the Baligou stream than in the Nanping stream ($p = 0.001$). A total of five fish species were identified across all sampling sites in the two streams: *Phoxinus oxycephalus* (Sauvage & Dabry de Thiersant, 1874), *Gnathopogon taeniellus* (Nichols, 1925), *Pseudorasbora parva* (Temminck & Schlegel, 1846), *Onychostoma macrolepis* (Bleeker, 1871), and *Rhinogobius giurinus* (Rutter, 1897). Notably, *p. oxycephalus* was the predominant species, and its CPUE was also significantly greater in the Baligou stream than in the Nanping stream ($p = 0.002$). Periphytic diatoms are the most important competitors of filamentous algae [40]. After identification, all periphytic algae were periphytic diatoms. The biomass of periphytic diatoms in Baligou was slightly greater than those in Nanping but the difference was not statistically significant ($p > 0.05$).

Table 1. Environmental factors and biotic parameters (mean values ± standard error, n = 50) in the two streams.

Classification	Environmental Factors and Biotic Parameters	Nanping	Baligou
Environmental factors	T (°C)	19.64 ± 3.34 a	22.82 ± 3.32 b
	pH	8.53 ± 0.26	8.59 ± 0.25
	DO (mg/L)	8.95 ± 1.01	9.03 ± 1.50
	EC (μs/cm)	409.86 ± 24.79	388.76 ± 27.32
	Hardness (mol/L)	0.81 ± 0.04	0.79 ± 0.12
	V (m/s)	0.13 ± 0.19	0.10 ± 0.16
	Dep (m)	0.40 ± 0.44	0.26 ± 0.24
	Light (μmol/m^2·s)	403.16 ± 493.37	641.66 ± 605.37
	Elevation (m)	718.90 ± 101.61	446.20 ± 94.49
	SRP (mg/L)	0.024 ± 0.06 a	0.004 ± 0.00 b
	DIN (mg/L)	7.82 ± 3.00 a	6.60 ± 1.17 b
	Fe^{2+} (mg/L)	0.11 ± 0.03	0.12 ± 0.09
	SiO_3^{2-} (mg/L)	216.15 ± 22.69	212.84 ± 27.02
	TC (mg/L)	43.71 ± 5.83	40.39 ± 4.66
	IC (mg/L)	36.64 ± 4.98	34.21 ± 3.67
	Filamentous algae biomass (g/m^2)	49.73 ± 70.47 a	35.95 ± 37.28 b
	S. communis biomass (g/m^2)	4.27 ± 16.85 a	12.48 ± 22.66 b
	C. glomerata biomass (g/m^2)	38.88 ± 68.86 a	11.06 ± 13.46 b
	PDA (cells/cm^2)	51897.45 ± 58567.75	60180.55 ± 68943.36
	PDB (mg/m^2)	603.01 ± 618.43	759.94 ± 918.36
	MBA (cells/m^2)	445.02 ± 514.47 a	198.65 ± 189.86 b
	MBB (g/m^2)	9.35 ± 28.55 a	0.89 ± 1.29 b
	Phoxinus oxycephalus CPUE (g/trap·d)	2171 ± 871.32 a	7794 ± 3913.26 b
	Total fish CPUE (g/ trap·d)	3159 ± 340.74 a	8177 ± 586.73 b

Note(s): a, b: Different letters denote significant differences ($p < 0.05$) between the two streams based on the independent T-tests. T: water temperature, DO: dissolved oxygen, EC: electrical conductivity, V: water velocity, Dep: water depth, Light: light intensity, SRP: soluble reactive phosphorus, DIN: dissolved inorganic nitrogen, Fe^{2+}: ferrite, SiO_3^{2-}: silicate, TC: total carbon, IC: inorganic carbon, PDA: periphytic diatom abundance, PDB: periphytic diatom biomass, MBA: macrobenthic abundance, MBB: macrobenthic biomass, CPUE: catch per unit effort of fish.

3.2. Distribution Patterns of Filamentous Algae Communities

During the observation period, filamentous algae were highly abundant in both the Nanping and Baligou streams, with average biomasses of 49.73 ± 70.47 g/m^2 and 35.95 ± 37.28 g/m^2, respectively (Table 1). Notably, the total biomass of filamentous algae in the Nanping Reservoir was significantly greater than that in the Baligou River (biomass: $p = 0.036$; biomass: $p = 0.028$). Five taxa were identified across all sampling sites in the two streams, namely, *C. glomerata* (Linnaeus) Kützing, 1843, *S. communis (Hassall)* Kützing, *1849*, *Zygnema* sp. (C. Agardh in Liljeblad, 1816), *Oedogonium* sp. (Link ex Hirn, 1900), and *Tribonema* sp. (Derbes and Solier, 1851). Among these, *C. glomerata* and *S. communis* were the most dominant taxa and were responsible for forming benthic algal blooms (Figure 3). In contrast, the remaining three species collectively accounted for less than 25% in these two areas (Figure 4). Interestingly, the biomass of *S. communis* was significantly greater in the Baligou stream than in the Nanping stream ($p = 0.027$) (Figure 4a), while the biomass of *C. glomerata* was significantly greater in the Nanping stream than in the Baligou stream ($p < 0.001$) (Figure 4b). In the Nanping stream, *C. glomerata* accounted for 78.1% of the total biomass and remained abundant and persistent throughout the observation period. Moreover, *S. communis* contributed 35.4% of the total biomass in the Baligou stream.

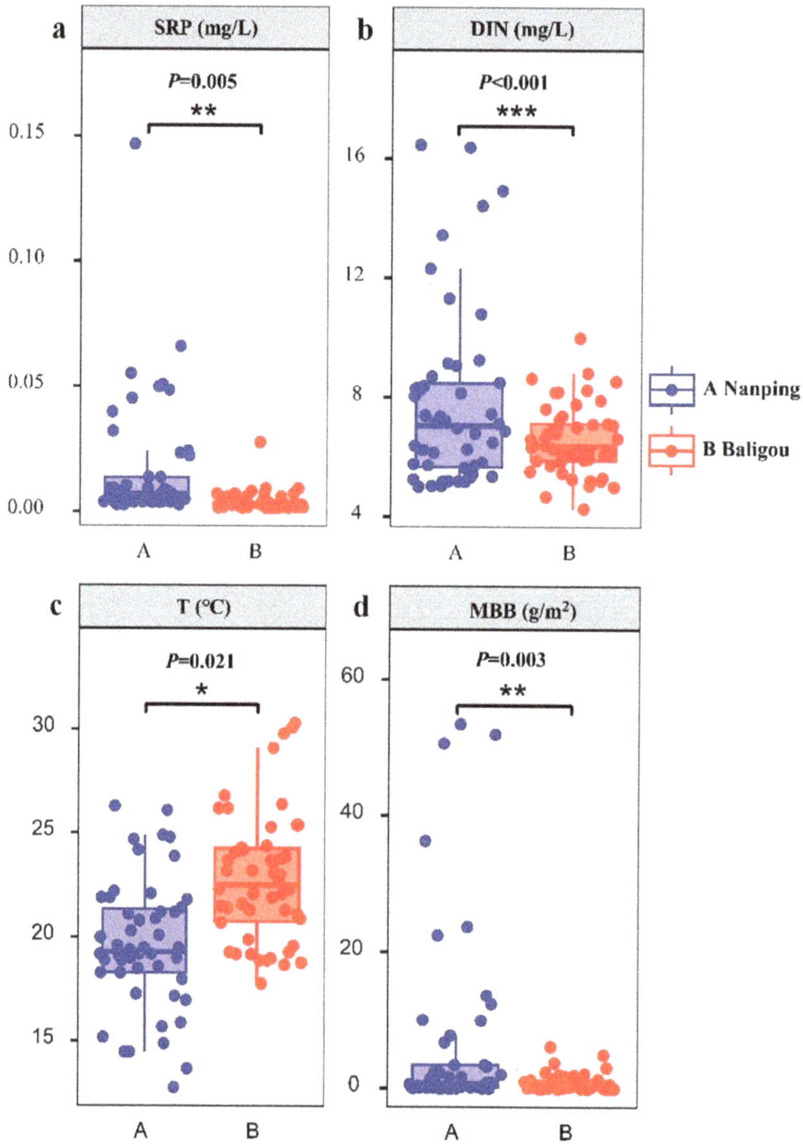

Figure 2. Environmental conditions exhibiting significant differences between the two streams. Boxes represent the interquartile range, central bars represent the median, and solid lines represent the data range. Corresponding letters denote means that do not statistically differ from one another. * Denote that statistically differ from the other ($p \leq 0.05$). ** indicates a significant difference ($p \leq 0.01$). *** indicates an extremely significant difference ($p \leq 0.001$). SRP: soluble reactive phosphorus, DIN: dissolved inorganic nitrogen, T: water temperature, MBB: macrobenthic biomass.

Figure 3. Morphology of the *C. glomerata* (**a**,**b**) and *S. communis* (**c**,**d**).

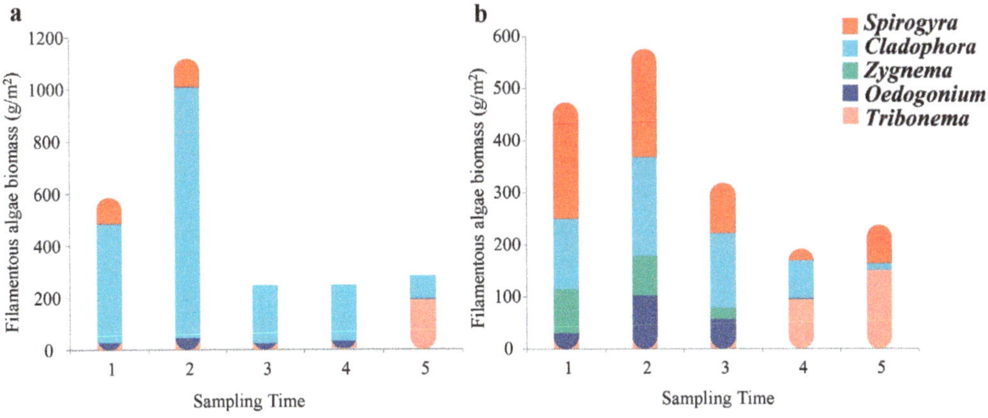

Figure 4. Taxa composition of filamentous algae communities in the Nanping (**a**) and Baligou (**b**) plots.

To assess the difference in community structure for filamentous algae in the two streams, PCoA was conducted, incorporating all temporal and spatial sampling sites. The first two axes explained 32.0% of the variance in the species composition of the filamentous algae. The analysis revealed that the filamentous algal species in the two streams formed distinct communities, indicating a significant difference in community structure ($p = 0.005$) (Figure 5).

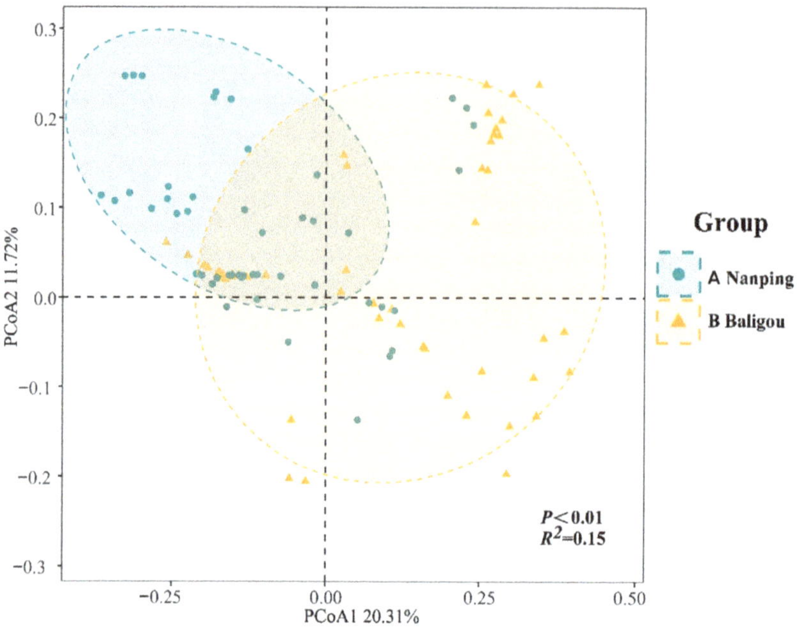

Figure 5. Principal coordinate analysis (PCoA) of filamentous algae communities in the two streams.

3.3. Determinants of the Distribution of Filamentous Algae Communities

RDA was conducted using 17 variables to assess the influence of abiotic and biotic factors on the biomass of filamentous algal species in the two streams. These variables included water temperature, depth, velocity, pH, DO, EC, TC, IC, SRP, DIN, light intensity, elevation, hardness, ferrite, silicate, periphytic diatom biomass, and macrobenthic biomass. The forward selection results revealed that temperature, depth, velocity, DIN, SRP, macrobenthic biomass, and periphytic diatom biomass accounted for 19.46% and 15.18%, respectively, of the variation in environmental variables along the first and second axes (Figure 6, R^2 = 0.35, p = 0.001). Nutrient levels were identified as key determinants influencing the species composition of filamentous algae. Additionally, abiotic factors such as water depth, velocity, and temperature played significant roles in the distribution of filamentous algae. Biotic factors, including macrobenthic biomass and periphytic diatom biomass, were also found to be important variables.

The VPA results for the 17 explanatory variables indicated that temperature, DIN concentration, SRP concentration, pH, depth, and velocity collectively accounted for more than 70% of the biomass of *S. communis* (Figure 7a). On the other hand, DIN, depth, macrobenthic biomass, SRP, and temperature were identified as determinants of the biomass of *C. glomerata* (Figure 7b). These findings clearly demonstrated the significant contribution of nutrient levels to the presence of both filamentous algal species, highlighting the importance of water trophic status in determining their biomass.

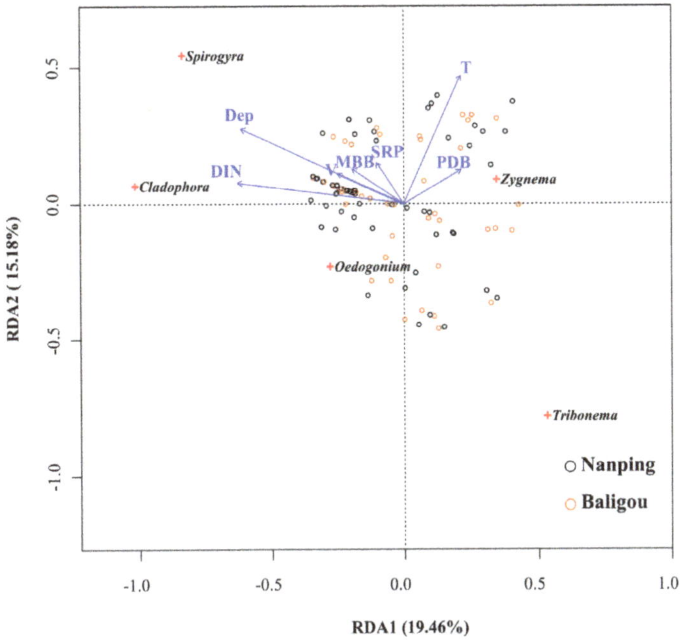

Figure 6. Redundancy analysis (RDA) of filamentous algal species composition and forward-selected environmental factors. T: water temperature, V: water velocity, Dep: water depth, SRP: soluble reactive phosphorus, DIN: dissolved inorganic nitrogen, PDB: periphytic diatom biomass, MBB: macrobenthic biomass.

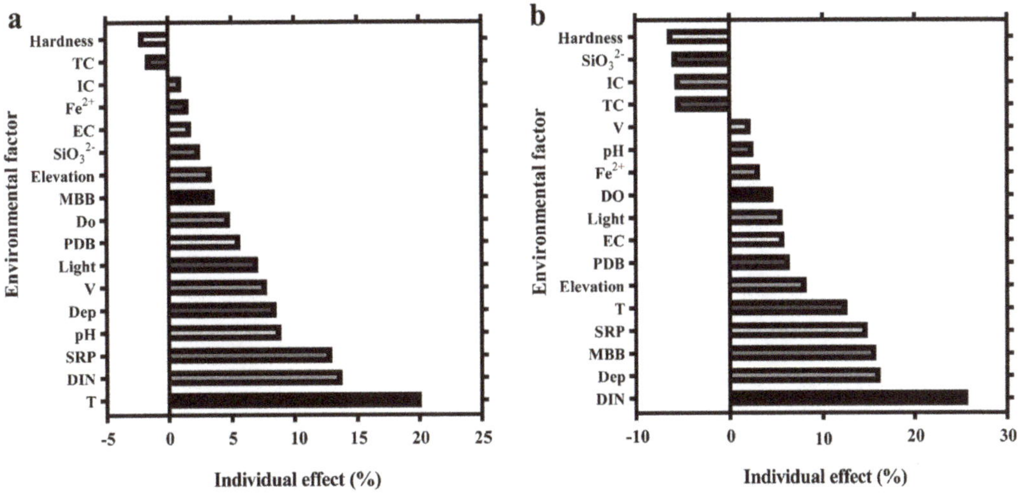

Figure 7. Contributions of the potential variables explaining the biomass of *S. communis* (**a**) and *C. glomerata* (**b**) in the VPA. T: water temperature, DO: dissolved oxygen, EC: electrical conductivity, V: water velocity, Dep: water depth, SRP: soluble reactive phosphorus, DIN: dissolved inorganic nitrogen, Fe^{2+}: ferrite, $SiO_3{}^{2-}$: silicate, TC: total carbon, IC: inorganic carbon, PDB: periphytic diatom biomass, MBB: macrobenthic biomass.

The best GLM results from the 17 explanatory variables are presented in Table 2. The determination coefficients (R^2) of 0.37 and 0.27, along with AIC values of 65.88 and 70.12 for *S. communis* and *C. glomerata*, respectively, indicated the valid performance of the GLMs. The Generalized Linear Model (GLM) analysis revealed that both *S. communis* and *C. glomerata* had positive correlations with water temperature, nutrient concentrations, and water depth. Moreover, the models revealed distinct differences for the two species. *S. communis* demonstrated a negative correlation with water velocity (Figure 8a), whereas *C. glomerata* showed a positive correlation with macrobenthic biomass (Figure 8b). A positive correlation observed between macrobenthic biomass and *C. glomerata* biomass suggested a potential mutually beneficial relationship between the two taxa. Notably, *C. glomerata* may serve as an important refuge and habitat for macrobenthos, while macrobenthos may help eliminate competitors for *C. glomerata* in benthic habitats. Macrobenthos increased the availability of light and nutrients for filamentous algae by feeding on periphytic diatoms, thereby promoting the growth of filamentous algae.

Table 2. Best GLMs for explaining the biomass of *S. communis* and *C. glomerata*.

Species	Adj-R^2	*AIC*	*BIC*	Best Model
S. communis biomass	0.38	65.88	69.92	Y = 0.343T + 0.308SRP + 0.249DIN + 0.22Dep − 0.187V + 0.643
C. glomerata biomass	0.32	70.12	75.38	Y = 0.365DIN + 0.322Dep + 0.238MBB + 0.198T + 0.152SRP + 1.782

Note(s): AIC: Akaike information criterion, BIC: Bayesian information criterion, T: water temperature, V: water velocity, Dep: water depth, SRP: soluble reactive phosphorus, DIN: dissolved inorganic nitrogen, MBB: macrobenthic biomass.

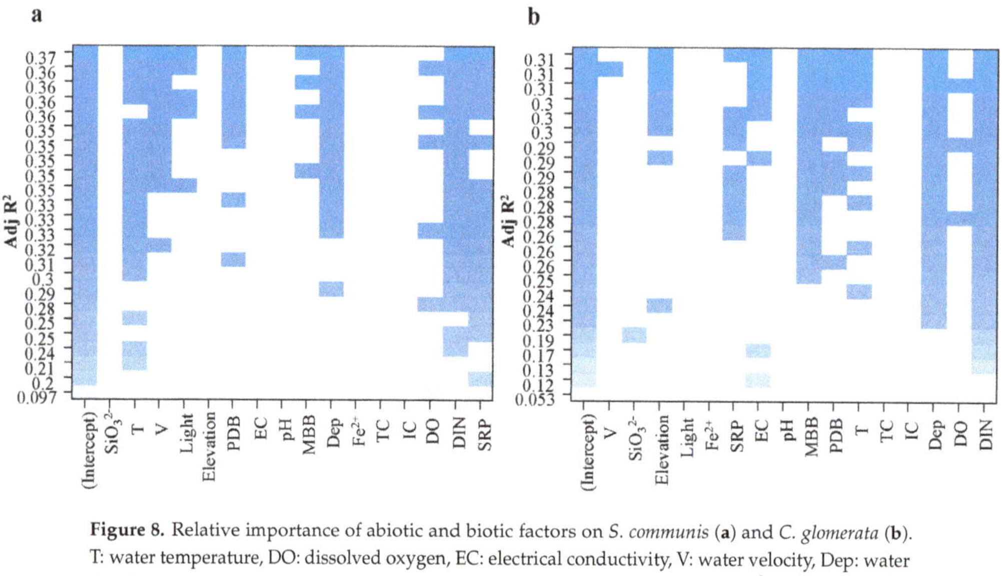

Figure 8. Relative importance of abiotic and biotic factors on *S. communis* (**a**) and *C. glomerata* (**b**). T: water temperature, DO: dissolved oxygen, EC: electrical conductivity, V: water velocity, Dep: water depth, SRP: soluble reactive phosphorus, DIN: dissolved inorganic nitrogen, Fe^{2+}: ferrite, $SiO_3{}^{2-}$: silicate, TC: total carbon, IC: inorganic carbon, PDB: periphytic diatom biomass, MBB: macrobenthic biomass.

The co-occurrence networks between filamentous algae and macrobenthos were further examined. Figure 9 demonstrates that the macrobenthic species associated with *S. communis* and *C. glomerata* were distinctly different. The results demonstrated that macrobenthos species associated with *S. communis* mostly exhibited negative effects, such as *Cordulegastridae* sp., *Polypedilum* sp., *Baetis* sp., and *Hydropsyche* sp., which are common consumers of *Spirogyra* [41]. Conversely, macrobenthic species closely related to *C. glomerata* were primarily herbivores with positive correlations, including *Radix swinhoei*

(Adams, 1866), *Hydropsyche* sp. (Pictet, 1834), and *Phryganea* sp. (Linnaeus, 1758). Due to the poor palatability of *Cladophora*, hardly any macrobenthos can feed on it. Competition from periphytic diatoms has become the most significant limiting factor for the growth of *Cladophora*. At sampling sites with abundant herbivorous macrobenthos, periphytic diatoms were rarely observed on *C. glomerata* (Figure 10a). In contrast, at sampling sites lacking herbivorous macrobenthos, a large amount of periphytic diatoms were found attached to *C. glomerata* (Figure 10b). These findings suggested that these herbivores may positively affect the growth of *C. glomerata* by consuming periphytic diatoms attached to filamentous algae.

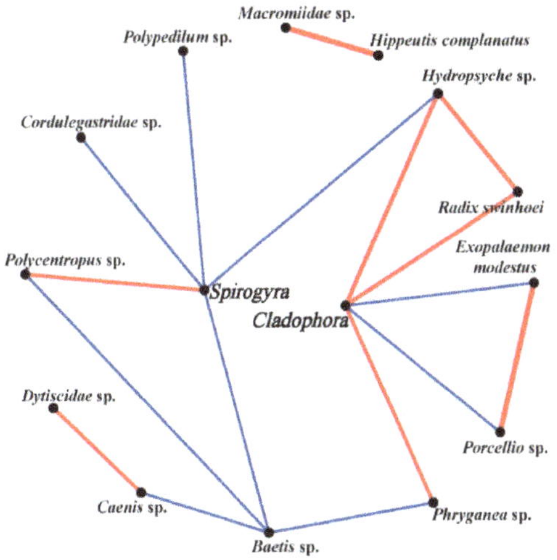

Figure 9. Co-occurrence networks between the filamentous algae and macrobenthic taxa. The red lines represent positive correlations, and the blue lines represent negative correlations.

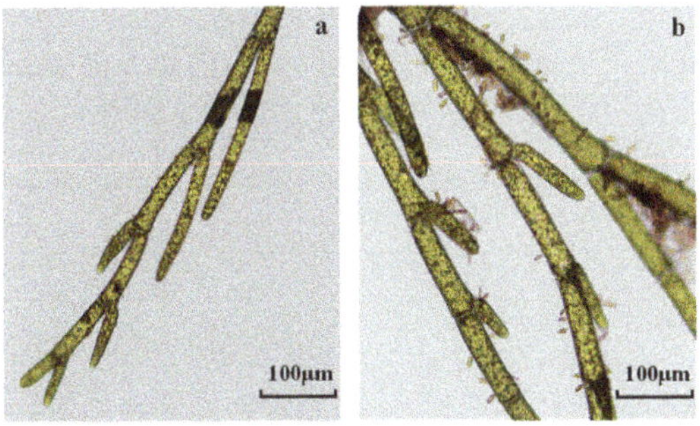

Figure 10. *C. glomerata* at the sites with (**a**) and without macrobenthos (**b**).

4. Discussion

Recent research has identified the frequent bloom of two common filamentous algae, *Cladophora* and *Spirogyra*, in streams [42]. While these streams have exhibited a transition from *Spirogyra* to *Cladophora*, the precise mechanisms underlying this shift remain unclear and lack definitive evidence. During the observed period, we found a low diversity but large biomass of filamentous algae in the southern Taihang catchment. Our study revealed that *S. communis* and *C. glomerata* were the two most prevalent filamentous algal species capable of causing blooms in this area. *C. glomerata* tended to dominate under high-nutrient conditions, while *S. communis* was more likely to thrive at slightly lower nutrient levels, reflecting their different ecological preferences.

Previous studies have shown that the coverage and biomass of filamentous algae generally increase in aquatic ecosystems experiencing eutrophication. For instance, Bosch et al., Higgins et al., and Olsen et al. have indicated this trend [43–45]. Additionally, Page et al. conducted a study in Conesus Lake, New York, and demonstrated a correlation between high filamentous algae coverage and elevated nitrogen and phosphorus loadings in streams [46]. Francoeur et al. [15] investigated the reappearance of filamentous algae blooms in the Great Lakes and found that the growth of filamentous algae was stimulated by mussel nutrient excretion and artificial fertilization with nitrogen and phosphorus. In our study, we observed a significantly greater total biomass of filamentous algae in Nanping compared to Baligou, confirming an increase in filamentous algae coverage with trophic level. Additionally, *C. glomerata* exhibited greater susceptibility to eutrophication than *S. communis* did, consistent with previous findings that *Spirogyra* is tolerant to oligotrophic conditions, while *Cladophora* prefers eutrophic conditions [47,48]. Hawes [49] noted that *Spirogyra* can store nitrogen and phosphorus, even at their low concentrations in water, allowing for the development of a large biomass of this taxon, highlighting its tolerance to nutrient-poor conditions. Han et al. [50] also suggested that internal reserves of inorganic nutrients likely contribute to the initial development of *Spirogyra*, even under low nutrient concentrations. Therefore, as nutrient levels increase, the filamentous algae community may shift from *Spirogyra* to *Cladophora*, which have lower palatability and greater biomass [51].

In addition to nutrient levels, water temperature is also believed to play a significant role in the distribution of filamentous algae, as different taxa have specific temperature preferences that affect their growth and survival [10,12,44,51]. It is thought that the growth of *Cladophora* is favored by relatively high water temperatures [44]. Adams and Stone [52] determined that the optimal temperature range for *Cladophora* was from 25 to 30 °C. Furthermore, Cambra-Sanchez and Aboal [53] reported that the *Spirogyra* typically appears in early spring and is subsequently succeeded by a vigorous growth of *Cladophora* in early summer, suggesting distinct temperature preferences for these two species. However, our findings showed that *S. communis* had a broader distribution in the Baligou, where temperatures were higher, compared to the Nanping stream. The positive effect of water temperature on the distribution of *S. communis*, as indicated in the GLM, also suggested that this species has a broad temperature niche.

Several studies have documented the preferences of these two species for other environmental factors. For instance, Margalef [54] noted that the filamentous algae of *Spirogyra* are replaced by *C. fracta* and *C. glomerata* mats in more permanent waters. *Spirogyra* can survive periodic drought, produce durable zygospores, and become the dominant genus in all types of temporary water [53,55]. Frossard et al. [56] reported that *Cladophora* attached to a solid substrate was more resistant to the effects of wind, waves, and current compared to weakly attached *Spirogyra*. In our study, the GLM results also indicated a preference for *S. communis* for slow flow. Additionally, *C. glomerata* was found to be more sensitive to water depth, confirming its preference for living in permanent waters that do not dry up in the summer.

We observed a positive correlation between *C. glomerata* and macrobenthos, which are larger aquatic organisms that can influence algal growth through grazing and nutrient recycling. It is well documented that grazers can remove periphytic diatoms from the surface of host plants, thus reducing the shading effect on the latter [57,58]. Ye et al. [59] demonstrated that grazing by the shrimp *Neocaridina denticulata sinensis* (Kemp, 1918) removed periphytic diatoms from the submerged plant *Vallisneria denseserrulata* (Makino, 1921), leading to a notable increase in biomass accumulation in the host plants. In spite of these studies have focused primarily on submerged macrophytes as host plants, the ecological effects of these macrobenthos are consistent. Some researchers believe that macrobenthos not only consume periphytic microalgae but also filamentous macroalgae [60]. Notably, a reduction in filamentous algae is often achieved through grazing by large herbivores rather than small species [61]. Pinowska [62] noted out that grazing and nutrient release by the small-sized species *Lymnaea turricula* stimulated the growth of *Cladophora* sp. The macrobenthic species related to *C. glomerata* in our study were mostly small herbivores, indicating that the presence of the macrobenthos may increase the growth of *Cladophora* mainly by reducing the biomass of periphytic diatoms. A high fish density in Baligou has a controlling effect on macrobenthic density, which promotes the development of periphytic diatoms. These diatoms compete strongly with *Cladophora*, ultimately leading to the suppression of *Cladophora* biomass (Figure 11 Left). In contrast, the low fish density in Nanping allows for the flourishing of macrobenthos, which controls periphytic diatoms, ultimately promoting the growth of *Cladophora* (Figure 11 Right).

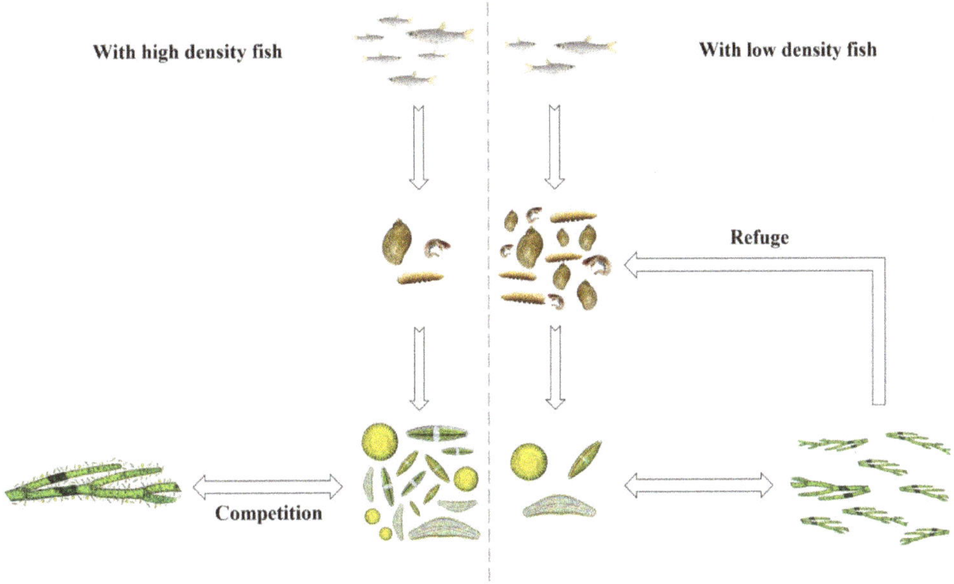

Figure 11. The mechanism diagram illustrating the variations in *Cladophora* biomass under two different fish population density conditions.

5. Conclusions

The findings of our study have important implications for river restoration. As nutrient levels rise, rivers undergo an initial increase in *Spirogyra* density, followed by *Cladophora*, often resulting in a shift from clear water to nuisance water with surface coverage. Many river restoration projects aim to reduce filamentous algal coverage as a primary goal. However, controlling *Cladophora* through biomanipulation is challenging due to its lower palatability to grazers than to other benthic algae [51]. Nevertheless, *Cladophora* provides substrates for periphytic diatom growth, creating the possibility of controlling it by leveraging the competition between *Cladophora* and its periphytic diatoms [63,64]. Food web management is often used to reduce the abundance of macrobenthos, which in turn decreases predation pressure on periphytic diatoms and can be an effective approach. Manipulating the benthic food web to diminish grazing control of periphytic diatoms may prove to be a crucial management tool, as *Cladophora* may be shaded by these periphytic diatoms. Increasing the population of benthivorous fishes, such as the indigenous species *Phoxinus oxycephalus*, in rivers may yield significant improvements with less effort. Nonetheless, managing trophic states in rivers remains the fundamental approach to water management.

Author Contributions: Conceptualization: B.Y., Y.Z., M.Z., Y.L., J.Z., X.W. (Xianfeng Wang), X.G., X.Z. and X.W. (Xiufen Wang); Data curation, X.L.; Formal analysis: Y.Z. and M.Z.; Funding acquisition: M.Z.; Methodology: B.Y., Y.Z., M.Z., X.L., Y.L., J.Z., X.W. (Xianfeng Wang), X.G., X.Z. and X.W. (Xiufen Wang); Resources: B.Y., X.L. and Y.L.; Software: Y.Z.; Writing—original draft: Y.Z.; Writing—review & editing: B.Y. All authors have read and agreed to the published version of the manuscript.

Funding: This research was funded by the China Agriculture Research System (CARS-50) and the National Natural Science Foundation of China (Grant No. U1904124). The article processing charge (APC) was funded by the China Agriculture Research System (CARS-50).

Data Availability Statement: The data generated during the present study are available from the corresponding author upon reasonable request.

Conflicts of Interest: The authors declare no conflict of interest.

References

1. Genkai-Kato, M.; Vadeboncoeur, Y.; Liboriussen, L.; Jeppesen, E. Benthic-planktonic coupling, regime shifts, and whole-lake primary production in shallow lakes. *Ecology* **2012**, *93*, 619–631. [CrossRef] [PubMed]
2. Lawniczak-Malińska, A.; Ptak, M.; Celewicz, S.; Choiński, A. Impact of lake morphology and shallowing on the rate of overgrowth in hard-water eutrophic lakes. *Water* **2018**, *10*, 1827. [CrossRef]
3. Kravtsova, L.S.; Izhboldina, L.A.; Khanaev, I.V.; Pomazkina, G.V.; Rodionova, E.V.; Domysheva, V.M.; Sakirko, M.V.; Tomberg, I.V.; Kostornova, T.Y.; Kravchenko, O.S.; et al. Nearshore benthic blooms of filamentous green algae in Lake Baikal. *J. Great Lakes Res.* **2014**, *40*, 441–448. [CrossRef]
4. Dodds, W.; Smith, V. Nitrogen, phosphorus, and eutrophication in streams. *Inland Waters* **2016**, *6*, 155–164. [CrossRef]
5. Oberholster, P.; Somerset, V.; Truter, J.; Botha, A.M. The interplay between environmental conditions and filamentous algae mat formation in two agricultural influenced South African rivers. *River Res. Appl.* **2017**, *33*, 388–402. [CrossRef]
6. Ruley, J.E.; Rusch, K.A. Development of a simplified phosphorus management model for a shallow, subtropical, urban hypereutrophic lake. *Ecol. Eng.* **2004**, *22*, 77–98. [CrossRef]
7. Timoshkin, O.A.; Moore, M.V.; Kulikova, N.N.; Tomberg, I.V.; Malnik, V.V.; Shimaraev, M.N.; Troitskaya, E.S.; Shirokaya, A.A.; Sinyukovich, V.N.; Zaitseva, E.P.; et al. Groundwater contamination by sewage causes benthic algal outbreaks in the littoral zone of Lake Baikal (East Siberia). *J. Great Lakes Res.* **2018**, *44*, 230–244. [CrossRef]
8. McCusker, M.; Dove, A.; Depew, D.; Howell, E.T. Factors affecting Cladophora growth in the eastern basin of Lake Erie: Analysis of a monitoring dataset (2012–2019). *J. Great Lakes Res.* **2023**, *49*, 790–808. [CrossRef]
9. Wehr, J.D.; Sheath, R.G.; Kociolek, J.P. Freshwater algae of North America: Ecology and classification. *J. Phycol.* **2003**, *39*, 624–625. [CrossRef]
10. Khanum, A. An ecological study of freshwater algal mats. *Bot. Bull. Acad. Sin.* **1982**, *23*, 89–104.
11. Pinn, E.; Jones, M. 20. Macroalgal mat development and associated changes in infaunal biodiversity. *Proc. Mar. Sci.* **2005**, *7*, 231–237. [CrossRef]
12. Pikosz, M.; Messyasz, B.; Gąbka, M. Functional structure of algal mat (*Cladophora glomerata*) in a freshwater in western Poland. *Ecol. Indic.* **2017**, *74*, 1–9. [CrossRef]
13. Pikosz, M.; Messyasz, B. Composition and seasonal changes in filamentous algae in floating mats. *Oceanol. Hydrobiol. Stud.* **2015**, *44*, 273–281. [CrossRef]
14. Çelekli, A.; Kapı, E.; Soysal, Ç.; Arslanargun, H.; Bozkurt, H. Evaluating biochemical response of filamentous algae integrated with different water bodies. *Ecotoxicol. Environ. Saf.* **2017**, *142*, 171–180. [CrossRef]
15. Francoeur, S.N.; Winslow, K.A.P.; Miller, D.; Peacor, S.D. Mussel-derived stimulation of benthic filamentous algae: The importance of nutrients and spatial scale. *J. Great Lakes Res.* **2017**, *43*, 69–79. [CrossRef]
16. Çelekli, A.; Bozkurt, H. Assessing biochemical responses of filamentous algae integrated with surface waters in Yavuzeli-Araban catchment. *Ecol. Eng.* **2021**, *159*, 106126. [CrossRef]
17. Wu, Z.; Zhao, X.; Ma, Y.; Zhao, X.; Zhao, T.; Yang, S.; Gao, L. Late Cenozoic geomorphology, geochronology and physiography of Yuntaishan in southern Taihang Mountain, North China. *Acta Geol. Sin.* **2010**, *84*, 230–239. [CrossRef]
18. Han, S.; Yang, Y.; Fan, T.; Xiao, D.; Moiwo, J.P. Precipitation-runoff processes in Shimen hillslope micro-catchment of Taihang Mountain, north China. *Hydrol. Process.* **2012**, *26*, 1332–1341. [CrossRef]
19. Chen, Y.-C. A Simple Method for Isolating Filaments as "Algal Seed Stock" from *Monostroma Latissimum* (Chlorophyta) Germlings, and Application for Mass Cultivation. *J. Phycol.* **2011**, *48*, 246–247. [CrossRef]
20. Guoxiang Liu, Z.H. *Flora Algarum Sinicarum Aquae Dulcis*; Science Press: Beijing, China, 2012; Volume 15, pp. 106–121.
21. Krammer, K.; Lange-Bertalot, H. Morphology and taxonomy of *surirella ovalis* and related taxa. *Diatom Res.* **1987**, *2*, 77–95. [CrossRef]
22. Lange-Bertalot, H.; Hofmann, G.; Werum, M.; Kelly, M.; Cantonati, M. *Freshwater Benthic Diatoms of Central Europe: Over 800 Common Species Used in Ecological Assessment*; Koeltz Botanical Books Schmitten-Oberreifenberg: Glashütten, Germany, 2017; Volume 942.

23. Tsukazaki, C.; Ishii, K.-I.; Saito, R.; Matsuno, K.; Yamaguchi, A.; Imai, I. Distribution of viable diatom resting stage cells in bottom sediments of the eastern Bering Sea shelf. *Deep. Sea Res. Part II Top. Stud. Oceanogr.* **2013**, *94*, 22–30. [CrossRef]
24. Kahlert, M.; Hasselrot, A.T.; Hillebrand, H.; Pettersson, K. Spatial and temporal variation in the biomass and nutrient status of epilithic algae in Lake Erken, Sweden. *Freshw. Biol.* **2002**, *47*, 1191–1215. [CrossRef]
25. Campos, C.A.; Kennard, M.J.; Júnior, J.F.G. Diatom and macroinvertebrate assemblages to inform management of Brazilian savanna's watersheds. *Ecol. Indic.* **2021**, *128*, 107834. [CrossRef]
26. Zilli, F.L. Distribution of benthic invertebrate biomass and secondary production in relation to floodplain connectivity in a large river system (Paraná River, Argentina). *Int. Rev. Hydrobiol.* **2013**, *98*, 284–293. [CrossRef]
27. Nelson, J. *Fishes of the World*, 3rd ed.; John Wiley&SonSINC: Hoboken, NJ, USA, 1994; Volume 3. [CrossRef]
28. Harley, S.J.; Myers, R.A.; Dunn, A. Is catch-per-unit-effort proportional to abundance? *Can. J. Fish. Aquat. Sci.* **2001**, *58*, 1760–1772. [CrossRef]
29. Gupta, C.; Langer, S.; Dhar, M. Variations in water quality parameters and their correlation with fish catch per unit effort of Bhini Stream, a tributary of River Ravi, Jammu and Kashmir, India. *Ecol. Environ. Conserv.* **2023**, *28*, 1967–1975. [CrossRef]
30. Norton, J.F. *Standard Methods for the Examination of Water and Sewage*; American Public Health Association: Washington, DC, USA, 2018; Volume 10.
31. Koutecký, P. MorphoTools: A set of R functions for morphometric analysis. *Plant Syst. Evol.* **2015**, *301*, 1115–1121. [CrossRef]
32. El-Horbaty, Y.S. A Monte Carlo permutation procedure for testing variance components in generalized linear regression models. *Comput. Stat.* **2023**, *39*, 2605–2621. [CrossRef]
33. Dutilleul, P.; Pelletier, B. A valid parametric test of significance for the average R^2 in redundancy analysis with spatial data. *Spat. Stat.* **2016**, *19*, 21–41. [CrossRef]
34. Lai, J.; Zou, Y.; Zhang, J.; Peres-Neto, P.R. Generalizing hierarchical and variation partitioning in multiple regression and canonical analyses using the rdacca. hp R package. *Methods Ecol. Evol.* **2022**, *13*, 782–788. [CrossRef]
35. Hamilton, B.H.; Nickerson, J.A. Correcting for endogeneity in strategic management research. *Strateg. Organ.* **2003**, *1*, 51–78. [CrossRef]
36. Hu, A.; Ju, F.; Hou, L.; Li, J.; Yang, X.; Wang, H.; Mulla, S.I.; Sun, Q.; Bürgmann, H.; Yu, C.P. Strong impact of anthropogenic contamination on the occurrence patterns of a riverine microbial community. *Environ. Microbiol.* **2017**, *19*, 4993–5009. [CrossRef] [PubMed]
37. Ma, B.; Wang, H.; Dsouza, M.; Lou, J.; He, Y.; Dai, Z.; Brookes, P.C.; Xu, J.; Gilbert, J.A. Geographic patterns of co-occurrence network topological features for soil microbiota at continental scale in eastern China. *ISME J.* **2016**, *10*, 1891–1901. [CrossRef] [PubMed]
38. Weerman, E.J.; Herman, P.M.J.; van de Koppel, J. Macrobenthos abundance and distribution on a spatially patterned intertidal flat. *Mar. Ecol. Prog. Ser.* **2011**, *440*, 95–103. [CrossRef]
39. Nieoczym, M.; Kloskowski, J. The role of body size in the impact of common carpCyprinus carpioon water quality, zooplankton, and macrobenthos in ponds. *Int. Rev. Hydrobiol.* **2014**, *99*, 212–221. [CrossRef]
40. Lange, K.; Liess, A.; Piggott, J.J.; Townsend, C.R.; Matthaei, C.D. Light, nutrients and grazing interact to determine stream diatom community composition and functional group structure. *Freshw. Biol.* **2010**, *56*, 264–278. [CrossRef]
41. Shimabukuro, E.M.; Henry, R. Macrophytes in tropical shallow lakes: An important food item to benthic entomofauna or an underused resource? *Entomol. Sci.* **2019**, *22*, 205–215. [CrossRef]
42. Filkin, N.R.; Sherwood, A.R.; Vis, M.L. Macroalgae from 23 Stream Segments in the Hawaiian Islands. *Pac. Sci.* **2003**, *57*, 421–431. [CrossRef]
43. Bosch, I.; Makarewicz, J.C.; Lewis, T.W.; Bonk, E.A.; Finiguerra, M.; Groveman, B. Management of agricultural practices results in declines of filamentous algae in the lake littoral. *J. Great Lakes Res.* **2009**, *35*, 90–98. [CrossRef]
44. Higgins, S.N.; Malkin, S.Y.; Todd Howell, E.; Guildford, S.J.; Campbell, L.; Hiriart-Baer, V.; Hecky, R.E. An ecological review of *cladophora glomerata* (Chlorophyta) in the laurentian great lakes. *J. Phycol.* **2008**, *3*, 1–7. [CrossRef]
45. Olsen, S.; Chan, F.; Li, W.; Zhao, S.; Søndergaard, M.; Jeppesen, E. Strong impact of nitrogen loading on submerged macrophytes and algae: A long-term mesocosm experiment in a shallow Chinese lake. *Freshwater Biology* **2015**, *60*, 1525–1536. [CrossRef]
46. Page, M.; Goldhammer, T.; Hilt, S.; Tolentino, S.; Brothers, S. Filamentous algae blooms in a large, clear-water lake: Potential drivers and reduced benthic primary production. *Water* **2022**, *14*, 2136. [CrossRef]
47. Chang, Y.-H.; Ku, C.-R.; Lu, H.-L. Effects of aquatic ecological indicators of sustainable green energy landscape facilities. *Ecol. Eng.* **2014**, *71*, 144–153. [CrossRef]
48. Gladyshev, M.; Gubelit, Y.I. Green tides: New consequences of the eutrophication of natural waters (invited review). *Contemp. Probl. Ecol.* **2019**, *12*, 109–125. [CrossRef]
49. Hawes, I. The seasonal dynamics of *Spirogyra* in a shallow, maritime Antarctic lake. *Polar Biology* **1988**, *8*, 429–437. [CrossRef]
50. Han, H.; Chen, Y.; Jørgensen, S.E.; Nielsen, S.N.; Hu, W. A system-dynamic model on the competitive growth between *Potamogeton malaianus* Miq. *and Spirogyra sp. Ecol. Model.* **2009**, *220*, 2206–2217. [CrossRef]
51. Dodds, W.K.; Gudder, D.A. The ecology of Cladophora. *J. Phycol.* **1992**, *28*, 415–427. [CrossRef]
52. Adams, M.S.; Stone, W. Field studies on photosynthesis of *Cladophora glomerata* (Chlorophyta) in Green bay, Lake Michigan. *Ecology* **1973**, *54*, 853–862. [CrossRef]
53. Cambra, J.; Aboal, M. Filamentous green algae of Spain: Distribution and ecology. *Limnetica* **1992**, *8*, 213–220. [CrossRef]

54. Margalef, R.; Kinne, O. *Our Biosphere*; Ecology Institute Oldendorf: Oldendorf/Luhe, Germany, 1997; Volume 10.
55. Gebler, D.; Szoszkiewicz, K. Response of aquatic plants to extreme alterations in river morphology. *Water* **2022**, *14*, 3746. [CrossRef]
56. Frossard, V.; Versanne-Janodet, S.; Aleya, L. Factors supporting harmful macroalgal blooms in flowing waters: A 2-year study in the Lower Ain River, France. *Harmful Algae* **2014**, *33*, 19–28. [CrossRef]
57. Asaeda, T.; Sultana, M.; Manatunge, J.; Fujino, T. The effect of epiphytic algae on the growth and production of *Potamogeton perfoliatus* L. in two light conditions. *Environ. Exp. Bot.* **2004**, *52*, 225–238. [CrossRef]
58. Ding, Y.-F.; Sun, S.-J.; Feng, J.; Cui, P.; Zhang, D.; Long, Z.-Y. An assessment of lake ecology on the basis of the macrobenthos multi-metric index (MMI) in 11 lakes in the western region of Jilin, China. *Water* **2021**, *13*, 235. [CrossRef]
59. Ye, J.; Tang, Y.; Zhang, X.; Zhong, P.; Liu, Z. Omnivorous shrimp *Neocaridina denticulata* sinensis enhances the growth of submerged macrophyte *Vallisneria denseserrulata*. *Knowl. Manag. Aquat. Ecosyst.* **2019**, *5*, 32. [CrossRef]
60. Yang, L.; He, H.; Guan, B.; Yu, J.; Yao, Z.; Zhen, W.; Yin, C.; Wang, Q.; Jeppesen, E.; Liu, Z. Mesocosm experiment reveals a strong positive effect of snail presence on macrophyte growth, resulting from control of epiphyton and nuisance filamentous algae: Implications for shallow lake management. *Sci. Total Environ.* **2020**, *705*, 135958. [CrossRef]
61. Fang, L.; Wong, P.K.; Lin, L.; Lan, C.; Qiu, J.W. Impact of invasive apple snails in Hong Kong on wetland macrophytes, nutrients, phytoplankton and filamentous algae. *Freshw. Biol.* **2010**, *55*, 1191–1204. [CrossRef]
62. Pinowska, A. Effects of snail grazing and nutrient release on growth of the macrophytes *Ceratophyllum demersum* and *Elodea canadensis* and the filamentous green alga *Cladophora* sp. *Hydrobiologia* **2002**, *479*, 83–94. [CrossRef]
63. Stevenson, R.J.; Stoermer, E. Luxury consumption of phosphorus by benthic algae. *BioScience* **1982**, *32*, 682–683. [CrossRef]
64. Huang, R.; Boney, A. Seasonal ecology of littoral epiphytic diatoms on Great Cumbrae Island. *Trans. Bot. Soc. Edinb.* **1985**, *44*, 309–322. [CrossRef]

water

MDPI

Article

Sedimentological, Geochemical, and Environmental Assessment in an Eastern Mediterranean, Stressed Coastal Setting: The Gialova Lagoon, SW Peloponnese, Greece

Maria Papakonstantinou, Spyros Sergiou, Maria Geraga, Amalia Prandekou, Xenophon Dimas, Elias Fakiris, Dimitris Christodoulou and George Papatheodorou *

Laboratory of Marine Geology and Physical Oceanography (Oceanus-Lab), Department of Geology, University of Patras, Rio, 26500 Patras, Greece; mariapap@upatras.gr (M.P.); sergiou@upatras.gr (S.S.); mgeraga@upatras.gr (M.G.); a.prandekou@ac.upatras.gr (A.P.); xendimas@upatras.gr (X.D.); fakiris@upatras.gr (E.F.); dchristo@upatras.gr (D.C.)

* Correspondence: gpapathe@upatras.gr

Abstract: This study describes the prevalent sedimentological and geochemical patterns and investigates the environmental status of the bottom of Gialova lagoon, a highly vulnerable coastal site of the EU's Natura 2000 network. For this task, lithological, geochemical, and microfaunal analyses of sediment samples were combined with a high-resolution bathymetric survey. Potential pollution was determined using geochemical-based (EF, I-geo, and PLI) and faunal (Foram-AMBI) indices. We find that sedimentation is mainly controlled by the bottom morphology, hydrodynamic variations, and biogenic productivity of the lagoon. The application of the multivariate factor analysis technique revealed four dominant factors explaining the geochemical processes occurring in the lagoon. The first factor, namely "terrigenous aluminosilicates associated with C_{org} vs. autochthonous biogenic carbonates", discriminates the deposition of detrital sediments, related to the high adsorption of heavy metals—versus bioclastic sediments. The "sulfides" factor represents an anoxic phase of the lagoon floor, whereas the "Mn-oxyhydroxides" factor indicates increased manganese content with several compounded trace elements. The "phosphate" factor reveals multiple sources of phosphorus in the lagoon. The lagoon bottom shows negligible to minor contamination in heavy metals, except Mo and Pb, which induce moderate pollution levels. The maximum contamination and environmental stress concern two small-sized, shallow basins within the lagoon.

Keywords: Gialova lagoon; geochemistry; heavy metals; pollution; environmental indices; factor analysis

1. Introduction

According to the definition of Kjerfve [1], coastal lagoons are shallow, usually shore-parallel oriented water bodies, separated from the open sea by a barrier and connected at least intermittently to the adjoining sea by one or more restricted inlets. Lagoons result from the interplay between coastal processes such as littoral drift and fluvial (deltaic) progradation, while their formation is mainly related to regional climate and landscape morphology. Due to their particular geomorphological setting, modern coastal lagoons are highly vulnerable environments, exposed to both natural and anthropogenic pressures. Natural drivers comprise wave action, tidal effects, freshwater inflows, and variations in sediment fluxes and sea level, while human-induced forces include agriculture, touristic activities, industrialization, and the composite impact of the rapidly growing climatic crisis, such as abrupt flooding events alternating with extended aridity, that leads to intensified coastal erosion [1–5].

A high number of coastal lagoons are distributed across the boundaries of the Mediterranean Sea. Given their significance in terms of human resources and biodiversity, they constitute socio-environmentally protected areas under EU-directives and protocols [6]. Previous studies have focused on the ecological status of several Mediterranean coastal lagoons, investigating the interrelationships between sedimentary processes and potential human-driven pollution and aiming to establish specific guidelines in terms of sustainability management and environmental monitoring [6–9]. Sediment investigation concerns the variation in the lithological, geochemical, and biological characteristics over the lagoon's bottom, thus leading to the discrimination of several sedimentary facies that reflect differentiations in the sediment sources, hydrodynamic regime, and hydrological conditions of the lagoon [1,10–13]. This is typically accomplished by standard techniques such as grain-size analysis, macro- and microscopic analysis of sediment texture and composition, bulk elemental analysis, organic carbon measurement, and targeted microfaunal analyses.

The determination of heavy metal concentration in the lagoon's sediments is a principal procedure for assessing the possible anthropogenic contamination, which is strongly related to the agricultural and industrial uses of the surrounding plain [14–23]. The distribution and accumulation of heavy metals depend on the sediments' grain size, the metal's chemical properties, the physiochemical parameters of the water column (Eh, pH, dissolved oxygen, and organic load), the biochemical state of the environment, and physical transport [18,23,24]. Heavy metals are mainly distinguished by their high toxicity to ecosystems, a wide range of sources, non-degradable pollution by biological processes, and their bioaccumulation behavior. Heavy metals get incorporated in sediments, accumulating predominately in the fine (<63 μm) size fraction [1,20], while maximum adsorption is observed in the fine-very-fine silt and clay fractions (<16 μm) [21,25].

The degree of sediment contamination from heavy metals and metalloids is widely determined in lagoonal environments using the following indices: the Enrichment Factor (EF), the Geoaccumulation Index (I-geo), and the Pollution Loading Index (PLI) [17,20,26]. These indices are mathematical formulas that compare the concentration of a certain metal (or multiple) in the potentially contaminated sediment with its natural background value. The most common metals and metalloids used for contamination assessment are: As, Cd, Co, Cr, Cu, Mn, Ni, Pb, V, and Zn [20,26]. Additionally, indices based on the faunal composition of the lagoon's bottom have been employed to evaluate human-related pressures. Most prominently, the Foram-AMBI index has been implemented as a foraminiferal biomonitoring tool in a wide range of aquatic ecosystems, such as deep-water environments of the North Atlantic [27], the Mediterranean Sea [28], and transitional waters in Europe like estuaries and lagoons [29].

The present study focuses on the Gialova lagoon (SW Peloponnese, Greece), which represents a typical example of a Mediterranean coastal lagoon and is marked as a wildlife refuge of high importance [30]. Even though it belongs to the Natura 2000 European community network as a Special Protection Area (SPA) and Site of Community Interest (SCI), the lagoon has experienced intense pressure from human activities during the last 70 years, including extensive drainage, agriculture, and touristic activities [30–33]. As a result, these activities led to the reduction in the lagoon's extent from 7.5 km^2 to 2.5 km^2 while also causing severe environmental deterioration, including frequent dystrophic events and a prominent increase in water salinity, which altered wetland habitats [30,32,34–36].

To date, the only information on the sedimentology and contamination degree of the modern lagoon's bottom comes from analyses of a set of samples collected back in 1995, published by Kontopoulos and Bouzos [31] and Avramidis et al. [32]. According to these works, the sedimentary characteristics show distinct zonation that follows the lagoon's hydrodynamic regime, while environmental pollution was found at low levels. Our paper expands on the investigation of the aforementioned studies, performing sedimentological (granulometry, composition), geochemical (bulk geochemistry, organic carbon), and microfaunal (benthic foraminifera) analyses on new bottom samples collected during a 2020 survey, combined with high-resolution bathymetry. The data presented here were treated with an integrated analytical scheme, comprising standard analytical methods with multivariate statistical techniques (hierarchical clustering and factor analysis), while potential pollution was determined using both geochemical-based (EF, I-geo, and PLI) and faunal (Foram-AMBI) indices. Consequently, this new study shares two main objectives: (a) present an updated and more detailed perspective of the current state of the lagoon in terms of sedimentary processes and pollution loads, and (b) evaluate the recent environmental evolution of the lagoon over the last ~30 years by comparing with the results of Kontopoulos and Bouzos [31] and Avramidis et al. [32] in order to establish spatiotemporal variations of a highly vulnerable and socio-environmentally protected coastal setting.

2. Materials and Methods

2.1. Study Area

The Gialova lagoon (Figure 1) is located in the SW Peloponnese, Greece, separated from the Navarino bay (Ionian Sea, Eastern Mediterranean) by a 3.3 km long and 0.15 km wide sand barrier. The interaction with the bay occurs through a narrow inlet that is formed on the barrier. The surrounding geological setting of the Gialova lagoon consists of Holocene alluvial deposits and dunes, Plio-Pleistocene conglomerates, marls, and sandstones, and Eocene to Oligocene turbidites, while the alpine bedrock consists of Upper Cretaceous to Eocene limestones [32]. Core sediment records from the Gialova lagoon and the surrounding wetland reveal that this coastal setting has been highly vulnerable and prone to climatic and environmental shifts over the last 5000 years, including subjection to extended dry or wet periods and exposure to high-energy tsunamigenic events [37–39].

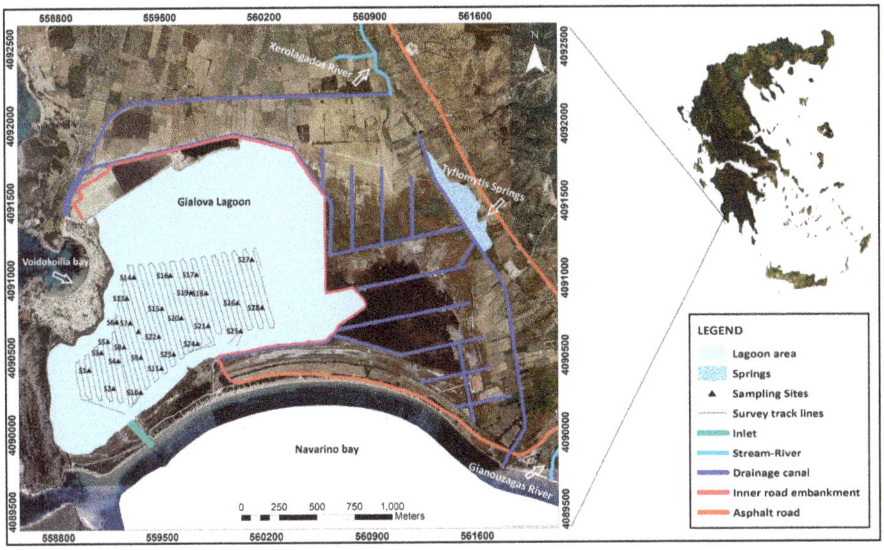

Figure 1. Overview of the Gialova lagoon and the main surrounding human-induced interventions. The basic geomorphological features are depicted, as are the sampling sites and the acoustic survey tracklines of the present study. Coordinate system: Hellenic Geodetic Reference System, 1987.

The lagoon is fed in freshwater and sediment by the Xirolagkados River and Tyflomyti springs (Figure 1). Before the 1950s, the Xirolagkados River was the main sediment supplier discharging in the northern sector of the lagoon, while Tyflomyti springs occurred in the eastern margin, pouring freshwater into the lagoon and the adjoining marshes of the Gianouzagas River plain [31,32]. Nevertheless, during the 1950s, the Xirolagkados River and Tyflomyti springs were diverted to Voidokoilia Bay and Navarino Bay, respectively. Currently, groundwater from the neighboring alluvial plain, together with the construction of two canals that have restored the connection with the Xirolagkados River and Tyflomyti springs since 1998, allow the input of freshwater into the lagoon [35,36].

The lagoon is quite shallow, while the hydrological regime shows intense seasonal variations, affected by the equilibrium between evaporation and precipitation/freshwater input [40]. Water temperature and salinity range between 15–32 °C and 30–70‰, respectively, the pH values between 7 and 9 and the dissolved oxygen levels between 4 and 11 mg/L, with the latter being higher in the winter season [40,41]. Concerning trophic state levels, the lagoon is characterized as eutrophic in April and hypereutrophic in August regarding phosphate content, whereas ammonium concentrations exhibit high values on an annual basis [40]. The high trophic levels are in line with the extended seagrass growth in the lagoon, as over 25% of its bottom is covered by submerged aquatic vegetation [42]. Notably, during the summer season, the eastern part of the lagoon becomes partially dried out.

2.2. Fieldwork

A combined acoustic (bathymetric) and sampling survey was conducted in the autumn of 2020. Bathymetric data were obtained using a high-resolution sonar (Lowrance—Elite-5 Ti) equipped on board an unmanned surface vehicle (USV) [43]. The imprinting of the lagoon's bathymetry was accomplished using a TotalScan-type sensor, while coverage was attained in the 20 m slant range, achieving 25% overlapping. Further, a 455 kHz operation pulse frequency was selected as the optimum choice due to the extremely shallow waters and high turbidity in the water column, most probably due to the presence of suspended solids (nutrients) and/or bottom sediment resuspension.

The design of track lines (Figure 1) was made on Mission Planner Software version 1.3.74, to support USV navigation on water. The USV platform was further equipped with a Global Positioning System (GPS—Magellan NAV 6500) for the positioning data and a high-resolution digital camcorder placed in a waterproof case for the data ground-truthing. For the position of USV during operation, real-time kinematic satellite navigation technology (RTK GPS) (Emlid Reach M2) was used, based onshore, with less than 10 cm accuracy. For the acquisition of the data, USV operated in water at 1 m/s speed (≈2 knots) for about nine working hours in total.

The sampling campaign was designed following the bathymetric survey. Twenty-eight (28) surficial sediment samples (Figure 1) were collected from pre-defined sampling sites, oriented by the backscatter data gained from the acoustic survey, covering the entire spectrum of the different backscatter levels. The samples were collected using a Van–Veen grab, corresponding to the upper ~20 cm of the lagoon's bottom, while positioning was determined with a Garmin GPS 60 device. Sub-samples for laboratory analyses (S1–S28) were taken from the central part of the grab to avoid contamination, sealed in polyethylene bags, and stored at 4 °C until the beginning of laboratory analyses. Systematic sediment sampling is considered a very effective ground-truthing technique in the acoustic survey.

2.3. Laboratory Analyses

2.3.1. Sedimentology

The sedimentological examination of the 28 collected samples included macro- and microscopic textural and compositional observations, together with grain size analysis. Macroscopic description is aimed at the general characterization of the sample, focusing primarily on features such as color (based on the Munsell Color Chart), lithology, and abundance of biogenic material. Microscopic observations were performed in the sand fraction of the samples (grain size > 63 μm, or <4Φ, see below) using a Leica MZ6 binocular stereo microscope after wet sieving and drying the retained fraction.

Grain size analysis was carried out by laser diffraction using a Malvern Mastersizer 2000 particle analyzer. Hydrogen peroxide (H_2O_2) treatment was applied before analysis to eliminate organic matter. The logarithmic (original) graphical measures of Folk and Ward, [44] were calculated with the GRADISTAT software version 8.0 [45], while sediment classification followed the nomenclature of Folk [46]. Accordingly, the five statistical parameters: median (Md), mean size (Mz), sorting (σi), skewness (Sk), and kurtosis (KG) of the grain size distribution as well as the size fractions are expressed in the logarithmic phi scale (Φ). Thus, sand, silt, and clay regard the size fraction as between −1Φ–4Φ, 4Φ–8Φ, and >8Φ, respectively [46].

The interrelationships between grain size and statistical parameters were examined in order to investigate the prevailing environmental conditions during sediment deposition [12,46]. We used the bivariate diagram of median (Md) versus sorting (σi) to differentiate between riverine, wave, and quiet water processes in a lagoonal setting [47] and the most appropriate of the discriminant functions proposed by Sahu [48], which suggests a higher influence between shallow marine versus fluvial (deltaic) environments: $Y_{sm:f} = 0.2852 \times Md - 8.7604 \times \sigma i^2 - 4.8932 \times Sk + 0.0482 \times KG$. When $Y < -7.4190$, a fluvial (deltaic) deposit is indicated, whereas when $Y > -7.4190$, shallow marine conditions prevail. The above proxies concern the statistical parameters obtained through the original graphical method of Folk and Ward [44] and have been widely used in the research of lagoonal and coastal settings [49–52].

2.3.2. Geochemistry

Bulk geochemistry and Total Organic Carbon (C_{org}) analyses in the 28 samples were both conducted at ALS Geochemistry Ltd. (Loughrea, Ireland), certified to ISO/IEC 17025:2017. Bulk geochemical analysis targets the elemental composition of the sediment. It was performed through a multi-element ultratrace method that combined four-acid digestion (HCl, HNO_3, $HClO_4$, and HF) with an inductively plasma—mass spectrometry (ICP—MS) finish. The four-acid digestion breaks down most silicate and oxide minerals, allowing for a "near-total" recovery of most analytes. The following forty-three (43) major and trace elements were determined: Al, Ca, Fe, K, Mg, Na, S, Ti (major, %) and Ag, As, Ba, Be, Cd, Ce, Co, Cr, Cs, Cu, Ga, Hf, In, La, Li, Mn, Mo, Nb, Ni, P, Pb, Rb, Sb, Sc, Sn, Sr, Ta, Th, Tl, U, V, W, Y, Zn, Zr (trace, ppm). Total organic carbon content (C_{org}) was determined using a LECO furnace carbon analyzer. Pre-treatment with HCl (25%) was necessary for removing inorganic carbonates.

2.3.3. Benthic Foraminifera

Benthic foraminifera tests were studied in 14 samples. Each sample was wet-sieved through 125 μm, then oven-dried at 50 °C. A number of 200–300 foraminiferal tests were used for microfaunal analysis. In cases where the dried residue exceeded the 200–300 tests of benthic foraminifera, it was split with a microsplitter in order to provide an adequate number. Then, each sample was weighed, and each foraminifera specimen was picked and identified up to the species level when possible based on the World Register of Marine Species database taxonomy (WoRMS). In addition, foraminiferal parameters have been calculated, like foraminiferal abundance (number of specimens per 1 g of dry sediment), relative foraminiferal abundance (the percent composition of a certain species relative to the total number of foraminifera), and Shannon-Weaver index ($Hs = -\sum_{i=1}^{R} p_i ln p_i$) as an estimation of the species diversity.

2.4. Contamination Assessement

For the assessment of the metal contamination levels and the possible anthropogenic impact on the sediments of Gialova lagoon, three (3) of the most commonly used pollution indices were employed: (a) Enrichment Factor (EF), (b) Geo-accumulation Index (I-geo), and (c) Pollution Loading Index (PLI) [17,26]. The global average shale values proposed by Turekian and Wedepohl [53] were used as natural background sediment reference concentrations. Additionally, for the implementation of the Foram-AMBI index, foraminiferal species were assigned to one of five ecological groups, according to the sensitivity of species to increasing stress [27–29]. The assignment of each species to an ecological category was mainly based on the list of species from intertidal and transitional waters in the Mediterranean Sea [29] and similar ecological areas that were found in the literature (e.g., [28]). All the above is summarized in Table 1, including basic information for each index.

Table 1. Environmental Indices used in the present study.

Index	Procedures of Calculation	Values	Description-Classification	Reference
EF	EF = $(Metal/RE)_{sample}/(Metal/RE)_{background}$. "RE" stands for the Reference Element. Aluminum (Al) was selected as the most suitable reference element, as it is mostly indicative of the natural, lithogenic fraction of the sediments [54].	EF ≤ 1 1 < EF ≤ 3 3 < EF ≤ 5 5 < EF ≤ 10 10 < EF ≤ 25 25 < EF ≤ 50 EF > 50	No enrichment Minor enrichment Moderate enrichment Moderate to severe enrichment Severe enrichment Very severe enrichment Extremely severe enrichment	[55]
I-geo	I-geo = $\log_2 (C_n/1.5B_n)$, where C_n is the measured content of element "n", and B_n is the background content of the "n" element.	Igeo ≤ 0 0 < Igeo < 1 1 < Igeo < 2 2 < Igeo < 3 3 < Igeo < 4 4 < Igeo < 5 Igeo ≥ 5	Unpolluted Unpolluted to moderately polluted Moderately polluted Moderately to heavily polluted Heavily polluted Heavily to extremely polluted Extremely polluted	[56]
PLI	PLI = $(CF_1 \times CF_2 \times CF_3 \times \ldots \times CF_n)^{1/n}$ where CF_{metals} is the ratio between the content of each metal to the background values in sediment; $CF_{metals} = C_{metal}/C_{background}$. The PLI ascribes an evaluation of the overall toxicity status of the sediments.	PLI < 1 PLI = 1 PLI > 1	Unpolluted sediments Baseline level of contamination Progressive deterioration of the environmental conditions and increasing pollution	[57]
Foram-AMBI	AMBI = {(0*%GI) + (1.5*%GII) + (3*%GIII) + (4.5*%GIV) + (6*%GV)}/100 Where GI-GV the relative abundance of each ecological group. Specifically, GI is the "sensitive species", GII "Indifferent species", GIII "3rd-order opportunistic species", GIV "2nd-order opportunistic species" and GV "1st-order opportunistic species"	0 < AMBI ≤ 1.2 1.2 < AMBI ≤ 3.3 3.3 < AMBI ≤ 4.3 4.3 < AMBI ≤ 5.5 5.5 < AMBI ≤ 7	High Ecological Quality Status (EcoQs) Good EcoQs Moderate EcoQs Poor EcoQs Bad EcoQs	[28,58]

2.5. Statistical Treatment—Hierarchical Clustering and Factor Analysis

Hierarchical clustering was performed on the grain size distribution (GSD) data using the statistical software SPSS v27 to identify possible similarities between the GSD curves and, consequently, classify the samples into distinct groups. Clustering was based on Ward's method with Squared Euclidean distance intervals.

Factor analysis is a multivariate statistical method that detects any interrelationships within a set of variables (R-mode) or objects (Q-mode) [59]. Factor analysis has been successfully used in the sedimentary research of Greek lagoons, unraveling certain patterns in the distribution of heavy metals, organic material, and textural components [16,20,60]. R-mode factor analysis was performed to investigate the interrelationships between the examined geochemical parameters (variables) and identify the processes that explain their spatial distribution as well as possible common sources [61,62], whereas Q-mode factor analysis was designed to investigate interrelationships between samples (objects) on the basis of geochemical properties (variables).

The aims of the Q-mode analysis, according to Reyment and Joreskog [59], are: (i) to find the minimum number of "end-members" needed to account for the compositional variation observed (ii) to identify the composition of the "end-member" in relation to the original variables; and (iii) to describe each sample in terms of the "end-member". To bring to fruition the aims of the Q-mode analysis, a plot of factor loadings for the first two factors was constructed, the loadings of which were normalized by being divided by their communalities.

In the current study, R-mode and Q-mode factor analysis were conducted via the statistical software program IBM SPSS v27 on the whole geochemical dataset, containing 28 samples × 44 variables (C_{org} and the 43 major and trace elements; see Supplementary File S3). The analysis was conducted following the steps described by Papatheodorou et al. [63].

3. Results and Discussion

3.1. Bathymetry

The acquired acoustic data showed that Gialova Lagoon is very shallow, with water depths up to 0.7 m. The detailed bathymetry showed that the lagoon floor is not flat, but the variation in topography forms three small-sized basins located in the eastern, central, and western parts of the examined area (Figure 2). The highest depths occur in the eastern part, while in the western basin, two smaller sub-basins are distinguished. The shallower depths were obtained along the borders and in the central part of the lagoon. Yet, it is worth mentioning the clear view of an elongated ridge running from east to southwest at the central part of the lagoon, which could possibly be related to fishing activities [30,33] (Figure 2). Moreover, traces of another ridge have been observed running the lagoon from north to south.

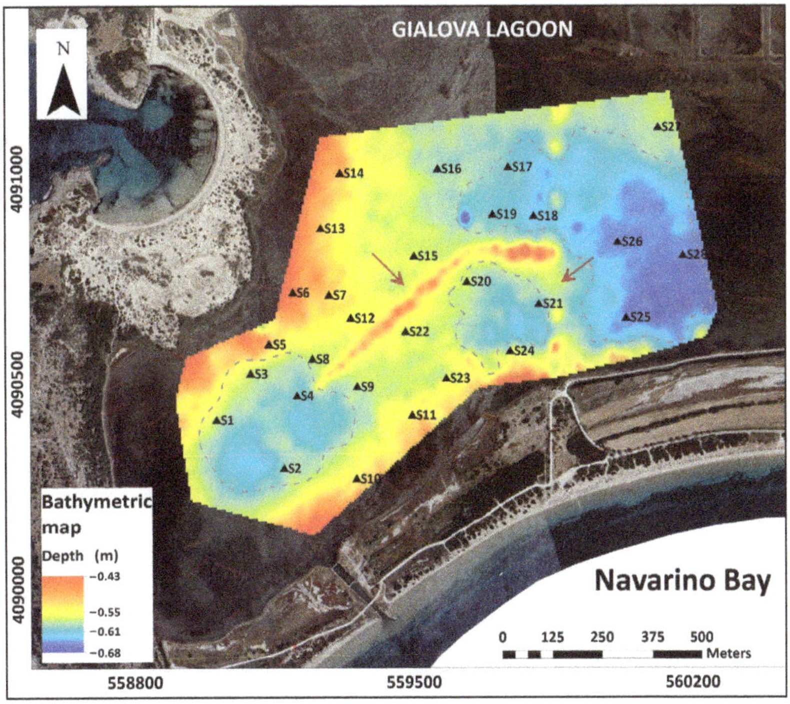

Figure 2. Bathymetric map of the Gialova lagoon. The sampling sites are also shown. The shallow basins are highlighted with dashed elliptic shapes, while elongated ridges are indicated with arrows.

3.2. Sedimentology

3.2.1. Grain Size Distribution

The grain size analyses of the collected bottom sediment samples (Supplementary File S1) showed that the average proportion of sand, silt, and clay is in the order of 7.50%, 63.17%, and 29.32%, respectively, while the mean size (Mz) ranged from 4.99 Φ to 7.28 Φ, with an average value of 6.67 Φ, implying that the lagoon's sediments are composed principally of silt-sized sediments.

The coarser sediments accumulate along the southwestern border of the lagoon, just north of the sandy barrier that separates the lagoon from Navarino Bay (Figure 3a). However, the distribution of the fine material does not exactly follow the morphology of the lagoon's floor. The high percentage of clay and the high values (Φ) of Mz (Figure 3b,c) are not located in the three shallow basins of the lagoon but slightly north of them.

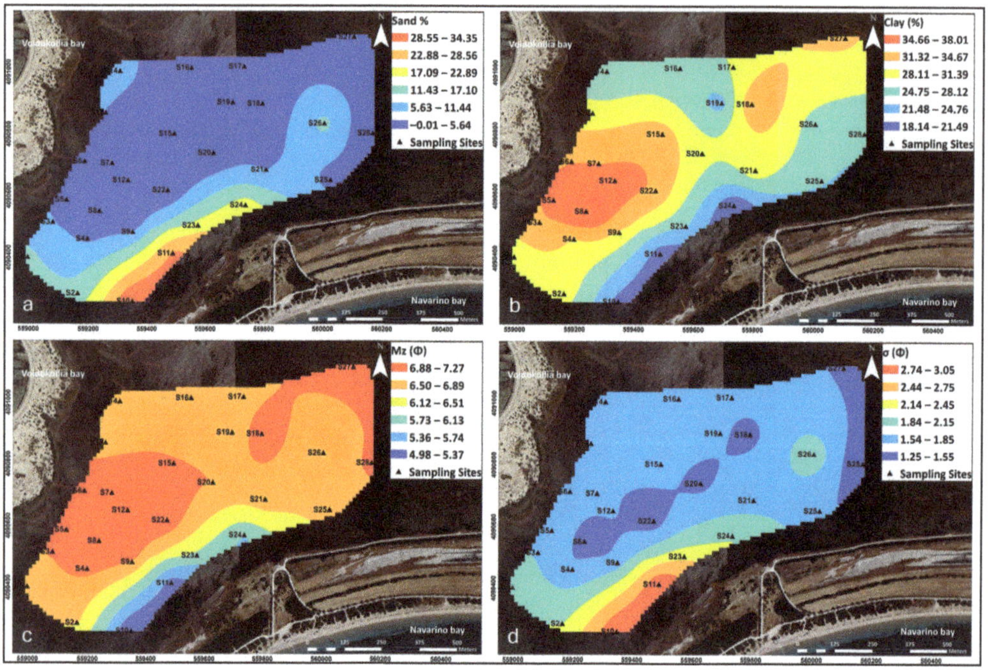

Figure 3. Spatial distribution of (**a**) Sand (%), (**b**) Clay (%), (**c**) Mz (Φ), and (**d**) σ (Φ) (Sorting) over the Gialova lagoon bottom.

The Gialova sediments are characterized as poorly to very poorly sorted, with σi values ranging from 1.25 Φ to 3.03 Φ, with an average value of 1.80 Φ (Figure 3d). Sorting values follow the trend of sand content, presenting the highest values along the southwestern border of the lagoon and diminishing towards the central part. Skewness (Sk) displays low variability and ranges between −0.36 Φ and 0.02 Φ with an average value of −0.17 Φ (Supplementary File S1), thus characterizing mainly coarse-skewed sediments. Finally, kurtosis fluctuates between platykurtic and mesokurtic values (KG: 0.68–1.06 Φ), with an average value of 0.91 Φ (Supplementary File S1).

3.2.2. Sand Composition

The microscopic examination led to the determination of three major classes that comprise the sand grains: (a) lithological components (detrital lithics and various mineral grains)—(b) biogenic components (fragments and/or individual specimens of benthic foraminifera, gastropods, ostracods, bivalves, etc.)—(c) organic components (wood fragments, algal constituents, and pyritized features). The relative proportions of lithogenic, biogenic, and organic components in the sandy material were determined on a semi-quantitative basis, as described by Chevillon [11]. The mean proportions of the above classes are 33%, 62%, and 5% for the lithogenic, biogenic, and organic grains, respectively (Supplementary File S1), suggesting that the sandy material of Gialova sediments is primarily of biogenic origin.

3.2.3. Insights on the Depositional Environment

The overall sedimentological work indicates a number of common features among the studied samples, which, in turn, reflect similar processes at the sampling sites. At first, all samples are fine-grained (dominant mud and silt) and accompanied by various seagrass residues and skeletal remains (Supplementary File S2, Table S1), thus suggesting a low-energy environment and adequate biogenic productivity. The sand content throughout the lagoon has a principally bioclastic composition (as mentioned in the previous section), comprising fragments and individuals of mollusk shells (gastropods and bivalves), ostracods, and benthic foraminifera. Bioclastic sand is commonly observed in lagoons and is closely related to increased organic productivity and seagrass growth [11,64–66]. The prevalent low-energy conditions are further supported by the bivariate Md—σi plot on the grain size data, while the discriminant function analysis indicates that the fluvial influence overcomes the effects of seawater intrusions throughout the lagoon [47,48] (Supplementary File S2, Figure S2).

Secondly, all samples share the same very dark gray color according to the Munsell Color Chart (5Y 3/1) implying similar eutrophic conditions throughout the lagoon [1]. This is further supported by the relatively high and weakly varying C_{org} values of the samples (1.35–2.02% range, mean value: 1.67%; Table 2) and agrees with the eutrophic character of the lagoon waters [40]. Thirdly, the large majority of the samples are characterized by coarse—skewed grain size distribution curves, likely reflecting a winnowing process along the lagoon bottom. This is a typical process in shallow lagoon environments and corresponds to the removal of the fine-grained distribution tail through re-suspension, induced by wind action [21,67,68].

Table 2. Mean, minimum, and maximum values of the major and trace elements and organic carbon (C_{org}) in the surface sediments of Gialova lagoon.

Element	Unit	Mean	Min	Max
Ag	ppm	0.03	0.02	0.05
Al	%	4.47	3.11	5.33
As	ppm	8.40	6.20	10.00
Ba	ppm	136.79	110.00	160.00
Be	ppm	1.22	0.87	1.50
Ca	%	11.18	7.48	16.05
Cd	ppm	0.13	0.10	0.16
Ce	ppm	37.99	35.40	40.80
Co	ppm	14.81	10.30	17.40
Cr	ppm	122.50	91.00	151.00
Cs	ppm	4.06	2.70	4.93
Cu	ppm	30.50	20.50	36.50
Fe	%	3.20	2.13	3.84
Ga	ppm	10.61	7.14	12.65
Hf	ppm	1.33	1.00	1.60
In	ppm	0.04	0.03	0.05
K	%	1.40	1.00	1.63
La	ppm	20.09	18.00	22.90
Li	ppm	43.01	31.20	50.20
Mg	%	1.97	1.27	2.25
Mn	ppm	1026.32	646.00	1150.00
Mo	ppm	6.32	3.26	8.90
Na	%	2.72	1.82	3.61
Nb	ppm	7.46	5.10	8.80
Ni	ppm	103.20	66.50	123.50
P	ppm	395.71	330.00	450.00
Pb	ppm	35.46	26.30	44.40
Rb	ppm	80.07	55.20	95.20
S	%	1.50	0.99	1.84

Table 2. *Cont.*

Element	Unit	Mean	Min	Max
Sb	ppm	0.49	0.36	0.57
Sc	ppm	10.60	7.00	12.70
Sn	ppm	1.62	1.10	2.00
Sr	ppm	713.00	434.00	1145.00
Ta	ppm	0.48	0.33	0.57
Th	ppm	5.91	4.54	7.55
Ti	%	0.22	0.16	0.26
Tl	ppm	0.42	0.31	0.51
U	ppm	2.94	2.10	3.90
V	ppm	69.18	46.00	83.00
W	ppm	0.85	0.60	1.00
Y	ppm	16.97	13.50	18.70
Zn	ppm	64.50	45.00	79.00
Zr	ppm	48.38	32.70	59.80
C_{org}	%	1.67	1.35	2.02

However, and despite the above-mentioned similarities, the detailed examination through hierarchical clustering of the grain size distribution (GSD) curves plus the textural and compositional evaluation of the samples revealed the presence of four (4) dominant sedimentary facies that describe the lagoon's bottom (Supplementary File S2, Figure S1, Table S1). These facies reflect the hydrodynamic variations within the lagoon, revealing different conditions between the western and eastern small basins (Facies 1 versus Facies 4) and the presence of a N-S-SW mixing zone (Facies 2 and 3). The four facies are thoroughly presented in Supplementary File S2.

3.3. Geochemistry

3.3.1. Elemental Concentrations in Gialova Lagoon Sediments

The concentration values of all measured major and trace elements, as well as organic carbon (C_{org}), are shown in Table 2. Additionally, the measured values per sample are presented in Supplementary File S3.

Herein, the spatial distribution of selected elements' concentrations (Al, As, Ca, Cd, Cr, Cu, Fe, Mn, Mo, Ni, Pb, P, S, Zn) and C_{org} is discussed. Among them, specific elements have been grouped and described together due to their highly similar spatial distribution. These groups are Al-Cr-Cu-Fe-Ni-Pb-S-Zn (Group A) and As-Cd-Mo (Group B). The distribution of the other elements (Ca, Mn, P, and C_{org}) is discussed separately.

The maximum concentrations of Group A elements are observed slightly north of the western shallow basin (Figure 4). Low concentrations are detected at the southwestern margin, close to the communication channel (inlet) with Navarino Bay, and at the northern part of the lagoon.

The spatial distribution pattern of Group B metals shows two areas of maximum concentrations: (i) at the western sub-basin of the shallow western basin and (ii) at the shallow area between the western and eastern basins of the lagoon (Figure 5). Low concentrations are measured at the eastern and northern margins of the lagoon and close to the inlet.

Manganese (Mn) shows a different spatial pattern compared to those of Groups A and B. The maximum concentrations are found at the eastern shallow basin and at the north of the western shallow basin (Figure 6a). Furthermore, low Mn concentrations are measured close to the inlet (Figure 6a).

The concentration of phosphorus (P) shows the highest values at the eastern shallow basin of the lagoon, while it is also increased along the northwestern sector (Figure 6b). On the other hand, low values are observed near the inlet and across a zone that extends from the southern part towards the central and northern sides of the lagoon.

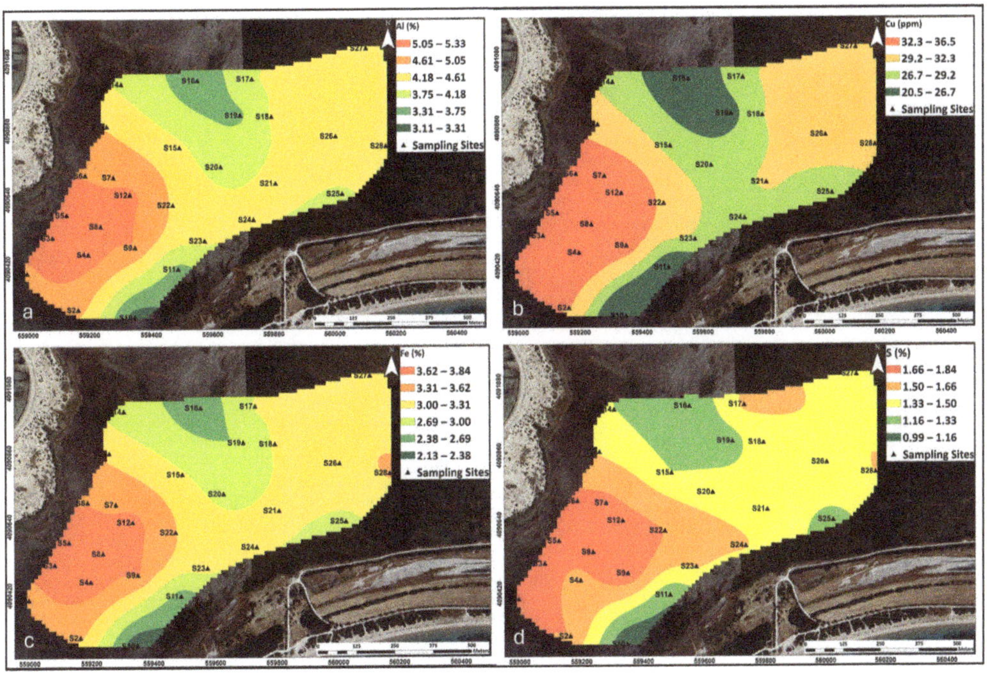

Figure 4. Spatial distribution of representative Group A elements: (**a**) Al (%), (**b**) Cu (ppm), (**c**) Fe (%) and (**d**) S (%) concentrations in Gialova lagoon bottom sediments.

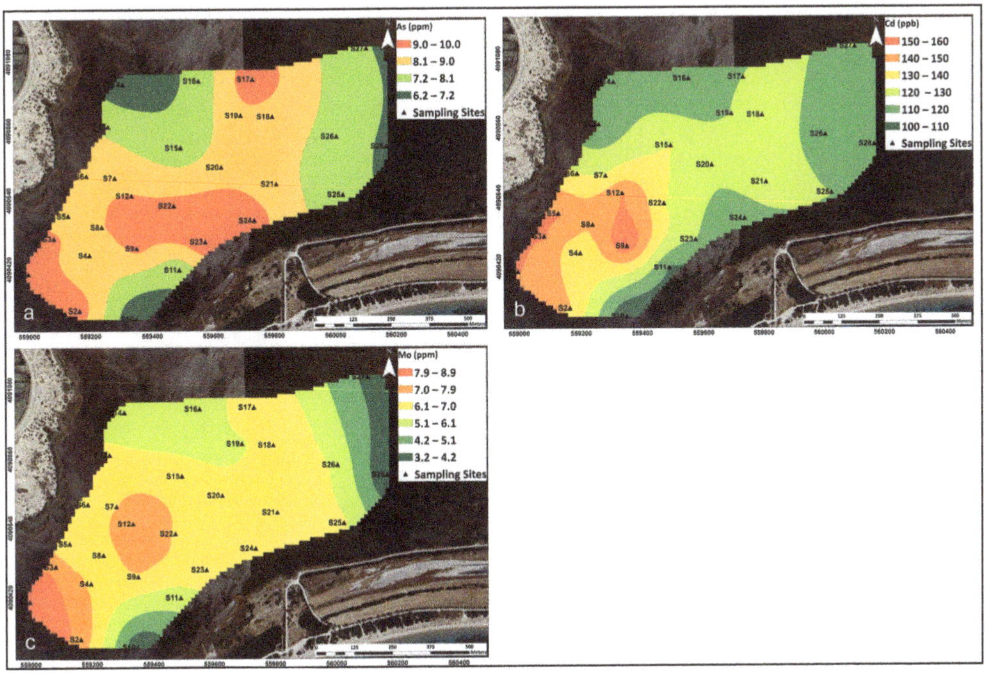

Figure 5. Spatial distribution of Group B metals: (**a**) As (ppm), (**b**) Cd (ppb), and (**c**) Mo (ppm) concentrations in Gialova lagoon bottom sediments.

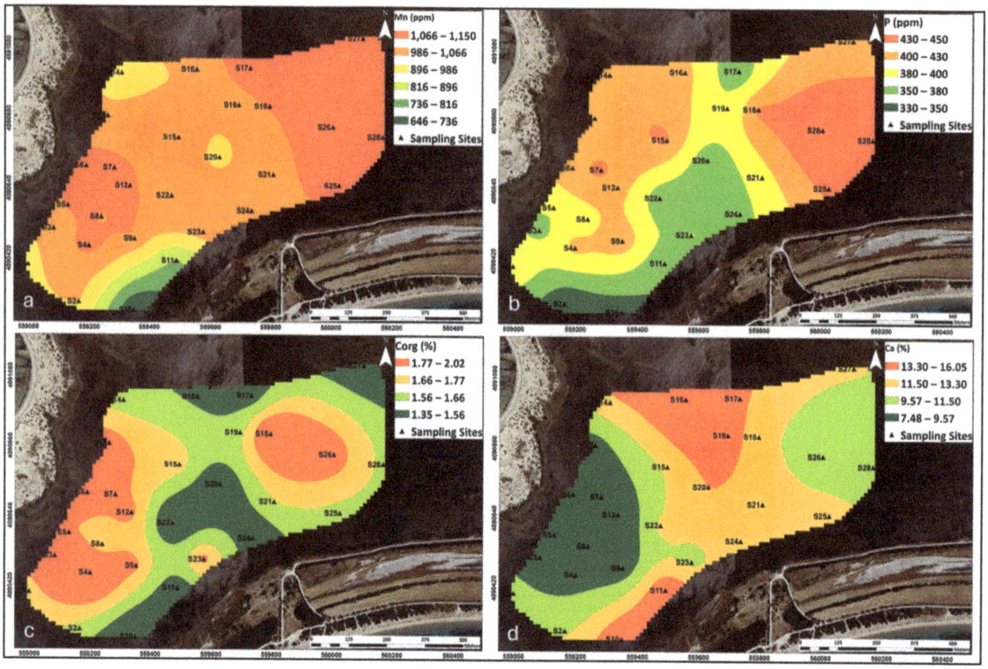

Figure 6. Spatial distribution of (**a**) Mn (ppm), (**b**) P (ppm), (**c**) C_{org} (TOC) (%), and (**d**) Ca (%) in Gialova lagoon bottom sediments.

The C_{org} values vary between 1.35 and 2.02% with a mean content of 1.67%, which is a typical range for Mediterranean coastal lagoon sediments [14,21,23,69]. The spatial distribution of C_{org} is shown in Figure 6c. The highest C_{org} percentages (>1.8%) were measured at the eastern sub-basin of the shallow western basin and at the western part of the eastern shallow basin, while low concentration values (<1.5%) were obtained along the sand-rich depositional zone behind the barrier as well as the central part and northern margin of the lagoon. This pattern highlights the enhanced adsorption of organic matter in the fine silt and clay fractions of the sediment. It also suggests that organic matter is primarily deposited in the two depressed morphological basins where the burial process prevents oxidation and that the oxygenation of the lagoon is ascribed to the saline water inflows from the adjacent Navarino Bay through the inlet.

Calcium (Ca) concentration in Gialova lagoon sediments is found between 7.48 and 16.05% (mean value: 11.18%) and shows an almost opposite distribution pattern with Group A elements (Figures 4 and 6d). Specifically, the maximum concentrations (>13%) are noticed in the northern sector and close to the inlet, whereas the minimum values (<9.5%) lie in the western shallow basin.

3.3.2. Geochemical Comparison with Other Greek Lagoons

The range of Cd, Cr, Cu, Fe, Mn, Ni, Pb, V, and Zn concentrations found in the present study are comparable with those from other lagoonal and coastal environments in Greece (Table 3) and are usually lower than the mean concentration for those elements. Exceptions are Mn and Pb, which present higher concentrations in Gialova lagoon (present study) in relation to their mean concentrations found in other lagoonal and coastal environments in Greece. In addition, the comparison of the mean concentrations of Cd, Cr, Cu, Fe, Ni, V, and Zn found in the present study is lower than the average shale values, but that of Mn and Pb is higher.

Table 3. Mean values of selected heavy metal concentrations in the Gialova lagoon compared to other Greek aquatic systems and average Shales' proportion.

Element	Cd	Cr	Cu	Fe	Mn	Ni	Pb	V	Zn	Reference
Element Unit	ppm	ppm	ppm	%	ppm	ppm	ppm	ppm	ppm	
Gialova lagoon	0.13	122.50	30.50	3.20	1026.32	103.20	35.46	69.18	64.50	This study
Gialova lagoon	0.81	118.20	57.51	4.76	1549	172.11	36.29	69.24	98.47	[32]
Messolonghi Lagoon	–	101.00	20.00	2.36	630.00	84.00	16.00	75.00	60.00	[16]
Koumoundourou Lake	–	58.00	21.00	0.58	155.00	28.00	53.00	23.00	83.00	[70]
Alikes Lagoon	–	251.67	29.17	3.59	630.00	134.17	9.88	109.33	68.89	[71]
Aetoliko Lagoon	–	140	88	4.17	837	75.6	–	–	122.2	[72]
Kleisova Lagoon	–	–	13.00	1.64	562.00	62.00	–	–	29.00	[60]
Rhodia Lagoon	–	231	37	2.83	867	124	36	112	72	[73]
Tsoukalio Lagoon	–	274	31	3.16	1191	131	26	108	76	[73]
Logarou Lagoon	–	302	44	4.74	922	221	25	153	105	[73]
Tsopeli Lagoon	–	295	48	3.98	665	168	26	129	100	[73]
Navarino Bay—upper sediments	–	–	66.00	–	–	151.00	–	–	352.00	[74]
Mean (min-max)		196.76 (58–302)	38.87 (13–88)	3.18 (0.58–4.76)	800.8 (155–1549)	119.99 (28–221)	28.52 (16–36.3)	97.32 (23–109.3)	81.45 (29–122.2)	
Average Shales	0.3	90	45	4.72	850	68	20	130	95	[53]

Moreover, when comparing the metal concentration values in Gialova sediments between 2020 (this study) and 1995 [32], there is an evident downslope trend for several metals, such as Cd, Cu, Mn, Ni, and Zn. This decline in metal concentrations could be partially attributed to the different analytical methods employed in the two studies; however, it could also indicate efficient modifications in the recent environmental conditions and/or changes in sedimentary processes (see Section 3.7).

3.4. Benthic Foraminifera

A total of 26 benthic foraminiferal species were identified in the samples. These are *Ammonia* spp., *Ammonia parkinsoniana*, *Ammonia perlucida*, *Ammonia tepida*, *Asterigerinata mammila*, *Bulimina marginata*, *Cassidulina* sp., *Cibicides* spp., *Elphidium crispum*, *Elphidium* sp., *Eponides* sp., *Gyroidina* sp., *Haynesina germanica*, *Lenticulina orbicularis*, *Melonis baleaanus*, *Miliolinella subrotunda*, *Nonion* spp., *Quinqueloculina auberiana*, *Quinqueloculina seminula*, *Quinqueloculina vulgaris*, *Quinqueloculina* spp., *Rosalina globularis*, *Spiroloculina* sp., *Triloculina trigonula*, *Rectuvigerina phlegeri*, and *Valvulineria* sp.

The relative abundance of recognized species varies from site to site, while only nine species show relative abundances greater than 2% in at least one sample (Supplementary File S4). The foraminifera microfauna is largely dominated by *A. tepida* (73.75% on average), followed by *A. parkinsoniana* (15.12% on average), and *Q. seminula* (3.23% on average). The foraminifera abundance varies from 510.7 indv/g in S13 to 5225.9 indv/g in S15, with an average value of 1492.5 indv/g. The Shannon–Weaver index (Hs) ranged from 0.65 (S28) to 1.25 (S13), with an average value of 0.85.

In addition, morphological abnormalities have been observed in most sampling sites, except for S13 and S28, while the highest abundance of deformed individuals was observed at S18 (6.6% of total fauna). The deformed assemblage is exclusively the species *Ammonia tepida*, which is 3.31% on average in the total fauna and 4.7% of the *Ammonia tepida* fauna. The abnormalities in hyaline specimens are likely associated with imperfect biomineralization [75] that can be attributed to the development of cavities in the wall or to disorganization in the pattern of crystallites [76]. Environmental factors such as pollution and/or environmental stress are potential factors that could contribute to the observed disorganization [76].

3.5. Geochemical Processes—Elemental Source Identification

The results of the R-mode factor analysis analysis led to the discrimination of certain geochemical processes along the bottom sediments and allowed the identification of possible sources from which the metals have been derived into the lagoon.

A four (4) factor model was utilized to describe the geochemical dataset without losing significant information. These factors explain about 92% of the total variance, and each variable shows communalities higher than 0.65 (Table 4). This means that the four-factor model expresses the analyzed variables efficiently [59].

Table 4. Communalities and varimax rotated R-mode factor loadings (R-mode) of the geochemical dataset. Variables with loadings ≥0.4 are highlighted in bold. High negative loadings are indicated in italics.

Variable	Communalities	Rotated Component Matrix			
		Factor 1	Factor 2	Factor 3	Factor 4
Ag	0.646	**0.731**	0.299	−0.141	0.046
Al	0.995	**0.927**	0.257	0.253	0.074
As	0.796	0.242	**0.736**	0.374	−0.236
Ba	0.947	**0.901**	0.249	0.268	0.028
Be	0.977	**0.943**	0.218	0.197	0.044
Ca	0.928	*−0.898*	−0.106	−0.191	−0.272
Cd	0.789	**0.650**	**0.605**	0.027	−0.010
Ce	0.856	**0.494**	−0.102	0.304	*−0.713*
Co	0.971	**0.862**	0.359	0.311	0.039
Cr	0.876	**0.929**	−0.038	−0.029	−0.101
Cs	0.993	**0.917**	0.294	0.248	0.058
Cu	0.982	**0.899**	0.292	0.274	0.110
Fe	0.995	**0.903**	0.264	0.319	0.095
Ga	0.994	**0.913**	0.275	0.273	0.102
Hf	0.918	**0.904**	0.231	0.210	0.060
In	0.906	**0.790**	0.339	0.362	0.186
K	0.990	**0.897**	0.280	0.277	0.177
La	0.881	**0.556**	0.162	**0.544**	*−0.500*
Li	0.935	**0.860**	0.380	0.221	0.048
Mg	0.958	**0.677**	0.186	**0.432**	**0.527**
Mn	0.911	0.297	0.144	**0.800**	**0.403**
Mo	0.926	0.369	**0.884**	0.036	−0.086
Na	0.757	0.265	−0.016	0.191	**0.806**
Nb	0.989	**0.902**	0.254	0.313	0.113
Ni	0.993	**0.894**	0.334	0.279	0.061
P	0.829	0.183	−0.169	0.212	**0.850**
Pb	0.973	**0.927**	0.239	0.236	−0.014
Rb	0.994	**0.910**	0.304	0.260	0.072
S	0.919	**0.780**	**0.403**	0.386	0.003
Sb	0.868	0.126	**0.883**	0.171	0.209
Sc	0.992	**0.914**	0.285	0.260	0.086
Sn	0.974	**0.928**	0.188	0.237	0.146
Sr	0.904	*−0.908*	−0.064	−0.246	−0.123
Ta	0.976	**0.911**	0.239	0.288	0.075
Th	0.897	**0.849**	0.105	**0.406**	0.006
Ti	0.996	**0.924**	0.254	0.264	0.090
Tl	0.972	**0.938**	0.267	0.129	0.059
U	0.830	0.387	**0.824**	−0.007	−0.027
V	0.975	**0.890**	0.361	0.219	0.067
W	0.945	**0.889**	0.204	0.322	0.093
Y	0.934	**0.485**	0.169	**0.813**	0.098
Zn	0.986	**0.925**	0.308	0.187	0.041
Zr	0.854	**0.733**	0.217	**0.519**	0.015
C_{org}	0.702	**0.597**	0.295	−0.033	**0.508**
Explained Variance (%)		61.506	12.784	10.215	7.378

Factor 1 explains the largest proportion (61.5%) of the total variance and can be considered a bipolar factor (Table 4). It displays high positive loadings for Al, Ba, Be, Co, Cr, Cs, Cu, Fe, Ga, Hf, In, K, Li, Nb, Ni, Pb, Rb, S, Sc, Sn, Ta, Th, Ti, Tl, V, W, and Zn and moderate positive loadings for Ag, Cd, Ce, La, Mg, Y, Zr, and C_{org}. In contrast, Ca and Sr show high negative loadings, constituting the negative pole of the factor.

Based on the loadings of the positive pole of Factor 1, this pole can be identified as the "terrigenous aluminosilicates" factor pole and represents the terrigenous inputs with a rather constant mineralogical sediment composition. Al, Fe, K, and Ti are major constituents of common silicate minerals and originate from the weathering release of parent materials in the local bedrock [77,78]. Additionally, Mg, Ni, Nb, and Rb provide good geochemical proxies for clay minerals, such as kaolinite, chlorite, illite, and smectite [79,80].

The terrigenous aluminosilicates factor pole exhibits a close relationship with C_{org}, as indicated by the moderate positive factor loading (Table 4). Several previous studies in shallow coastal areas and lagoonal environments have shown that clay minerals and organic matter build aggregates and flocs that absorb trace metals and then sink to form a "fluffy layer" [81,82]. Although the "fluffy layer" was not observed and sampled during the Gialova survey, its existence cannot be excluded. Moreover, the low-moderate (0.39) loading of U on Factor 1 could be related to the adsorption and complexion of U in humic organic material and poorly crystalline clay minerals [83,84]. It is also important to mention that although most of the elements with high positive loadings on Factor 1 are mostly of terrigenous origin, some of them could also be absorbed onto the aluminosilicate/C_{org} particles from anthropogenic sources. Based on the above, the positive pole of Factor 1 can be considered a "terrigenous aluminosilicate associated with C_{org}" factor, which additionally represents the potential adsorption of trace metals.

The high factor loading of S (0.780) combined with Fe (0.903) and many other heavy metals may outline acid-volatile sulfides (AVSs), a geochemical phase that contributes to the binding of metals [85]. AVS is an operationally defined reactive sulfide fraction that mainly comprises dissolved (hydrogen) sulfides and mackinawite (FeS), which may differ largely in their composition and capacity for precipitating trace metals in different aquatic environments [85]. It should also be mentioned that the existence of more stable forms of sulfides, like elemental sulfur and pyrite (FeS_2), cannot be excluded for Gialova lagoon. Consequently, the contribution of sulfur in the first factor further supports that a portion of metals could be absorbed by C_{org} and sulfides and do not hold solely lattice-held positions in the aluminosilicate minerals.

The high negative loadings of Ca and Sr in this factor imply an antipathetic relationship between the autochthonous biogenic carbonates and the terrigenous aluminosilicates. The element assemblages of Ca and Sr behave similarly to each other due to their similar ionic radius (rSr = 0.113 nm vs. rCa = 0.099 nm) and are the dominant components of most bioclastic materials in sediments [77,86]. In the present study, Ca presents a strong correlation with Sr and, contrariwise, a high anti-correlation with all typical terrigenous elements (Al, Ti, K, Fe, Rb, etc.). Notably, the macro- and microscopic examination of sediments suggests that the calcium carbonate component is principally of biogenic origin in Gialova lagoon, thus indicating that the primary production rates overcome the detrital intrusions of the distant limestone bedrock. Hence, the Ca and Sr loadings comprise the negative pole of Factor 1, associated with biogenic sedimentation.

Overall, the inverse relation between the main element assemblages of Factor 1 is the dominant geochemical process of the surface sediments of Gialova Lagoon, as indicated by the very high variance explained by the analysis. A similar "terrigenous aluminosilicates vs. autochthonous biogenic carbonates" bipolar factor was also determined in the geochemical data of other lagoons in western Greece (Messolonghi lagoon, [16]; Prokopos lagoon, [20]), thus highlighting the competitive contrast between detrital and biogenic sedimentation in lagoonal environments of the wider region.

The positive pole of Factor 1 (terrigenous aluminosilicates) shows high score values at the two basins of the lagoon and in the area north of the western one (Figure 7a). This pattern follows the distribution of clay and sediment mean size (Mz) (Figure 3b,c), which further indicates the potential adsorption of trace metals and C_{org} into the finer-grained fractions of the sediments [21,87]. The increased metal and C_{org} loads at the two basins hint at increased eutrophication levels and seagrass growth [88,89] but are not related to significant $CaCO_3$ deposition.

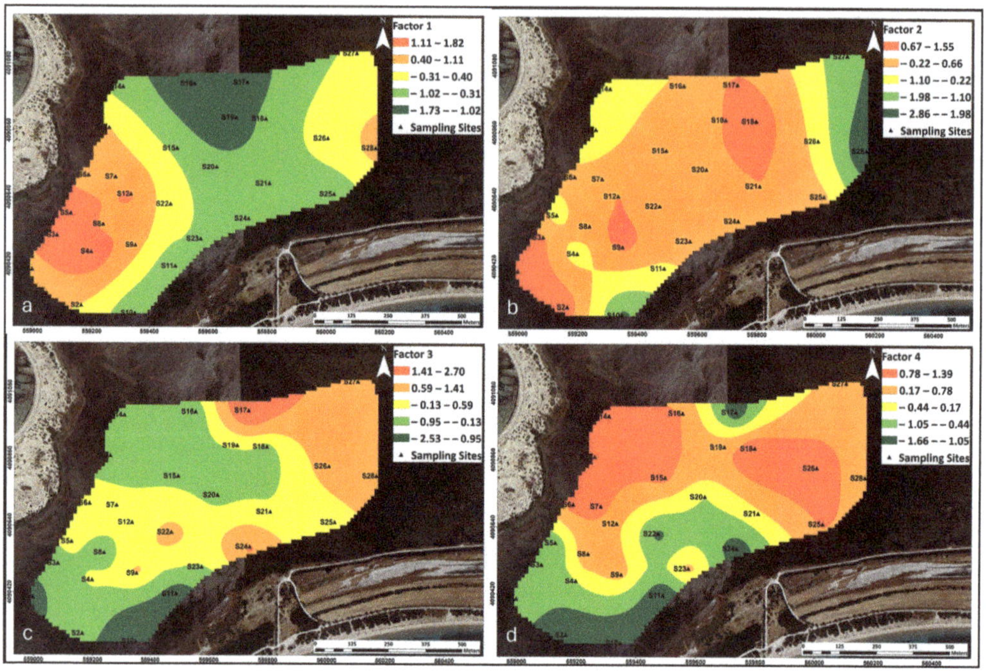

Figure 7. Spatial distribution of Factors 1–4 scores (**a**–**d**).

Regarding the negative pole of Factor 1 (autochthonous biogenic carbonates), it shows high values at the southern, central, and especially northern parts of the Gialova lagoon, forming a N-S-SW distribution zone (Figure 7a). At these parts of the lagoon, sediments of Facies 2 and 3 prevail, characterized by increased sand (mostly biogenic) and the common presence of seagrass residues (Supplementary File S2, Figure S1). These sediments display mainly platykurtic and bimodal grain size distributions, which could be explained by one of the following: (a) an elevated proportion of the coarser-grained fractions; (b) mixing processes; or (c) both.

The second factor (**Factor 2**) accounts for 12.8% of the total variance of the variables (Table 4) and shows high positive loadings on Mo, Sb, and U and moderate positive loadings on As, Cd and S.

The loading profile and the moderately positive loading of S suggest that this factor can be considered a solely "sulfide" factor. Mo, As, and Sb show intense correlation against S, as indicated by the high to moderate loadings, representing an anoxic phase in the surface sediments of the lagoon [24,90]. Divalent metals, such as Cd, have a high affinity with sulfides in sediments, especially in anoxic conditions [91]. Sulfides act as good scavengers of divalent metals that are trapped in the solid phase. Redox-sensitive metals like Mo and U show a similar distribution pattern, suggesting that redox processes are responsible for the deposition of both of these elements in the lagoon [90,92]. Moreover, Mo may be related to algal blooms because this element is essential for cyanobacterial uptake of nitrogen [93,94]. It is thought that algal remnants with increased concentrations of Mo are transported to the seafloor attached to particulate matter.

The spatial distribution of Factor 2 scores (Figure 7b) shows that the influence of sulfides takes place in a wider area at the central part of the lagoon. Maximum values of Factor 2 scores are observed at the western sub-basin of the western shallow basin and at the western part of the eastern shallow basin, indicating the enhanced influence of sulfides to those areas.

Factor 3 accounts for 10.2% of the total variance and shows high positive loadings on Mn and Y while having moderate loadings on La, Zr, Mg, and Th (Table 4). Manganese (Mn) is one of the main geochemical phases in the lagoon sediments and an important scavenger of heavy metals. Based on the loading profile, Factor 3 can be considered a "Mn-hydroxide" factor. The factor loading of C_{org} is negligible, indicating the competition between C_{org} and Mn-hydroxides for scavenging heavy metals.

The high to moderate loadings of Y, La, Zr, Mg, and Th on Factor 3 probably represent the adsorption and/or co-precipitation of these five elements into amorphous Mn oxyhydroxides. The Yttrium (Y), Lanthanum (La), and generally the Rare Earth Elements (REEs) distribution in surface aquatic sediments is mainly controlled by scavenging processes, in particular by Mn–oxides [95,96]. Y, La, Zr, Mg, and Th also show moderate—to—high loadings on Factor 1 (terrigenous aluminosilicates), which suggests the terrigenous (lattice-held) fraction of those elements and/or the limited scavenging from the aluminosilicates [96].

The spatial pattern of the Factor 3 scores implies that the influence of the Mn-oxyhydroxides is more intense at the eastern shallow basin and in the area between the two shallow basins of the lagoon (Figure 7c).

Factor 4 is a bipolar factor and accounts for 7.4% of the total variance of the data. It exhibits high positive loadings on P and Na and moderate positive loadings on Mg, Mn, and C_{org} (Table 4). The negative pole of the third factor presents moderate negative loadings on Ce and La (Table 4).

Phosphorus (P) has a key role in aquatic environments and is considered not only as a primary nutrient for aquatic ecology but also as the most critical limiting nutrient for aquatic productivity [97]. Sedimentary total phosphorus (TP) can be obtained in fractions, namely inorganic-P (IP), organic-P (OP), P bound to Al, Fe, and Mn-oxyhydroxides (Mn-P), and calcium-bound P (Ca-P) [98]. In the case of the Gialova lagoon, the positive correlation of P to Na, Mn and C_{org} suggests at least three P-fractions in the lagoon sediments: (a) inorganic-P in the form of sodium polyphosphate compounds [99], (b) Mn-oxyhydroxides, and (c) organic-P (OP). Based on the above, the positive pole of the fourth factor can be considered a "phosphate" factor of both natural (OP, Mn-P) and anthropogenic (phosphate and polyphosphate fertilizers) origin.

Mg, as mentioned before, also shows a high positive loading on Factor 1, suggesting that Mg concentrations in the lagoon are controlled by multiple processes, such as the "terrigenous aluminosilicates" distribution (Factor 1) and the sea salt influence and/or chlorophyll a degradation (Factor 4). Magnesium, being the metallic part of chlorophyll, can be found in association with phytoplankton and macrophytic debris, which contribute to the organic matter content of the sediments. The Mg released through chlorophyll degradation during cellular senescence and death [100] could replenish some of the Mg content of the sediments [16]. The moderately positive loading of the C_{org} further suggests this interpretation. Moreover, Mg has been used in magnesium-fortified phosphate fertilizers [101], and as such, an additional, anthropogenic source is possible.

The high loadings of Ce and La on the negative pole of Factor 4 probably represent the "dissolved" form, which includes both truly dissolved and/or colloidal and/or the labile (hydroxylamine) form [102].

The "phosphates" factor pole has an important contribution at the northwestern and eastern parts of the lagoon, as indicated by the spatial distribution of the factor scores (Figure 7d). The northwestern part is directly affected by the Xirolagkados River mouth and discharges, so it is most probable that the high Factor 4 scores at this area indicate increased accumulation of materials related to phosphate fertilizers rather than organic-P compounds (low C_{org} values at the northwestern part; Figure 6c). On the other hand, the co-variation with the Factor 3 scores at the eastern basin suggests a predominant coupling of P to Mn-oxyhydroxides (Mn-P).

3.6. Contamination Assessment (Environmental Indices)

3.6.1. Heavy Metal Pollution Indices (EF, I-geo, PLI)

The application of the metal-related pollution indices regarded a total of 14 metals (Ag, As, Cd, Co, Cr, Cu, Mn, Mo, Ni, Pb, Sn, U, V, and Zn), which are typically used for contamination assessment in environmental studies due to their increased toxicity levels and hazardous impact on ecosystems and human health [103–105]. Concerning Enrichment Factor (EF) and I-geo indices, the results are presented and discussed in terms of the average values of the above metals in the lagoon sediments. The total results per sample are presented in Supplementary File S5 for both indices.

Based on the EF variation, the Gialova lagoon sediments are characterized by relatively low metal pollution levels (Table 5). More particularly, the sediments present no enrichment (EF $\leq$ 1) in four metals (Ag, Cd, Sn, V), minor enrichment (1 < EF $\leq$ 3) in eight metals (As, Co, Cr, Cu, Mn, Ni, U, Zn), and moderate enrichment (3 < EF $\leq$ 5) in only two metals (Mo, Pb). Regarding the I-geo values, the lagoonal sediments are classified as unpolluted (I-geo $\leq$ 0) in nearly all metals except Mo and Pb, which cause up to moderate pollution (0 < Igeo < 1) (Table 5). The examined metals display the same order of contamination degree for both EF and I-geo as follows: Sn < Cd < Ag < V<As < Cu < Zn < Co < U<Mn < Cr < Ni < Pb < Mo (Table 5).

Table 5. Contamination assessment of the Gialova lagoon sediments based on the Enrichment Factor (EF) and Geoaccumulation Index (I-geo) values for the pollution-indicator metals concentration.

Metal	Enrichment Factor (EF)					Geoaccumulation Index (I-geo)				
	Min	Max	Mean	Class	State	Min	Max	Mean	Class	State
Sn	0.45	0.52	0.48	EF $\leq$ 1	No enrichment	−3.03	−2.17	−2.49	I-geo $\leq$ 0	Unpolluted
Cd	0.64	0.87	0.76	EF $\leq$ 1	No enrichment	−2.17	−1.49	−1.84	I-geo $\leq$ 0	Unpolluted
Ag	0.55	1.13	0.86	EF $\leq$ 1	No enrichment	−2.39	−1.07	−1.68	I-geo $\leq$ 0	Unpolluted
V	0.87	1.00	0.95	EF $\leq$ 1	No enrichment	−2.08	−1.23	−1.51	I-geo $\leq$ 0	Unpolluted
As	0.93	1.48	1.17	1 < EF $\leq$ 3	Minor enrichment	−1.65	−0.96	−1.23	I-geo $\leq$ 0	Unpolluted
Cu	1.15	1.25	1.21	1 < EF $\leq$ 3	Minor enrichment	−1.72	−0.89	−1.16	I-geo $\leq$ 0	Unpolluted
Zn	1.17	1.28	1.21	1 < EF $\leq$ 3	Minor enrichment	−1.66	−0.85	−1.16	I-geo $\leq$ 0	Unpolluted
Co	1.32	1.47	1.40	1 < EF $\leq$ 3	Minor enrichment	−1.47	−0.71	−0.95	I-geo $\leq$ 0	Unpolluted
U	1.03	1.69	1.43	1 < EF $\leq$ 3	Minor enrichment	−1.40	−0.51	−0.93	I-geo $\leq$ 0	Unpolluted
Mn	1.77	2.69	2.18	1 < EF $\leq$ 3	Minor enrichment	−0.98	−0.15	−0.32	I-geo $\leq$ 0	Unpolluted
Cr	2.13	3.43	2.45	1 < EF $\leq$ 3	Minor enrichment	−0.57	0.16	−0.15	I-geo $\leq$ 0	Unpolluted
Ni	2.52	2.79	2.71	1 < EF $\leq$ 3	Minor enrichment	−0.62	0.28	0.00	I-geo $\leq$ 0	Unpolluted
Pb	3.03	3.38	3.17	3 < EF $\leq$ 5	Moderate enrichment	−0.19	0.57	0.23	0 < Igeo < 1	Unpolluted to moderately polluted
Mo	2.18	5.49	4.36	3 < EF $\leq$ 5	Moderate enrichment	−0.26	1.19	0.66	0 < Igeo < 1	Unpolluted to moderately polluted

PLI values range from 0.56 to 0.95, with a mean value of 0.82 suggesting unpolluted sediments throughout the lagoon. According to the spatial distribution of the PLI values (Figure 8), the western shallow basin and the area north of it, as well as the western part of the eastern shallow basin, exhibit the highest levels, whereas the lowest values are noted at the aforementioned N-S-SW zone, where biogenic carbonates prevail.

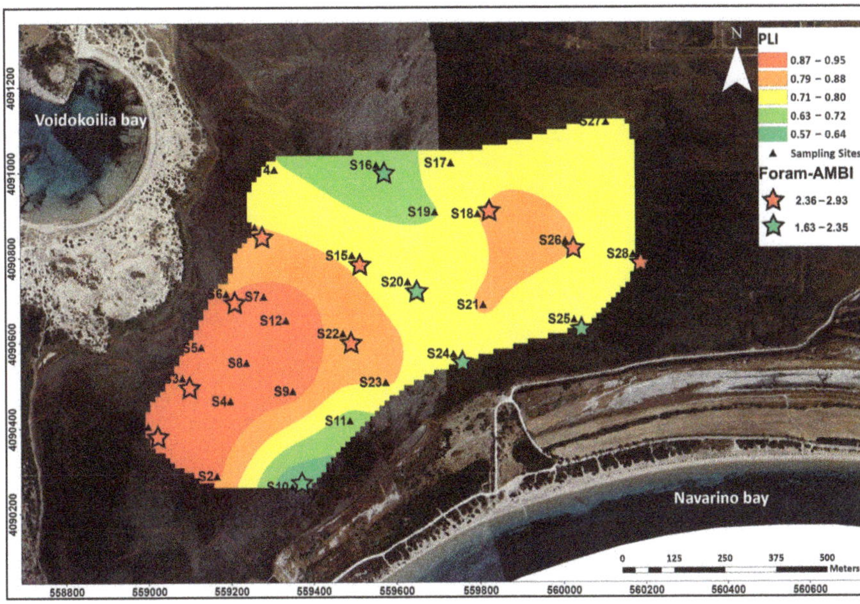

Figure 8. Spatial distribution of the PLI values in Gialova lagoon bottom sediments and Foram-AMBI values. The value-range for each index indicates the contamination degree and environmental stress, as presented in Table 1.

3.6.2. Q-mode Factor Analysis

Using the Q-mode factor analysis, we evaluated the interrelationships among the 28 sediment samples in terms of their geochemical composition and defined certain sediment clusters that provide significant information on the contamination assessment. Two factors account for more than 99.9% of the information among the studied samples (Supplementary File S6, Table S1). All samples display very high communalities, while according to the Kaiser–Meyer–Olkin (KMO) Test (0.837), the dataset is suitable for factor analysis. Therefore, the 2-factor model is considered efficient for the description and explanation of the samples' investigation.

By plotting the factor loadings on the two factor axes (Supplementary File S6, Figure S1), we portray the relationships between the samples according to the entire spectrum of their geochemical parameters. The samples occurring nearest the two axes are "end-member" samples (Clusters A and B). The other samples (Cluster C) can all be considered mixtures of the former two. Q-Factor 1 represents sediment samples with the highest elemental (except Ca and Sr) and C_{org} concentrations, as indicated by the chemical composition of their end-members (Cluster A) (Supplementary File S6, Table S2). On the other hand, the end-members of Q-Factor 2 (Cluster B) are characterized by the highest Ca and Sr concentrations among the whole geochemical dataset (Supplementary File S6, Table S2). The sediment samples of Cluster C, which is located at the midpoint of the Q-mode factor plot, show intermediate element concentrations compared to the end-member samples (Supplementary File S6, Table S2).

The two end-member clusters link the samples with the highest scores of the positive (Cluster A) and negative (Cluster B) poles of the aforementioned "terrigenous aluminosilicates associated with C_{org} vs. biogenic carbonates" R-mode factor, which represents the leading geochemical pattern in the Gialova lagoon (see Section 3.5). The Cluster A samples are located in the eastern sub-basin of the western shallow basin and in the area north of it (Figure 9). The area of Cluster A sediments matches very well with the maximum PLI values (Figure 8), suggesting that Cluster A sediments represent the elevated metal concentration area of the lagoon. On the other hand, the Cluster B sediments are located at the northern and southwestern boundaries of the lagoon, close to the Xirolagkados River mouth and the sand barrier/inlet, respectively (Figure 9). The area of Cluster B sediments coincides with the lowest PLI values of the lagoon (Figure 8) and the biogenic-rich sediments. Based on the above, the Q-factor plot can be considered to portray the entire spectrum of sediments in the lagoon in terms of elevated and low metal concentrations. The correct assignment of the lagoon sediments confirmed the adequacy of the Q-mode factor analysis as a classification method for the overall geochemistry of the lagoon.

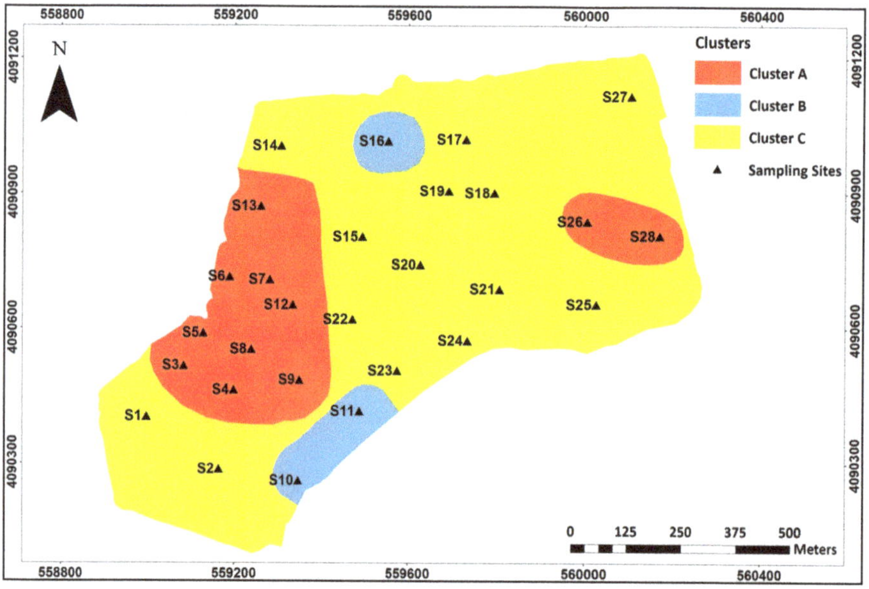

Figure 9. Spatial pattern of Gialova lagoon sediment contamination degree from lowest (Cluster B) to intermediate (Cluster C) and highest (Cluster A) levels.

3.6.3. Microfaunal Index (Foram-AMBI)

The Foram-AMBI values within the studied area exhibited a range from 2.84 (S28) to 1.63 (S24), averaging at 2.36. Employing the classification proposed by [58] for the AMBI index, the entirety of the lagoon demonstrated a "Good" ecological quality status. Notably, the most adverse (>2.36) conditions were predominantly observed in the western and eastern basins, whereas the N-S-SW zone, where biogenic carbonates prevail, showcased a comparatively higher ecological quality status (Figure 8). Consequently, a notable correlation between Foram-AMBI and the Pollution Loading Index (PLI) is observed, thus further emphasizing the utility of AMBI as an indicator for environmental quality assessment (Figure 8). Additionally, the Foram-AMBI values exhibited a negative correlation with the diversity index, suggesting that areas with better environmental conditions are preferred for multi-diverse microfauna (Supplementary File S4).

3.7. Depositional and Environmental Variations of the Last 30 Years

In this section, we compare our sedimentological and geochemical findings with the ones derived back in 1995 to address how and to which degree the environmental conditions within the Gialova lagoon have changed during a period of 25 years (until the 2020 sampling survey), referred to as the last 30 years approximately. The initial sedimentological results of the 1995 survey were first described by Kontopoulos and Bouzos [31], while the later work of Avramidis et al. [32] provided further details both on the sedimentology and geochemistry of the lagoon bottom, including contamination assessment via EF and I-geo indices. Nevertheless, we note at this point that the different methodological approaches in grain size (sieving and pipette vs. laser diffraction), bulk geochemistry (3-acids and ICP-OS vs. 4-acids and ICP-MS), and organic carbon (titration vs. combustion method) might affect the comparisons between the two surveys, and therefore we keep the comparison efforts under a more general perspective, thus avoiding overinterpretation.

Modern sediments are remarkably less abundant in sandy material compared to 1995, leading to grain size fining and much improved sorting values (Supplementary File S7, Table S1). Similarly to the present study, Kontopoulos and Bouzos [31] also defined the primarily biogenic composition of the sandy material throughout the lagoon. Consequently, a reduction in sand content is most likely associated with a decrease in biogenic productivity. This is further supported by the overall C_{org} reduction in the lagoon (Supplementary File S7, Table S1) and also by the fact that today carbonate material prevails in a relatively limited N-S-SW zone, whereas in 1995 it presented a widespread distribution of high values [31]. This modification could be the outcome of the construction of the two canals in 1998, which restored the freshwater inflows into the lagoon, favoring important rearrangements in the subaqueous ecosystems and physiochemical properties of the water column [35,36]. The large influence of the restored freshwater inflows is also confirmed by the discriminant function analysis on the sedimentological data, suggesting that the fluvial effects are more pronounced than the signals of seawater intrusions (Supplementary File S2, Figure S2b).

Another difference between the two surveys regards the heavy metal pollution levels of the lagoon sediments, which presents a clear reduction in the later survey (Supplementary File S7, Table S2). A possible explanation for this reduction could be the installation of a Wastewater Treatment Plant (WWTP) in 2016 at about 4 km north of the lagoon, affecting the Xirolagkados River drainage basin [106]. Notably, it has been documented that the operation of WWTPs leads to a gradual decrease in heavy metal accumulation in lagoonal and nearshore sediments [107,108], following a series of chemical treatments in the wastewater and sewage sludge [109,110]. However, we cannot exclude the possibility that the reduction in metal-related pollution results from changes in the hydrologic/hydrodynamic regime of the lagoon, namely due to the aforementioned construction of the two drainage canals.

4. Conclusions

In the present study, the compilation of acoustic, sedimentological, geochemical, and microfaunal data enabled us to assess the environmental conditions of the shallow Gialova lagoon and identify the dominant sedimentological and geochemical patterns. The environmental status of the lagoon has been achieved through the application of geochemical—and microfaunal—based indices and the statistical treatment of the datasets.

Although the Gialova lagoon is an extremely shallow environment, a detailed bathymetric map was obtained using an unmanned surface vehicle (USV). The lagoon floor can be considered flat, with three small basins forming in the central, western, and eastern parts.

The lagoon bottom shows rather low contamination with heavy metals except Mo and Pb, which induce moderate pollution levels. Comparing with the findings of a previous study conducted in 1995, we find that over the last ~30 years, biogenic productivity is currently at lower levels, most likely due to rearrangements in the subaqueous ecosystems and the physiochemical properties of the water column induced by the construction of two drainage canals in 1998. Moreover, we note a clear reduction in heavy metal pollution. It should be emphasized that the different methodologies used in the two studies cast doubt on the above comparisons.

The most significant sedimentary process in the lagoon regards detrital versus biogenic sedimentation in a low-energy environment. Terrigenous aluminosilicates and autochthonous biogenic carbonates are the principal geochemical phases of the lagoon, followed by sulfides, Mn-hydroxides, and phosphates. Organic matter seems to build aggregates and flocs with clay minerals and shows a close relationship with the sulfides. Certain metals (Cd, Zr, Mg, and Y) have been shown to be absorbed by more than one scavenger, suggesting that two different processes control their distribution in the surface sediments of the lagoon. Aluminosilicate-rich sediments predominately accumulate at the western and eastern parts of the lagoon, where two small-sized basins prevail, while biogenic carbonates prevail in a N-S-SW zone shaped by riverine flow (northern part) and seawater intrusions through the inlet (southwestern part). The terrigenous sediments are finer-grained (fine to very fine silt) and more abundant in C_{org} than the biogenic deposits, while they adsorb the largest proportion of trace metal loads. Accordingly, the maximum contamination and environmental stress concern the two shallow basins.

Overall, the findings of this study can contribute to the development of a robust socio-environmental management plan for a highly vulnerable coastal site of the EU's Natura 2000 network. In addition, the present work updates and promotes scientific knowledge on the natural versus anthropogenic processes that shape such settings under the threat of growing climatic instability across the eastern Mediterranean.

Supplementary Materials: The following supporting information can be downloaded at: https://www.mdpi.com/article/10.3390/w16162312/s1, Supplementary File S1: Sedimentological Results; Supplementary File S2: Sedimentary facies and Depositional Environment; Supplementary File S3: Geochemical Results; Supplementary File S4: Microfaunal Results; Supplementary File S5: Environmental Indices (EF and I-geo); Supplementary File S6: Q-mode Factor Analysis; and Supplementary File S7: Sedimentological and Geochemical comparisons.

Author Contributions: Conceptualization, G.P., M.P. and S.S.; methodology, M.P., S.S., A.P. and E.F.; validation, G.P. and M.G.; formal analysis, S.S., E.F. and G.P.; investigation, M.P., S.S., E.F., X.D., D.C. and A.P.; data curation, G.P., S.S. and D.C.; writing—original draft preparation, M.P. and S.S.; writing—review and editing, M.P., S.S., M.G. and G.P.; visualization, M.P., S.S. and E.F.; supervision, G.P. and M.G.; project administration, G.P. All authors have read and agreed to the published version of the manuscript.

Funding: This research received no external funding.

Data Availability Statement: Data is available upon request.

Acknowledgments: The authors would like to thank everyone involved in the fieldwork survey. Special thanks are given to Katerina Mavrogonatou and Christina Fourkalidi for their efforts in microfaunal analyses and to Nikos Mavromatis for the employment of the USV. Finally, we thank the three anonymous reviewers for their critical reviews that led to the improvement of the manuscript. The publication fees of this manuscript have been financed by the Research Council of the University of Patras.

Conflicts of Interest: The authors declare no conflicts of interest.

References

1. Kjerfve, B. *Coastal Lagoon Processes*; Elsevier Science B.V.: Amsterdam, The Netherlands, 1994; ISBN 0444882588.
2. Sousa, A.; García-Murillo, P.; Morales, J.; García-Barrón, L. Anthropogenic and Natural Effects on the Coastal Lagoons in the Southwest of Spain (Doñana National Park). *ICES J. Mar. Sci.* **2009**, *66*, 1508–1514. [CrossRef]
3. Chapman, P.M. Management of Coastal Lagoons under Climate Change. *Estuar. Coast. Shelf Sci.* **2012**, *110*, 32–35. [CrossRef]
4. Madkour, H.; Abdelhalim, M.A.K.; Branch, R.S.; Obirikorang, K.A.; El-taher, A. Environmental Implications of Surface Sediments from Coastal Lagoons in the Red Sea Coast. *J. Environ. Biol.* **2015**, *36*, 1421–1427.
5. Inda, H.; García-Rodríguez, F.; del Puerto, L.; Stutz, S.; Lopes Figueira, R.C.; de Lima Ferreira, P.A.; Mazzeo, N. Discriminating between Natural and Human-Induced Shifts in a Shallow Coastal Lagoon: A Multidisciplinary Approach. *Anthropocene* **2016**, *16*, 1–15. [CrossRef]

6. Quintana, X.D.; Boix, D.; Gascón, S.; Sala, J. *Management and Restoration of Mediterranean Coastal Lagoons in Europe*; Càtedra d'Ecosistemes Litorals Mediterrànis i LIFE Pletera: Torroella de Montgrí, Spain, 2018; Volume 220.
7. Pérez-Ruzafa, A.; Marcos, C.; Pérez-Ruzafa, I.M. Mediterranean Coastal Lagoons in an Ecosystem and Aquatic Resources Management Context. *Phys. Chem. Earth* **2011**, *36*, 160–166. [CrossRef]
8. Lillebø, A.I.; Stålnacke, P.; Gooch, G.D.; Krysanova, V.; Bielecka, M. Pan-European Management of Coastal Lagoons: A Science-Policy-Stakeholder Interface Perspective. *Estuar. Coast. Shelf Sci.* **2017**, *198*, 648–656. [CrossRef]
9. Soria, J.; Pérez, R.; Sòria-Pepinyà, X. Mediterranean Coastal Lagoons Review: Sites to Visit before Disappearance. *J. Mar. Sci. Eng.* **2022**, *10*, 347. [CrossRef]
10. El-Sammak, A.A. Textural Facies of Recent Sediments North of Sinai, Southeastern Mediterranean, Egypt. *J. Coast. Res.* **1995**, *11*, 904–910.
11. Chevillon, C. Skeletal Composition of Modern Lagoon Sediments in New Caledonia: Coral, a Minor Constituent. *Coral Reefs* **1996**, *15*, 199–207. [CrossRef]
12. Boggs, S. *Principles of Sedimentology and Stratigraphy*, 4th ed.; Pearson Prentice Hall: Upper Saddle River, NJ, USA, 2006; ISBN 9788490225370.
13. Molinaroli, E.; Sarretta, A.; Ferrarin, C.; Masiero, E.; Specchiulli, A.; Guerzoni, S. Sediment Grain Size and Hydrodynamics in Mediterranean Coastal Lagoons: Integrated Classification of Abiotic Parameters. *J. Earth Syst. Sci.* **2014**, *123*, 1097–1114. [CrossRef]
14. Christophoridis, A.; Stamatis, N.; Orfanidis, S. Sediment Heavy Metals of a Mediterranean Coastal Lagoon: Agiasma, Nestos Delta, Eastern Macedonia (Greece). *Transitional Waters Bull.* **2007**, *1*, 33–43. [CrossRef]
15. Prudêncio, M.I.; Gonzalez, M.I.; Dias, M.I.; Galan, E.; Ruiz, F. Geochemistry of Sediments from El Melah Lagoon (NE Tunisia): A Contribution for the Evaluation of Anthropogenic Inputs. *J. Arid Environ.* **2007**, *69*, 285–298. [CrossRef]
16. Karageorgis, A.P.; Sioulas, A.; Krasakopoulou, E.; Anagnostou, C.L.; Hatiris, G.A.; Kyriakidou, H.; Vasilopoulos, K. Geochemistry of Surface Sediments and Heavy Metal Contamination Assessment: Messolonghi Lagoon Complex, Greece. *Environ. Earth Sci.* **2012**, *65*, 1619–1629. [CrossRef]
17. Hernández-Crespo, C.; Martín, M. Determination of Background Levels and Pollution Assessment for Seven Metals (Cd, Cu, Ni, Pb, Zn, Fe, Mn) in Sediments of a Mediterranean Coastal Lagoon. *Catena* **2015**, *133*, 206–214. [CrossRef]
18. Maanan, M.; Saddik, M.; Maanan, M.; Chaibi, M.; Assobhei, O.; Zourarah, B. Environmental and Ecological Risk Assessment of Heavy Metals in Sediments of Nador Lagoon, Morocco. *Ecol. Indic.* **2015**, *48*, 616–626. [CrossRef]
19. Vasileiadou, K.; Pavloudi, C.; Kalantzi, I.; Apostolaki, E.T.; Chatzigeorgiou, G.; Chatzinikolaou, E.; Pafilis, E.; Papageorgiou, N.; Fanini, L.; Konstas, S.; et al. Environmental Variability and Heavy Metal Concentrations from Five Lagoons in the Ionian Sea (Amvrakikos Gulf, W Greece). *Biodivers. Data J.* **2016**, *4*, e8233. [CrossRef]
20. Katsaros, D.; Panagiotaras, D.; Kontopoulos, N.; Avramidis, P. Sediments Characteristics and Heavy Metals Distribution of a Very Shallow Protected Coastal Lagoon, Prokopos Lagoon, Mediterranean Sea Western Greece. *Fresenius Environ. Bull.* **2017**, *26*, 6093–6103.
21. Zonta, R.; Botter, M.; Cassin, D.; Bellucci, L.G.; Pini, R.; Dominik, J. Sediment Texture and Metal Contamination in the Venice Lagoon (Italy): A Snapshot before the Installation of the MOSE System. *Estuar. Coast. Shelf Sci.* **2018**, *205*, 131–151. [CrossRef]
22. El Zrelli, R.; Yacoubi, L.; Wakkaf, T.; Castet, S.; Grégoire, M.; Mansour, L.; Courjault-Radé, P.; Rabaoui, L. Surface Sediment Enrichment with Trace Metals in a Heavily Human-Impacted Lagoon (Bizerte Lagoon, Southern Mediterranean Sea): Spatial Distribution, Ecological Risk Assessment, and Implications for Environmental Protection. *Mar. Pollut. Bull.* **2021**, *169*, 112512. [CrossRef]
23. Zoidou, M.; Sylaios, G. Ecological Risk Assessment of Heavy Metals in the Sediments of a Mediterranean Lagoon Complex. *J. Environ. Health Sci. Eng.* **2021**, *19*, 1835–1849. [CrossRef]
24. De Lacerda, L.D. Chapter 8 Biogeochemistry of Heavy Metals in Coastal Lagoons. In *Elsevier Oceanography Series*; Elsevier: Amsterdam, The Netherlands, 1994; Volume 60, pp. 221–241. [CrossRef]
25. Ramiro Pastorinho, M.; Telfer, T.C.; Nogueira, A.J.A.; Soares, A.M.V.M.; Ranville, J.F. An Evaluation of Trace Metal Distribution, Enrichment Factors and Risk in Sediments of a Coastal Lagoon (Ria de Aveiro, Portugal). *Environ. Earth Sci.* **2012**, *67*, 2043–2052. [CrossRef]
26. Tnoumi, A.; Angelone, M.; Armiento, G.; Caprioli, R.; Crovato, C.; De Cassan, M.; Montereali, M.R.; Nardi, E.; Parrella, L.; Proposito, M.; et al. Heavy Metal Content and Potential Ecological Risk Assessment of Sediments from Khnifiss Lagoon National Park (Morocco). *Environ. Monit. Assess.* **2022**, *194*, 356. [CrossRef] [PubMed]
27. Alve, E.; Korsun, S.; Schönfeld, J.; Dijkstra, N.; Golikova, E.; Hess, S.; Husum, K.; Panieri, G. Foram-AMBI: A Sensitivity Index Based on Benthic Foraminiferal Faunas from North-East Atlantic and Arctic Fjords, Continental Shelves and Slopes. *Mar. Micropaleontol.* **2016**, *122*, 1–12. [CrossRef]
28. Jorissen, F.; Pia, M.; Almogi-labin, A.; Barras, C.; Bergamin, L.; Bicchi, E.; El, A.; Ferraro, L.; Mcgann, M.; Morigi, C.; et al. Developing Foram-AMBI for Biomonitoring in the Mediterranean: Species Assignments to Ecological Categories. *Mar. Micropaleontol.* **2018**, *140*, 33–45. [CrossRef]
29. Bouchet, V.M.P.; Frontalini, F.; Francescangeli, F.; Sauriau, P.G.; Geslin, E.; Martins, M.V.A.; Almogi-Labin, A.; Avnaim-Katav, S.; Di Bella, L.; Cearreta, A.; et al. Relative Abundances of Benthic Foraminifera in Response to Total Organic Carbon in Sediments: Data from European Intertidal Areas and Transitional Waters. *Mar. Pollut. Bull.* **2021**, *164*, 112071. [CrossRef]

30. Maneas, G.; Makopoulou, E.; Bousbouras, D.; Berg, H.; Manzoni, S. Anthropogenic Changes in a Mediterranean Coastal Wetland during the Last Century-The Case of Gialova Lagoon, Messinia, Greece. *Water* **2019**, *11*, 350. [CrossRef]
31. Kontopoulos, N.; Bouzos, D. The Sedimentology of the Modern Lagoons in the Western Peloponnesus. *Bull. Geol. Soc. Greece* **2011**, *44*, 1–18. [CrossRef]
32. Avramidis, P.; Iliopoulos, G.; Kontopoulos, N.; Panagiotaras, D.; Barouchas, P.; Nikolaou, K.; Papadopoulou, P. Depositional Environments, Sediment Characteristics, Palaeoecological Analysis and Environmental Assessment of an Internationally Protected Shallow Mediterranean Lagoon, Gialova Lagoon—Navarino Bay, Greece14. *Earth Environ. Sci. Trans. R. Soc. Edinb.* **2015**, *105*, 189–206. [CrossRef]
33. Bray, L.; Faulwetter, S.; Kaberi, H.; Karageorgis, A.P.; Kastanidi, E.; Katsiaras, N.; Pavlidou, A.; Providakis, N.; Sigala, K.; Voutsinas, E.; et al. Assessing Pressure Drivers on the Benthic Ecosystem in the Coastal Zone of Western Messinia, Greece. *Estuar. Coast. Shelf Sci.* **2022**, *274*, 107935. [CrossRef]
34. Koutsoubas, D.; Arvanitidis, C.; Dounas, C.; Drummond, L. Community Structure and Dynamics of the Molluscan Fauna in a Mediterranean Lagoon (Gialova Lagoon, SW Greece). *Belg. J. Zool.* **2000**, *130*, 131–138.
35. Chatzigeorgiou, G.; Reizopoulou, S.; Maidanou, M.; Naletaki, M.; Orneraki, E.; Apostolaki, E.; Arvanitidis, C. Macrobenthic Community Changes Due to Dystrophic Events and Freshwater Inflow: Changes in Space and Time in a Mediterranean Lagoon (Gialova Lagoon, SW Greece). *Estuar. Coast. Shelf Sci.* **2011**, *94*, 111–121. [CrossRef]
36. Manzoni, S.; Maneas, G.; Scaini, A.; Psiloglou, B.E.; Destouni, G.; Lyon, S.W. Understanding Coastal Wetland Conditions and Futures by Closing Their Hydrologic Balance: The Case of the Gialova Lagoon, Greece. *Hydrol. Earth Syst. Sci.* **2020**, *24*, 3557–3571. [CrossRef]
37. Willershäuser, T.; Vött, A.; Hadler, H.; Ntageretzis, K.; Emde, K.; Brückner, H. Holocene Palaeotsunami Imprints in the Stratigraphical Record and the Coastal Geomorphology of the Gialova Lagoon near Pylos (Southwestern Peloponnese, Greece). *Z. Geomorphol.* **2015**, *59*, 215–252. [CrossRef]
38. Emmanouilidis, A.; Katrantsiotis, C.; Norström, E.; Risberg, J.; Kylander, M.; Sheik, T.A.; Iliopoulos, G.; Avramidis, P. Middle to Late Holocene Palaeoenvironmental Study of Gialova Lagoon, SW Peloponnese, Greece. *Quat. Int.* **2018**, *476*, 46–62. [CrossRef]
39. Katrantsiotis, C.; Kylander, M.E.; Smittenberg, R.; Yamoah, K.K.A.; Hättestrand, M.; Avramidis, P.; Strandberg, N.A.; Norström, E. Eastern Mediterranean Hydroclimate Reconstruction over the Last 3600 Years Based on Sedimentary N-Alkanes, Their Carbon and Hydrogen Isotope Composition and XRF Data from the Gialova Lagoon, SW Greece. *Quat. Sci. Rev.* **2018**, *194*, 77–93. [CrossRef]
40. Papakonstantinou, M. Side Scan Sonar Bottom Relief Interpretation and Ecological Evaluation of Gialova Lagoon, Pref. of Messinia, Greece, as Well as Land Cover Land Use (LCLU) Digital Mapping in the Wider Protected Area. Master's Thesis, University of Patras, Patras, Greece, 2015.
41. Bouzos, D.; Kontopoulos, N.; Avramidis, P. Oceanographic Observations in the Lagoon of Gialova, (SW Peloponnese). In Proceedings of the 6th Pan-Hellenic Geographical Conference of the Hellenic Geographical Society, Thessaloniki, Greece, 3–6 October 2002; pp. 247–254.
42. Papakonstantinou, M.; Papatheodorou, G.; Christodoulou, D.; Papastergiadou, E. Seasonal Variations of Water Quality and Aquatic Macrophytes Survey Based on Side Scan Sonar, in a Shallow Coastal Lagoon (Gialova, SW Peloponnesus, Greece). In Proceedings of the 16th International Conference on Environmental Science and Technology, Rhodes, Greece, 4–7 September 2019. [CrossRef]
43. Papakonstantinou, M.; Fakiris, E.; Geraga, M.; Papatheodorou, G. Exploring Shallow Lagoons with USV Mapping Technology. *Hydro Int.* **2023**, *27*, 30–34.
44. Folk, R.L.; Ward, W. Brazos River Bar: A Study in the Significance of Grain Size Parameters. *J. Sediment. Pet.* **1957**, *27*, 3–26. [CrossRef]
45. Blott, S.J.; Pye, K. Gradistat: A Grain Size Distribution and Statistics Package for the Analysis of Unconsolidated Sediments. *Earth Surf. Process. Landf.* **2001**, *26*, 1237–1248. [CrossRef]
46. Folk, R.L. *Petrology of Sedimentary Rocks*; Hemphill Publishing Company: Austin, TX, USA, 1974.
47. Stewart, H. Sedimentary Reflections of Depositional Environment in San Miguel Lagoon, Baja California, Mexico. *Am. Assoc. Pet. Geol. Bull.* **1958**, *42*, 2587–2618. [CrossRef]
48. Sahu, B.K. Depositional Mechanisms from the Size Analysis of Clastic Sediments. *SEPM J. Sediment. Res.* **1964**, *34*, 73–83. [CrossRef]
49. Farhat, H.I. Textural Features and Transportation Mode of Nile Delta Coastal Lagoon Surficial Sediments (Lake Burullus, Ramsar Site). *SN Appl. Sci.* **2019**, *1*, 1013. [CrossRef]
50. Tnoumi, A.; Khalidi, K.E.; Zourarah, B.; Angelone, M.; Armiento, G.; Caprioli, R.; Crovato, C.; De Cassan, M.; Montereali, M.R.; Proposito, M.; et al. Studies on the Textural Characteristics of Sediments From Khnifiss Lagoon (Southern Coast of Morocco). *Int. J. Adv. Res. Eng. Technol.* **2020**, *11*, 2299–2311. [CrossRef]
51. Botziolis, C.; Bourli, N.; Zoumpouli, E.; Papadopoulou, P.; Dimopoulos, N.; Kovani, A.; Zelilidis, P.; Aspioti, D.C.; Iliopoulos, G.; Zelilidis, A. The Knowledge and Application of Sedimentary Conditions of Shallow Marine and Tidal Waters of Ionian Islands, Greece: Implications for Therapeutic Use. *Geosciences* **2024**, *14*, 48. [CrossRef]

52. Azidane, H.; Michel, B.; Bouhaddioui, M.E.; Haddout, S.; Magrane, B.; Benmohammadi, A. Grain Size Analysis and Characterization of Sedimentary Environment along the Atlantic Coast, Kenitra (Morocco). *Mar. Georesour. Geotechnol.* **2021**, *39*, 569–576. [CrossRef]
53. Turekian, K.K.; Wedepohl, K.H. Distribution of the Elements in Some Major Units of the Earth's Crust. *Geol. Soc. Am. Bull.* **1961**, *72*, 175–192. [CrossRef]
54. Birch, G.F. An Assessment of Aluminum and Iron in Normalisation and Enrichment Procedures for Environmental Assessment of Marine Sediment. *Sci. Total Environ.* **2020**, *727*, 138123. [CrossRef] [PubMed]
55. Chen, C.W.; Kao, C.M.; Chen, C.F.; Dong, C. Di Distribution and Accumulation of Heavy Metals in the Sediments of Kaohsiung Harbor, Taiwan. *Chemosphere* **2007**, *66*, 1431–1440. [CrossRef] [PubMed]
56. Muller, G. Schwermetalle in Den Sedimenten Des Rheins Veranderungen Selt 1971. *Umschau* **1979**, *79*, 778–783.
57. Tomilson, D.L.; Wilson, J.G.; Harris, C.; Jeffrey, D. Problems in the Assessment of Heavy Metal Levels in Estuaries and the Formation of a Pollution Index. *Helgol. Meeresunters.* **1980**, *33*, 566–575.
58. Borja, A.; Franco, J.; Pérez, V. A Marine Biotic Index to Establish the Ecological Quality of Soft-Bottom Benthos within European Estuarine and Coastal Environments. *Mar. Pollut. Bull.* **2000**, *40*, 1100–1114. [CrossRef]
59. Reyment, R.A.; Joreskog, K.G. *Applied Factor Analysis in the Natural Sciences*; Cambridge University Press: Cambridge, UK, 1996.
60. Papatheodorou, G.; Hotos, G.; Geraga, M.; Avramidou, D.; Vorinakis, T. Heavy Metal Concentrations in Sediments of Klisova Lagoon (Southeast Mesolonghi-Aetolikon Lagoon Complex), W. Greece. *Fresenius Environ. Bull.* **2002**, *11*, 951–956.
61. Hua, Z.; Yinghui, J.; Tao, Y.; Min, W.; Guangxun, S.; Mingjun, D. Heavy Metal Concentrations and Risk Assessment of Sediments and Surface Water of the Gan River, China. *Pol. J. Environ. Stud.* **2016**, *25*, 1529–1540. [CrossRef]
62. Selvaraj, K.; Ram Mohan, V.; Szefer, P. Evaluation of Metal Contamination in Coastal Sediments of the Bay of Bengal, India: Geochemical and Statistical Approaches. *Mar. Pollut. Bull.* **2004**, *49*, 174–185. [CrossRef] [PubMed]
63. Papatheodorou, G.; Demopoulou, G.; Lambrakis, N. A Long-Term Study of Temporal Hydrochemical Data in a Shallow Lake Using Multivariate Statistical Techniques. *Ecol. Model.* **2006**, *193*, 759–776. [CrossRef]
64. Madsen, J.D.; Chambers, P.A.; James, W.F.; Koch, E.W.; Westlake, D.F. The Interaction between Water Movement, Sediment Dynamics and Submersed Macrophytes. *Hydrobiologia* **2001**, *444*, 71–84. [CrossRef]
65. Kontopoulos, N.; Avramidis, P. A Late Holocene Record of Environmental Changes from the Aliki Lagoon, Egion, North Peloponnesus, Greece. *Quat. Int.* **2003**, *111*, 75–90. [CrossRef]
66. De Falco, G.; Molinaroli, E.; Conforti, A.; Simeone, S.; Tonielli, R. Biogenic Sediments from Coastal Ecosystems to Beach-Dune Systems: Implications for the Adaptation of Mixed and Carbonate Beaches to Future Sea Level Rise. *Biogeosciences* **2017**, *14*, 3191–3205. [CrossRef]
67. Valia, H.; Cameron, B. Skewness as a Paleoenvironmental Indicator. *SEPM J. Sediment. Res.* **1977**, *47*, 784–793. [CrossRef]
68. Martins, L.R. Recent Sediments and Grain-Size Analysis. *Gravel* **2003**, *1*, 90–105.
69. Magni, P.; De Falco, G.; Como, S.; Casu, D.; Floris, A.; Petrov, A.N.; Castelli, A.; Perilli, A. Distribution and Ecological Relevance of Fine Sediments in Organic-Enriched Lagoons: The Case Study of the Cabras Lagoon (Sardinia, Italy). *Mar. Pollut. Bull.* **2008**, *56*, 549–564. [CrossRef]
70. Karageorgis, A.P.; Katsanevakis, S.; Kaberi, H. Use of Enrichment Factors for the Assessment of Heavy Metal Contamination in the Sediments of Koumoundourou Lake, Greece. *Water Air Soil Pollut.* **2009**, *204*, 243–258. [CrossRef]
71. Panagiotaras, D.; Papoulis, D.; Kontopoulos, N.; Avramidis, P. Geochemical Processes and Sedimentological Characteristics of Holocene Lagoon Deposits, Alikes Lagoon, Zakynthos Island, Western Greece. *Geol. J.* **2012**, *47*, 372–387. [CrossRef]
72. Dassenakis, M.; Krasakopoulou, E.; Matzara, B. Chemical Characteristics of Aetoliko Lagoon, Greece, after an Ecological Shock. *Mar. Pollut. Bull.* **1994**, *28*, 427–433. [CrossRef]
73. Karageorgis, A.P. Geochemical Study of Sediments from the Amvrakikos Gulf Lagoon Complex, Greece. *Transitional Water Bull.* **2007**, *1*, 3–8. [CrossRef]
74. Varnavas, S.P.; Panagos, A.G.; Laios, G. Trace Elements in Surface Sediments of Navarino Bay, Greece. *Environ. Geol. Water Sci.* **1987**, *10*, 159–168. [CrossRef]
75. Geslin, E.; Debenay, J.P.; Lesourd, M. Abnormal Wall Textures and Test Deformation in Ammonia (Hyaline Foraminifer). *J. Foraminifer. Res.* **1998**, *28*, 148–156.
76. Frontalini, F.; Buosi, C.; Da Pelo, S.; Coccioni, R.; Cherchi, A.; Bucci, C. Benthic Foraminifera as Bio-Indicators of Trace Element Pollution in the Heavily Contaminated Santa Gilla Lagoon (Cagliari, Italy). *Mar. Pollut. Bull.* **2009**, *58*, 858–877. [CrossRef] [PubMed]
77. Rothwell, R.G.; Croudace, I.W. Twenty Years of XRF Core Scanning Marine Sediments: What Do Geochemical Proxies Tell Us? In *Micro-XRF Studies of Sediment Cores. Developments in Paleoenvironmental Research*; Croudace, I., Rothwell, R., Eds.; Springer: Dordrecht, The Netherlands, 2015; Volume 17, ISBN 978-94-017-9848-8.
78. Veizer, J.; Mackenzie, F.T. *Evolution of Sedimentary Rocks*, 2nd ed.; Elsevier Ltd.: Amsterdam, The Netherlands, 2014; Volume 9, ISBN 9780080983004.
79. Clark, M.W.; Davies-Mcconchie, F.; McConchie, D.; Birch, G.F. Selective Chemical Extraction and Grainsize Normalisation for Environmental Assessment of Anoxic Sediments: Validation of an Integrated Procedure. *Sci. Total Environ.* **2000**, *258*, 149–170. [CrossRef]

80. Aksu, A.E.; Yaşar, D.; Uslu, O. Assessment of Marine Pollution in Izmir Bay: Heavy Metal and Organic Compound Concentrations in Surficial Sediments. *Turk. J. Eng. Environ. Sci.* **1998**, *22*, 387–415.
81. Löffler, A.; Leipe, T.; Emeis, K.C. The "Fluffy Layer" in the Pomeranian Bight (Western Baltic Sea): Geochemistry, Mineralogy and Environmental Aspects. *Meyniana* **2000**, *52*, 85–86.
82. Glasby, G.P.; Szefer, P.; Geldon, J.; Warzocha, J. Heavy-Metal Pollution of Sediments from Szczecin Lagoon and the Gdansk Basin, Poland. *Sci. Total Environ.* **2004**, *330*, 249–269. [CrossRef] [PubMed]
83. Langmuir, D. Uranium Solution-Mineral Equilibria at Low Temperatures with Applications to Sedimentary Ore Deposits. *Geochim. Cosmochim. Acta* **1978**, *42*, 547–569. [CrossRef]
84. Borovec, Z. Trace Element Levels in Sediments of the Czech Part of the Elbe River. *GeoJournal* **1996**, *40*, 299–309. [CrossRef]
85. Rickard, D.; Morse, J.W. Acid Volatile Sulfide (AVS). *Mar. Chem.* **2005**, *97*, 141–197. [CrossRef]
86. Milliman, J.D. *Marine Carbonates. Recent Sedimentary Carbonates Part 1. Marine Carbonates*; Springer: Berlin, Germany, 1974.
87. Chen, Y.-M.; Gao, J.-b.; Yuan, Y.-Q.; Ma, J.; Yu, S. Relationship between Heavy Metal Contents and Clay Mineral Properties in Surface Sediments: Implications for Metal Pollution Assessment. *Cont. Shelf Res.* **2016**, *124*, 125–133. [CrossRef]
88. Aires, T.; Muyzer, G.; Serrão, E.A.; Engelen, A.H. Seaweed Loads Cause Stronger Bacterial Community Shifts in Coastal Lagoon Sediments than Nutrient Loads. *Front. Microbiol.* **2019**, *9*, 3283. [CrossRef] [PubMed]
89. Fielding, J.J.; Croudace, I.W.; Kemp, A.E.S.; Pearce, R.B.; Cotterill, C.J.; Langdon, P.; Avery, R. Tracing Lake Pollution, Eutrophication and Partial Recovery from the Sediments of Windermere, UK, Using Geochemistry and Sediment Microfabrics. *Sci. Total Environ.* **2020**, *722*, 137745. [CrossRef] [PubMed]
90. Huerta-Diaz, M.A.; Morse, J.W. Pyritization of Trace Metals in Anoxic Marine Sediments. *Geochim. Cosmochim. Acta* **1992**, *56*, 2681–2702. [CrossRef]
91. Bertolin, A.; Frizzo, P.; Rampazzo, G. Sulphide Speciation in Surface Sediments of the Lagoon of Venice: A Geochemical and Mineralogical Study. *Mar. Geol.* **1995**, *123*, 73–86. [CrossRef]
92. Prange, A.; Kremling, K. Distribution of Dissolved Molybdenum, Uranium and Vanadium in Baltic Sea Waters. *Mar. Chem.* **1985**, *16*, 259–274. [CrossRef]
93. Dellwig, O.; Beck, M.; Lemke, A.; Lunau, M.; Kolditz, K.; Schnetger, B.; Brumsack, H.J. Non-Conservative Behaviour of Molybdenum in Coastal Waters: Coupling Geochemical, Biological, and Sedimentological Processes. *Geochim. Cosmochim. Acta* **2007**, *71*, 2745–2761. [CrossRef]
94. Mori, C.; Beck, M.; Striebel, M.; Merder, J.; Schnetger, B.; Dittmar, T.; Pahnke, K.; Brumsack, H.J. Biogeochemical Cycling of Molybdenum and Thallium during a Phytoplankton Summer Bloom: A Mesocosm Study. *Mar. Chem.* **2021**, *229*, 103910. [CrossRef]
95. Bau, M.; Koschinsky, A. Oxidative Scavenging of Cerium on Hydrous Fe Oxide: Evidence from the Distribution of Rare Earth Elements and Yttrium between Fe Oxides and Mn Oxides in Hydrogenetic Ferromanganese Crusts. *Geochem. J.* **2009**, *43*, 37–47. [CrossRef]
96. Brito, P.; Prego, R.; Mil-Homens, M.; Caçador, I.; Caetano, M. Sources and Distribution of Yttrium and Rare Earth Elements in Surface Sediments from Tagus Estuary, Portugal. *Sci. Total Environ.* **2018**, *621*, 317–325. [CrossRef] [PubMed]
97. Moss, B.; Jeppesen, E.; Søndergaard, M.; Lauridsen, T.L.; Liu, Z. Nitrogen, Macrophytes, Shallow Lakes and Nutrient Limitation: Resolution of a Current Controversy? *Hydrobiologia* **2013**, *710*, 3–21. [CrossRef]
98. Thin, M.M.; Sacchi, E.; Setti, M.; Re, V. A Dual Source of Phosphorus to Lake Sediments Indicated by Distribution, Content, and Speciation: Inle Lake (Southern Shan State, Myanmar). *Water* **2020**, *12*, 1993. [CrossRef]
99. Khourchi, S.; Delaplace, P.; Bargaz, A. Polyphosphate Fertilizer Use Efficiency Strongly Relies on Soil Physicochemical Properties and Root-Microbial Activities. *Geoderma* **2023**, *429*, 116281. [CrossRef]
100. Louda, J.W.; Li, J.; Liu, L.; Winfree, M.N.; Baker, E.W. Chlorophyll-a Degradation during Cellular Senescence and Death. *Org. Geochem.* **1998**, *29*, 1233–1251. [CrossRef]
101. Lu, Z.; Wang, Y.; Degryse, F.; Huang, C.; Hou, C.; Wu, L.; Jiang, R.; McLaughlin, M.J.; Zhang, F. Magnesium-Fortified Phosphate Fertilizers Improve Nutrient Uptake and Plant Growth without Reducing Phosphorus Availability. *Pedosphere* **2022**, *32*, 744–751. [CrossRef]
102. Leybourne, M.I.; Johannesson, K.H. Rare Earth Elements (REE) and Yttrium in Stream Waters, Stream Sediments, and Fe-Mn Oxyhydroxides: Fractionation, Speciation, and Controls over REE + Y Patterns in the Surface Environment. *Geochim. Cosmochim. Acta* **2008**, *72*, 5962–5983. [CrossRef]
103. Briffa, J.; Sinagra, E.; Blundell, R. Heavy Metal Pollution in the Environment and Their Toxicological Effects on Humans. *Heliyon* **2020**, *6*, e04691. [CrossRef] [PubMed]
104. Abd Elnabi, M.K.; Elkaliny, N.E.; Elyazied, M.M.; Azab, S.H.; Elkhalifa, S.A.; Elmasry, S.; Mouhamed, M.S.; Shalamesh, E.M.; Alhorieny, N.A.; Abd Elaty, A.E.; et al. Toxicity of Heavy Metals and Recent Advances in Their Removal: A Review. *Toxics* **2023**, *11*, 580. [CrossRef]
105. Das, S.; Sultana, K.W.; Ndhlala, A.R.; Mondal, M.; Chandra, I. Heavy Metal Pollution in the Environment and Its Impact on Health: Exploring Green Technology for Remediation. *Environ. Health Insights* **2023**, *17*, 1–10. [CrossRef]
106. Prochaska, C.; Zouboulis, A. A Mini-Review of Urban Wastewater Treatment in Greece: History, Development and Future Challenges. *Sustainability* **2020**, *12*, 6133. [CrossRef]

107. Sarbu, N.D.; Mihai, D.; Panter, Z.; Salcianu, R.; Mudura, R.; Mircea, L.C.; Konvalina, P. Research on the Efficiency of Wastewater Treatment Facility in Heavy Metal Removal from Leachate. *Rom. Biotechnol. Lett.* **2020**, *25*, 1768–1775. [CrossRef]
108. Gkaragkouni, A.; Sergiou, S.; Geraga, M.; Christodoulou, D.; Dimas, X.; Papatheodorou, G. Metal Pollution Chronology and Ecological Risk Assessment in Marine Sediments of Perama—Salamina Strait, Saronikos Gulf, Greece. *Reg. Stud. Mar. Sci.* **2024**, *76*, 103584. [CrossRef]
109. Qasem, N.A.A.; Mohammed, R.H.; Lawal, D.U. Removal of Heavy Metal Ions from Wastewater: A Comprehensive and Critical Review. *NPJ Clean Water* **2021**, *4*, 36. [CrossRef]
110. Geng, H.; Xu, Y.; Zheng, L.; Gong, H.; Dai, L.; Dai, X. An Overview of Removing Heavy Metals from Sewage Sludge: Achievements and Perspectives. *Environ. Pollut.* **2020**, *266*, 115375. [CrossRef]

water

Article

Estimation of Phytoplankton Primary Productivity in Qinghai Lake Using Ocean Color Satellite Data: Seasonal and Interannual Variations

Xuan Ban [1,*,†], Yingchao Dang [2,†], Peng Shu [3], Hongfang Qi [4], Ying Luo [4], Fei Xiao [1], Qi Feng [1] and Yadong Zhou [1]

1 Key Laboratory for Environment and Disaster Monitoring and Evaluation of Hubei Province, Innovation Academy for Precision Measurement Science and Technology, Chinese Academy of Sciences, Wuhan 430077, China; xiaof@apm.ac.cn (F.X.); fengqi@apm.ac.cn (Q.F.); zhouyadong@apm.ac.cn (Y.Z.)

2 Hubei Key Laboratory of Three Gorges Project for Conservation of Fishes, Chinese Sturgeon Research Institute, China Three Gorges Corporation, Yichang 443100, China; dang_yingchao@ctg.com.cn

3 State Key Laboratory of Water Resources and Hydropower Engineering Science, Wuhan University, Wuhan 430072, China; shupeng@whu.edu.cn

4 Key Laboratory of Breeding and Protection of Gymnocypris przewalskii in Qinghai Province, Rescue Center of Gymnocypris przewalskii, Xining 810016, China; qhf1970@163.com (H.Q.); lyquncr@163.com (Y.L.)

* Correspondence: banxuan@apm.ac.cn; Tel.: +86-13476825599

† These authors contributed equally to this work.

Abstract: Estimation of primary production in Qinghai Lake is crucial for the aquatic ecosystem management in the northeastern Qinghai–Tibet Plateau. This study used the Vertically Generalized Production Model (VGPM) with ocean color satellite data to estimate phytoplankton primary productivity (PP) in Qinghai Lake during the non-freezing period from 2002 to 2023. Field data from 2018 and 2023 were used to calibrate and verify the model. The results showed a seasonal trend in chlorophyll-a and PP, with the lowest values in May and peaks from June to September. Qinghai Lake was identified as oligotrophic, with annual mean chlorophyl-a of 0.24–0.40 μg/L and PP of 40–369 mg $C/m^2/day$. The spatial distribution of PP was low in the center of the lake and high near the shores and estuaries. An interesting periodic increasing trend in PP every 2 to 4 years was observed from 2002 to 2023. This study established a remote sensing method for PP assessment in Qinghai Lake, revealing seasonal and interannual variations and providing a useful example for monitoring large saline mountain lakes.

Keywords: Qinghai Lake; Vertically Generalized Production Model; chlorophyll-a; phytoplankton primary production; remote sensing

1. Introduction

Aquatic ecosystems are highly dependent on phytoplankton for primary production (PP). In large lakes, phytoplankton often accounts for more than 95% of PP [1]. Qinghai Lake, located in the northeastern part of the Qinghai–Tibet Plateau, is the largest and highest inland saline lake in China. Primary productivity plays a key role in maintaining the balance of the aquatic ecosystem in Qinghai Lake, and its assessment is importance for effective aquatic ecosystem management of the lake [2]. The growth in various aquatic organisms in Qinghai Lake basically depends on the energy accumulated by phytoplankton photosynthesis, so phytoplankton can be regarded as the primary energy for fisheries in Qinghai Lake [3]. Since 2001, Qinghai Province has implemented a comprehensive lake closure and stocking strategy for nearly two decades, with a fishing ban policy to protect a special fish, the naked carp (*Gymnocypris przewalskii*), in Qinghai Lake [4]. PP estimates provide important guidance for the conservation and management of fisheries in Qinghai Lake and for the development of a comprehensive lake closure and no-fishing policy.

The traditional monitoring methods of PP are the light and dark bottle method based on scattered sample points from in situ observation; however, these methods cannot be used to analyze the dynamic monitoring of long time series on a large spatial scale [5]. The area of Qinghai Lake is about 4500 km^2. Most studies on ecological parameters (i.e., chlorophyll-a, dissolved oxygen, temperature, nitrogen, etc.) are mainly based on a small number of field samples from cruise surveys. These methods are time-consuming and labor-intensive, and are also unable to enable spatial distribution estimation over a long time sequence [6–8]. Remote sensing monitoring technology, with its advantages of wide spatial coverage and high temporal continuity, is a powerful tool for monitoring lake environments at large spatial scales [9]. PP monitoring and assessment of Qinghai Lake by remote sensing is currently one of the most efficient methods, but it has not been reported so far [10,11]. Ocean color (OC) inversion products obtain environmental parameters from multiple satellite data (the Sea-viewing Wide Field-of-view Sensor (SeaWiFS) and Moderate Resolution Imaging Spectroradiometer (MODIS)) to obtain long time series (2002–present) of environmental parameters to quantify and analyze global or regional ecological and environmental changes in water bodies. This approach has the advantages of high temporal resolution (revisit period of 1 day), numerous effective data points, and wider coverage than other satellite data [12]. The standard OC algorithm is mainly a set of algorithms developed based on datasets collected from coastal or open ocean waters, including the optical properties of saline alpine lakes [12]. Therefore, OC products are also expected to be applied to the study of the PP assessment of long time series in Qinghai Lake.

Several empirical, analytical, and bio-optical models have been applied to estimate marine PP [13–18]. Among them, empirical models are usually driven by basic remote sensing products, supported by measured data, to estimate the spatial and temporal distribution of biological parameters related to lake ecology [11]. For example, the Vertically Generalized Production Model (VGPM) proposed by Behrenfeld and Falkowski is one of the most-used models to estimate PP based on products such as chlorophyll-a and water temperature from satellite remote sensing inversions [13]. The VGPM is a vertically integrated model that combines empirical relationships with phytoplankton photosynthetic mechanisms and relies on optical principles. It has the advantage of having few input variables and can be driven by satellite remote sensing data products to estimate PP [14]. Due to the intricate challenges posed by the analysis of water quality parameters in inland lakes, encompassing optical complexity, adjacency effects, atmospheric correction issues, and diminished water quality and transparency, accurately estimating phytoplankton productivity (PP) in lakes remains a more daunting task compared to the marine environment [15]. Although we have considerable knowledge of photosynthetic processes and ocean optics, which has been extensively applied in marine research, there are still limitations in using satellite data to monitor primary productivity (PP) in lakes based on observational data [16]. It has been found that re-parameterization of the original inverse relationship based on in situ data can eliminate systematic biases associated with the algorithm [16–18]. Therefore, it is imperative to develop remote sensing estimation models for primary productivity in specific areas. Qinghai Lake is an ideal lake for use of the VGPM method because of its good water quality and high transparency. This study attempts to develop an improved VGPM model applicable to Qinghai Lake using OC products and in situ observational data, and applies this model to estimate chlorophyll-a and PP levels during the non-freezing period of Qinghai Lake over 20 years. This study aims to provide a theoretical basis for the long-term monitoring of PP and dynamic assessment of spatial and temporal variation patterns, and to provide scientific and technological support for the evaluation of ecological restoration effects and fishery management of Qinghai Lake.

2. Materials and Methods

2.1. Study Area

Qinghai Lake is located between 99°36′~100°16′ E longitude and 36°32′~37°15′ N latitude with an elevation of about 3200 m. It has a total area of about 4500 km^2, a perimeter of about 360 km, and an average water depth of about 19 m. It is the largest inland saltwater lake on plateaus in China [2]. The Qinghai Lake region has a highland continental climate with abundant light and low precipitation. The lake region receives little precipitation throughout the year, with most of it falling from May to September. The water temperature varies seasonally. In summer, there is an obvious isothermal layer in the lake, with the highest temperature of 22.3 °C in August. The lower layer of the water has a lower temperature, with an average temperature of 9.5 °C. In autumn, the lake water is stirred up by windy weather, so the phenomenon of water temperature stratification disappears. In winter, the lake water temperature is subject to the isothermal inversion phenomenon. In January, the upper layer of the lake water under the ice is −0.9 °C, while the lower layer is 3.3 °C [19]. The sources of lake water recharge in Qinghai Lake are river water, rainwater, and groundwater. Among the major sources of recharge are five tributary rivers, namely Buha River, Shaliu River, Quanji River, Wuha Alam River, and Hargai River (Figure 1) [20]. In Qinghai Lake, the freezing season typically begins in November and lasts until April of the following year with ice cover, while the non-freezing season lasts from May to October [21,22]. As it is impossible to obtain the water color parameters by remote sensing when the lake surface is frozen, our study period is from May to October in the non-freezing period.

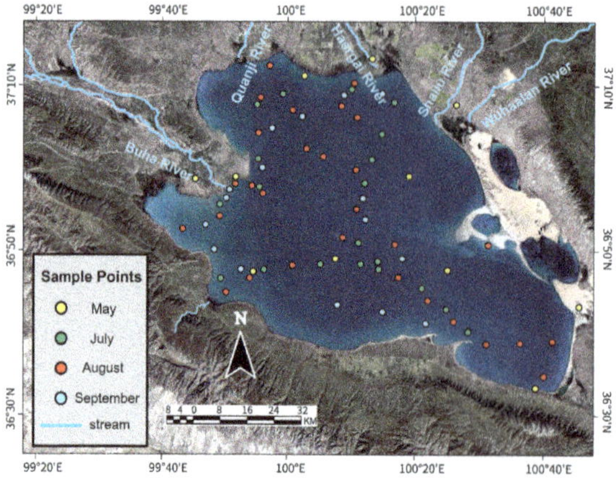

Figure 1. The distribution of sample points and location of the tributary rivers of Qinghai Lake.

2.2. Field Data

Field data were collected from 2018 to 2023, during the non-freezing period. Water temperature and chlorophyll-a were measured at 0.5 m underwater using a portable water quality multi-parameter analyzer (CTD, OCEAN SEVEN310, Idronaut, Brugherio, Italy) (Table 1). Approximately 30 sampling points were measured once throughout the Qinghai Lake, and a total of 257 sample points were measured in this study. The PP was measured with a light–dark bottle device in each sampling site [5].

Table 1. The information of the sampling data in the surface water (<0.5 m underwater) of Qinghai Lake.

Month	Date	Time	Number of Samples	Chlorophyll-a ($\mu g \cdot L^{-1}$)		Surface Water Temperature (°C)	Average Density of Phytoplankton ($ind \cdot L^{-1}$)	Biomass of Phytoplankton ($mg \cdot L^{-1}$)
				Range	Average	Average	Average	Average
May	14–15 May 2018	7:30~19:50	12	0.030~0.245	0.089	8.8	2400	0.0093
July	1–2 July 2018	8:00~19:38	20	0.070~0.788	0.404	15	21,400	0.0823
August	7–10 August 2018	8:10~19:15	29	0.070~0.730	0.258	17.5	50,730	0.238
September	26–27 September 2018	6:40~19:02	19	0.020~1.229	0.42	13.4	39,470	0.126
August	2–5 August 2019	6:30~18:30	30	0.134~0.594	0.295	13	49,268	0.221
September	19–20 September 2019	7:20~17:50	24	0.196~0.605	0.364	10	38,456	0.119
June	30 June–2 July 2021	8:45~16:20	25	0.198~0.475	0.296	11	14,399	0.044
August	19–23 August 2021	8:50~17:50	24	0.034~0.536	0.266	15	56,687	0.154
June	23–24 June 2022	7:06~17:16	23	0.211~0.719	0.354	15.5	44,106	0.673
May	25–26 May 2023	7:10~17:13	22	0.04~1.642	0.258	8.4	42,061	0.695
August	10–13 August 2023	7:10~18:30	29	0.14~3.23	1.404	17.4	52,693	1.857

A 1 L water sample was collected using a polymethylmethacrylate bottle for analyzing the phytoplankton. Water samples were collected from three layers at each sampling point 0.5 m below the water surface, and 15 mL of Lugol's solution was added to the sample for retention. Species classification and biomass count analysis (wet weight) were performed in the laboratory. A quantity of 1 L of water sample was concentrated to 10–25 mL to count the density and biomass of phytoplankton after standing and precipitation for more than 24 h in the laboratory. The concentrated sample was shaken well and then 0.1 mL was transferred to the phytoplankton counting chamber. Counting was performed under a 10 × 40 microscope, with usually 30–50 fields per slide, and the number of individuals or cells was counted by species. Each sample was counted twice and averaged. The results were valid only if the difference between the two counts and their average were less than 10%; otherwise, the third slide was counted.

The formula for calculating the number of phytoplankton of each water sample is as follows:

$$N = \frac{A}{a \cdot n} \cdot \frac{\mu}{V} \cdot p$$

where N is the number of phytoplankton in 1 L of water (#. L^{-1}); A is the area of the counting chamber (mm^2); a is the area of each visual field (mm^2); n is the number of individuals counted in each slide; μ is the volume of the concentrated water sample (mL); V is the volume of the counting chamber (mL); and p is the number of phytoplankton counted by each slide.

The counting results were converted into biomass (wet weight) by the cell volume method. The average of the counting results from the three layers of each sampling point was taken as the result of that sampling point.

2.3. Vertically Generalized Production Model (VGPM)

The VGPM model was developed to estimate marine primary productivity based on parameters such as temperature and chlorophyll-a samples in water quality classes I and II of the Environmental Quality Standards for Surface Water of China (GB3838-2002) [13,23]. The water quality of Qinghai Lake was classified as class I with high transparency, which is suitable for the use of this model to calculate PP. The parameters can be obtained using remote sensing monitoring and other resources, and the equation is:

$$PP = 0.66125 \times P^{B}_{ept} \times \frac{E_0}{E_0 + 4.1} \times Z_{eu} \times C_{opt} \times D_{irr} \quad (1)$$

where PP is the primary productivity (mg $C/m^2/day$) integrated from the surface layer to the euphotic layer; P^{B}_{ept} is the maximum rate of carbon fixation in the water column (mg C/mg Chl·h); E_0 is the photosynthetically available radiation (PAR) at the water surface ($\mu mol/m^2/day$); Z_{eu} is the depth of the true photosphere (m); D_{irr} is the photoperiod (h); and C_{opt} is the chlorophyll-a concentration (Chl-a) (μg/L) at the water surface.

2.3.1. Calculation of the Maximum Rate of Carbon Fixation in the Water Column

P^B_{ept} can be considered as a function of the temperature in the water column and is calculated as follows [15]:

$$P^B_{ept} = \left\{ \begin{array}{c} 1.13(T \leq -1.0) \\ 4.00(T \geq 28.5) \\ P^B_{ept}(-1.0 < T < 28.5) \end{array} \right\} \quad (2)$$

T (°C) is the surface temperature.

$$P^B_{ept} = 1.2956 + 2.749 \times 10^{-1}\mathrm{T} + 6.17 \times 10^{-2}\mathrm{T}^2 - 2.05 \times 10^{-2}\mathrm{T}^3 + 2.462 \times 10^{-3}\mathrm{T}^4 - 1.348 \times 10^{-4}\mathrm{T}^5 + 3.4132 \times 10^{-6}\mathrm{T}^6 - 3.27 \times 10^{-8}\mathrm{T}^7$$

2.3.2. Estimation of Photoperiod and Photosynthetically Available Radiation

The photoperiod D_{irr} is calculated from the average monthly sunshine hours from national weather station websites (http://data.cma.cn/ (accessed on 31 December 2023)). The spectral component of solar radiation that is effective for plant photosynthesis is called PAR, and has a wavelength range of 380–710 nm and largely coincides with visible light [24]. The photosynthetically available radiation E_0 is derived from remotely sensed products from the ocean color satellite MODIS.

2.3.3. Calculation of Euphotic Zone

The euphotic zone (Z_{eu}) is obtained from the diffuse attenuation coefficient Kd_{PAR} of photosynthetically available radiation obtained from OC products, which is calculated by the equation [24,25] below:

$$Z_{eu} = \frac{4.605}{Kd_{PAR}} \quad (3)$$

$$Kd_{PAR} = 0.896Kd_{490}{}^{0.873}, (r^2 = 0.98, n = 81, P < 0.001) \quad (4)$$

Since the diffuse attenuation coefficient of irradiance at a wavelength of 490 nm correlates well with the diffuse attenuation coefficient of photosynthetically available radiation Kd_{PAR}, Kd_{PAR} can be calculated from Kd_{490} [24,25].

2.4. Estimating Phytoplankton Primary Productivity in Qinghai Lake

2.4.1. Setting up the Model and Data Sources

In this study, an improved VGPM model was constructed to estimate the PP of Qinghai Lake. The model construction involved the following two steps: (1) Extracting the MODIS satellite data from the OC Level-2 products of NASA Goddard Space Flight Center (GSFC, https://Oceancolor.gsfc.nasa.gov/ (accessed on 31 December 2023)) in the non-freezing period (May to October) of Qinghai Lake from 2002 to 2023, with a total of 32,120 images at a spatial resolution of 1 km and a temporal resolution of daily satellite revisits. The data were pre-processed, specifically including batch clipping, matching, and reprojection in ArcGIS using a Python scripting program with a uniform WGS 84 coordinate system and UTM projection. (2) The relevant parameters required for the VGPM model to calculate PP (Equation (1)) were extracted from the above data products. The chlorophyll-a concentration (Copt) product was obtained using the standard blue-to-green band ratio algorithm in the NASA OC4 Algorithm products [26]. Surface water temperature (T) is a product obtained from MODIS mid- and far-infrared bands, inverted using the proto-algorithm algorithm, and validated by AVHRR satellite sensors and situ data. The photosynthetically available radiation (E_0) and the diffuse attenuation coefficient of photosynthetically available radiation at a wavelength of 490 nm ($Kd_{(490)}$) were OC data products from the Visible Infrared Imaging Radiometer Suite (VIIRS) [26]. Photoperiod was obtained from the nearest Republican National Reference Weather Station closest to Qinghai Lake (http://data.cma.cn (accessed on 31 December 2023)). These parameters were used as input parameters to the VGPM model shown in Equations (1)–(4) for the estimation of PP.

2.4.2. Calibration and Validation of VGPM Model

In this study, the survey data from 2018 to 2023 (mainly including chlorophyll-a concentration (C_{opt}) and surface water temperature (T)) were compared with the spatio-temporal synchronized data products extracted by OC to verify the accuracy of the model and calibrate the model. The final model-estimated PP values from the model can be compared with the measured mean phytoplankton densities, and the reasonableness of the model-estimated PP values can be verified by analyzing their correlations. First, the applicability of the OC data products in Qinghai Lake was verified by comparing the errors between the simulated and measured values of water temperature and chlorophyll-a from the model inversion. A total of 179 out of 256 samples in Qinghai Lake were selected to establish a regression equation with the temperature products of OC. The remaining 77 samples were used to validate the temperature values. The comparative analysis of the measured water temperature and the OC data product showed a correlation coefficient of 0.88, indicating that the OC water temperature data product could explain 88% of the measured data, with an average relative error of 5% and an absolute error range of −1.76~1.64 °C (Figure 2a). The water temperature error was within an acceptable range, so the OC water temperature data products could be directly substituted into the VGPM model for PP estimation.

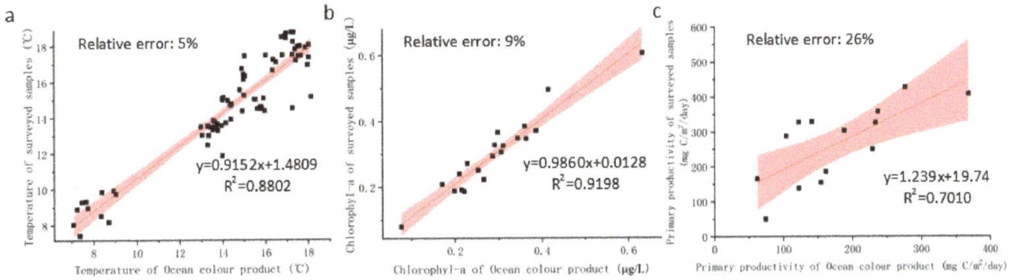

Figure 2. The comparison of the measured and ocean color remote sensing data in Qinghai Lake. (**a**) is the comparative analysis of the measured and the OC derived water temperature; (**b**) is the comparative analysis of the measured and the OC derived rectified chlorophyll-a; (**c**) is the comparative analysis of the measured and VGPM modelled primary productivity.

The results of the comparative analysis between the chlorophyll-a data products of OC and the measured values showed the chlorophyll-a product values of OC are much higher than the measured values and the correlation is low, so the chlorophyll-a products of OC cannot be directly used to calculate the PP values, and the chlorophyll-a values from the OC products need to be rectified with measured data before applying the VGPM model of Qinghai Lake (Figure 2b). The rectification process is as follows: (1) A total of 193 out of 256 samples were selected to establish a regression equation with chlorophyll-a products of OC; 43 measured samples collected between 2018 and 2023 were selected to rectify the chlorophyll-a products; and the remaining 20 samples were used to validate the rectified chlorophyll-a values. The validation results showed that the correlation coefficient between the calibrated chlorophyll-a and the measured values was 0.92, with a relative error of 9% and an absolute error range of −0.07 to 0.04 μg/L (Figure 2b), and the error of the calibrated chlorophyll-a values was within an acceptable range. A total of 14 matched pairs of PP from the surveyed data were compared with simultaneous VGPM-modeled PP results; the correlation coefficient was 0.70, with a relative error of 26% (Figure 2c), indicating that the VGPM model for Qinghai Lake has a good accuracy for predicting PP.

3. Results

3.1. Seasonal Variation of Chlorophyll-a during the Non-Freeze Period in Qinghai Lake

The 2019 results were used as an example to analyze the spatial distribution of chlorophyll-a in Qinghai Lake. The center of the lake had a low concentration of chlorophyll-a, while there was a much higher concentration near the lakeshore and in the estuary regions (Figure 3). From 2002 to 2023, the seasonal variation of chlorophyll-a showed a fluctuating trend with the low values occurring in May and the high values occurring in June, then decreasing from July to October (Figure 4). In May, the aquatic plants begin to proliferate as the thaw occurs after the freezing period. Meanwhile, the chlorophyll-a shows a lower value in the box plots with an interquartile distance of 0.07 g/L, indicating that chlorophyll-a concentration fluctuated less from year to year than other months of the year (interquartile range was 0.17–0.23 μg/L). Regarding the medians, the lowest median chlorophyll-a was 0.18 μg/L in May; from June to September, chlorophyll-a showed a gradual increase from 0.20 μg/L in June to 0.39 μg/L in September, and started to decrease to 0.32 μg/L in October. Concerning the minima, the minima of chlorophyll-a varied little from May to September, ranging from 0.11 to 0.16 μg/L, and the lowest minima occurred in October, at 0.06 μg/L. In terms of maxima, the lowest maxima chlorophyll-a was 0.79 μg/L in May, the maxima from June to October showed a trend of decreasing and then increasing, and the highest maxima also occurred in October, at 0.91 μg/L. The maximum value of 0.91 μg/L in October indicates that October is the month with the largest interannual fluctuation of chlorophyll-a. The multi-year monthly mean values of chlorophyll-a concentration during the non-freezing period of Qinghai Lake from 2002 to 2023 ranged from 0.24 to 0.40 μg/L, which is typical for an oligotrophic lake.

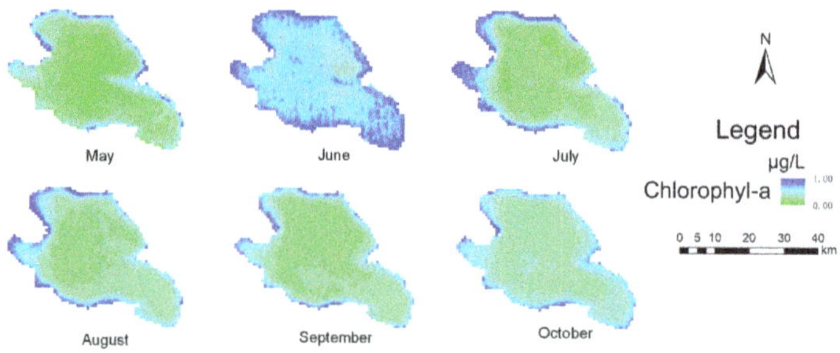

Figure 3. The spatial distribution of chlorophyll-a (MODIS Aqua) in Qinghai Lake from May to October, 2019.

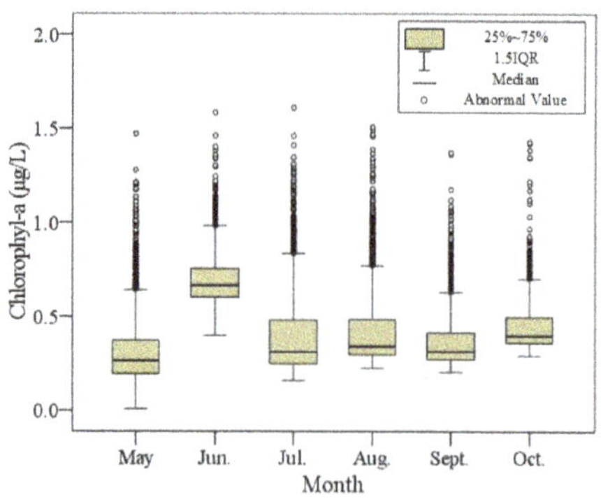

Figure 4. The boxplots of chlorophyll-a (MODIS Aqua) in Qinghai Lake from 2002 to 2023 in the non-freezing period. The five lines from the bottom to the top of the box plot represent the minimum, lower quartile, median, upper quartile, and maximum values, respectively.

3.2. Seasonal Changes in PP during the Non-Freezing Period of Qinghai Lake

The results estimated by MODIS Aqua data source were used to analyze the seasonal variation pattern of PP during the non-freezing period of Qinghai Lake. The spatial distribution of PP was basically consistent with chlorophyll-a, and the spatial distribution showed a small value in the center of the lake and a large value near the lakeshore and estuary areas (Figure 5, taking 2019 as an example). The seasonal variation trend also showed a fluctuation of first increasing and then decreasing, with the low value occurring in May and October and the high value occurring in July to September (Figure 6). The box plots of PP values were smaller in May and September, with interquartile distances of 87.37 mg C/m^2/day and 94.86 mg C/m^2/day, respectively, indicating that the PP values fluctuated less annually in May and September than in other months. With regard to the median, the mean PP was lowest in May at 16.76 mg C/m^2/day, and the PP from June to September showed a gradual increase from 92.77 mg C/m^2/day in June to 135.27 mg C/m^2/day in September, and then started to decrease to 119.38 mg C/m^2/day in October. The minimum value of PP also increased first and then decreased from May to October, ranging from 10.07 mg C/m^2/day to 28.30 mg C/m^2/day. The maximum value in May was the smallest among all the months, being 223.83 mg C/m^2/day. The maximum value from June to October showed a trend of first decreasing and then increasing, and the maximum value also appeared in July, which was 397.50 mg C/m^2/day. This indicates that July is the month with the highest seasonal value of PP, and it is also the month with the largest interannual variation in PP. The multi-year monthly average values of PP during the non-freezing period of Qinghai Lake from 2002 to 2023 ranged from 40 to 369 mg C/m^2/day.

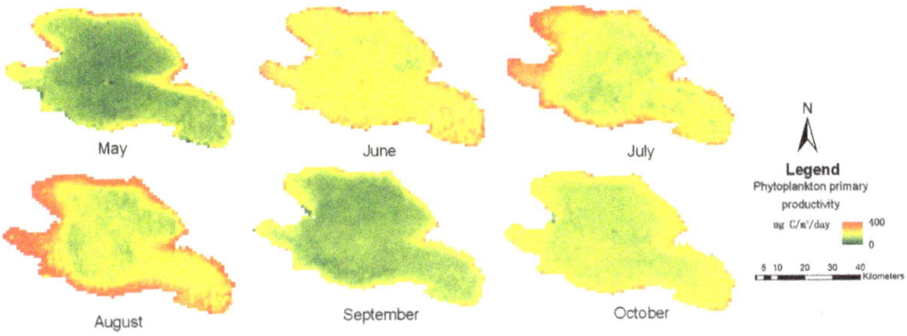

Figure 5. The spatial distribution of phytoplankton primary productivity in Qinghai Lake from May to October, 2019.

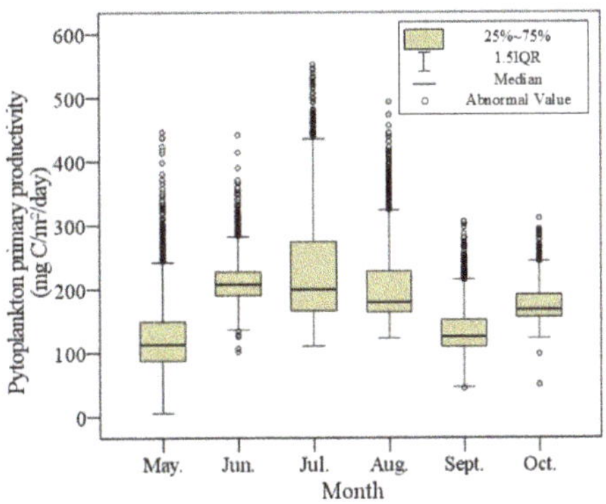

Figure 6. The boxplots of phytoplankton primary productivity (MODIS Aqua) in Qinghai Lake from 2002 to 2023 in the non-freezing period. The five lines from the bottom to the top of the box plot represent the minimum, lower quartile, median, upper quartile, and maximum values, respectively.

3.3. Interannual Variation of PP during the Non-Freezing Period of Qinghai Lake

The PP value of Qinghai Lake in the non-freezing period from 2002 to 2023, which is in the period of "closed lake for breeding fish", was estimated in our study. Estimated PP values in August are most representative for analysis of the interannual variation trend of PP values in Qinghai Lake from 2002 to 2023. This is because the only months in which the OC data product had no missing data records for Qinghai Lake were August and October; the month of August also coincides with a critical life stage for naked carp, and the PP could be related to fish resources through the food web [27]. The results show that the PP values near the lakeshore are larger than those near the center of the lake in all years in terms of spatial distribution. In particular, the PP values near the mouth of Buha River and Quanji River in the western region are significantly larger than those in other regions (Figure 7). The PP values showed a very significant interannual fluctuation trend, with values ranging from about 20 to 400 mg C/m^2/day. Among them, the PP values were less than 100 mg C/m^2/day from 2002 to 2004, which were smaller than those of other years (the green area in Figure 7). While the PP values increased significantly to 200–400 mg C/m^2/day from 2005 to 2007 (the color changes from green to yellow in the graph), the PP values decreased again to less than 100 mg C/m^2/day from 2008 to 2009. The PP values increased significantly in 2010, especially in the northwest area near the Buha and Quanji River regions. From 2011 to 2014, PP values began to decrease again, and most areas of the lake fluctuated in the range of 100 to 200 mg C/m^2/day. The PP values of the whole lake increased significantly again to about 300 mg C/m^2/day in 2015; in particular, the value for the area near the river estuary was larger than that for other areas. PP then decreased to below 200 mg C/m^2/day in 2016~2018, and PP increased to the highest value of about 400 mg C/m^2/day in 2019. Overall, the range of PP values in Qinghai Lake from 2002 to August 2019 seems to be from about 20 to 400 mg C/m^2/day, showing a cyclical fluctuating upward trend, with PP values decreasing and then increasing significantly at intervals of 2–4 years, reaching the highest value in 2019, then decreasing from 2020 to 2023 (Figure 8).

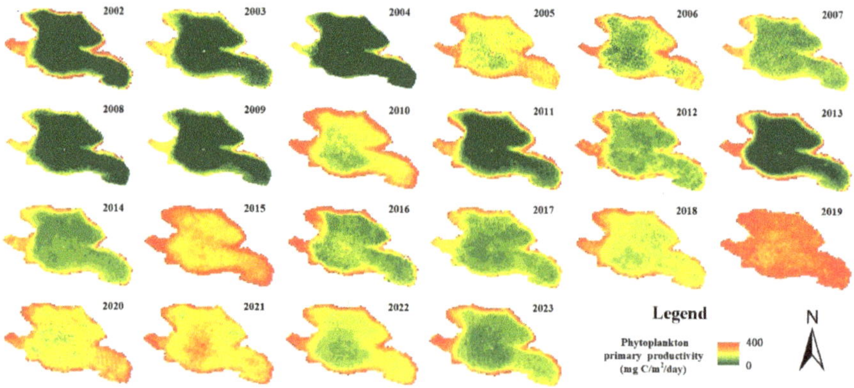

Figure 7. Spatial and temporal distribution map of phytoplankton primary productivity (PP) in Qinghai Lake from 2002 to 2023 (August).

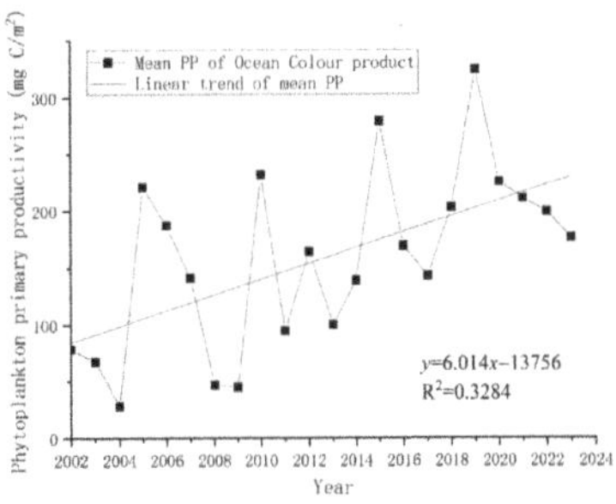

Figure 8. Trend of mean of phytoplankton primary productivity (PP) in Qinghai Lake from 2002 to 2023 in August.

4. Discussion

4.1. Limitation and Improvement of Remote Sensing Estimation Method for Phytoplankton Primary Productivity in Qinghai Lake

Due to the large difference between chlorophyll-a from OC products and in situ samples, the chlorophyll-a from OC products cannot be directly used for calculation in the VGPM model; therefore, we used the measured data to rectify the OC products. After rectification, the relative error between the calibrated chlorophyll-a and the measured value was less than 9%, which could be substituted into the VGPM formula to calculate the PP value of Qinghai Lake. The reason for this error is, on the one hand, that the measured sample points do not coincide with the time of the remote sensing satellite products. The time period of the sampling point measurement is from about 6:30 a.m. to 19:30 p.m., while the MODIS satellite transit time is 6:00 a.m. and 15:00 p.m. Due to the large area of Qinghai Lake (about 4500 km^2), the measurement work required several days to obtain a large amount of sample data for the average distribution of the whole lake. Therefore, it is impossible to collect the sample points of the whole lake within the satellite transit time period, and only the satellite data products close to the sampling time point were selected for verification. On the other hand, because the spatial resolution of the MODIS satellite OC product is 1 km, the actual sampling point value and the remote sensing satellite extraction value have a spatial error of 1 km. However, because the MODIS satellite has a long time series and high-temporal-resolution data, combined with the large area of Qinghai Lake, enough sampling points (more than 4000 data points) can be extracted for interpolation, and the error caused by the spatial resolution is within the acceptable range. In contrast, other satellites (Landsat, Sentinel series satellites, etc.) have high spatial resolution, but low temporal resolution because of their long revisit period; thus, they have few images per month (cloudy weather above Qinghai Lake), and it is impossible to obtain enough valid sample points to cover the whole lake. Therefore, in order to estimate the PP results over a long time series for each month of the whole Qinghai Lake, MODIS satellite imagery is the most suitable for use in PP estimation.

Although the error is caused by the time and space issues described above, it is an absolute error, and does not affect the relative value from the same satellite data source. Therefore, the results of the spatial and temporal variation trend of chlorophyll-a and PP values reflected by the model results are credible. In addition, due to the limitations of remote sensing satellite products, there are many dates without effective data. In this study, the number of days with effective data of chlorophyll-a in each month of the OC data products from 2002 to 2023 was counted (if the data points of the obtained remote sensing data products can cover more than 50% of the lake area, this day is considered as providing effective remote sensing data) (Figure 9). The results show that the number of days with effective data is higher in August and October than in May and July, so the accuracy of the remote sensing data products is higher in August and October than in other months. The PAR in the OC data product cannot be compared with the results of the VGPM model because it was not measured in our study. In the future, this parameter will be measured synchronously to better verify the model.

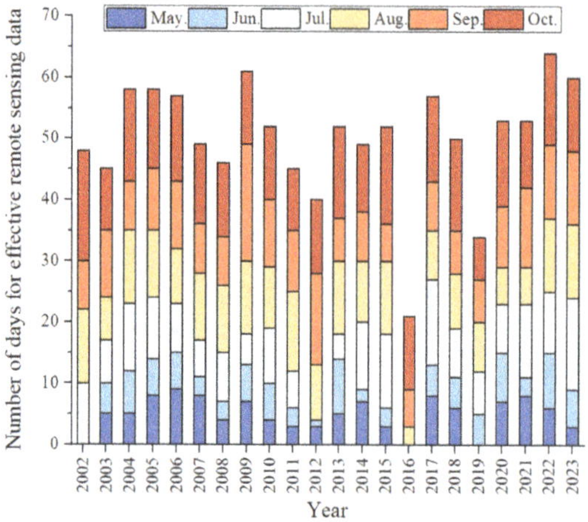

Figure 9. Statistical results of the effective days of chlorophyll-a in OC data products in each month from 2002 to 2023.

4.2. Validation Analysis of Estimation Results of Phytoplankton Primary Productivity in Qinghai Lake

In other lakes at a high altitude and having high salinity, water quality and transparency are similar to those of Qinghai Lake, and these lakes thus highly suitable for the assessment of their long-term and large-scale primary productivity of plankton using the method proposed in this study. Other results show that chlorophyll-a and PP are relatively low in these similar ecosystems, which are oligotrophic lakes [28–30]. Studies have shown that the primary productivity of phytoplankton is mainly affected by temperature, salinity, sunshine duration, nutrients, wind, and the growth cycle of phytoplankton [28–30]. Factors such as water temperature and light cycle have been considered in this model. Therefore, comparison of the measured phytoplankton with the PP trend simulated by this model was considered to verify the reliability of the model (taking 2018 as an example). The average biomass and density of phytoplankton at 20 sampling sites monitored from May to October in 2018 were compared with the simulated PP. The results showed that the average density of phytoplankton in May is the smallest in the surveyed months, and has a similar trend to the simulated PP values in Haixin Mountain, Sankuaishi, and Heima Estuary (Figure 10). In August, the average density of phytoplankton in Heimahekou, Haixinshan, and Erlangjian is largest among the surveyed months, and the simulated PP values also had a similar trend [31]. The results of the VGPM model showed that the PP values reached their peak in July, and the measured average density of phytoplankton was the highest in August and September. Since the PP is also affected by light and temperature, and the average temperature of Qinghai Lake from May to October is highest in July, temperature has the greatest impact on PP [32]. The variation trend and distribution of the average density of phytoplankton is basically consistent with the simulated PP; therefore, the simulated PP from the VGPM model can well reflect the spatial and temporal distribution of phytoplankton primary productivity in Qinghai Lake.

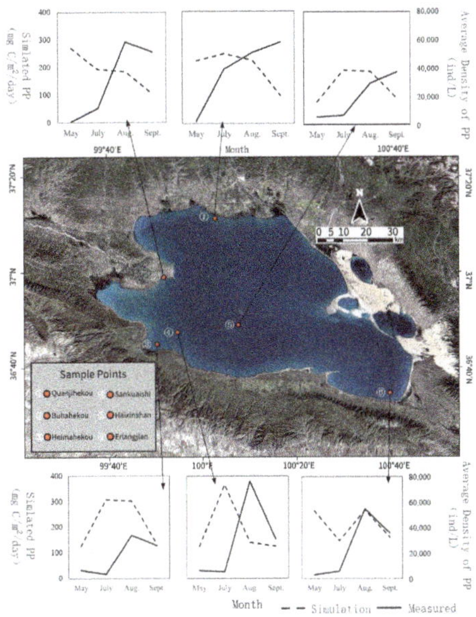

Figure 10. Contrast of average density of the phytoplankton and simulated PP in Qinghai Lake.

4.3. The Change Trend of Chlorophyll-a and PP in Qinghai Lake in the Past Twenty Years and Its Causes

The seasonal variation in chlorophyll-a and PP values from 2002 to 2023 showed a trend of first increasing and then decreasing in most years. This phenomenon may be caused by the life stage of phytoplankton and the feeding of naked carp in Qinghai Lake. Naked carp is an anadromous spawning fish that migrates from the lake to the tributaries of Qinghai Lake to spawn. Its peak spawning period is around mid-June, and adult fish mostly migrate to the tributaries. Thus, fewer adults remain in the lake, which is one of the reasons for the increase in PP in July and August [2,27]. During this period, the phytoplankton just start to grow from May to September, and there are fewer fish in the lake to feed on it, causing the PP value to peak in August. After August, the adult naked carp and their larvae migrate into the lake and start to feed on the plankton in Qinghai Lake, causing the PP value to decrease. In September and October, due to the feeding of naked carp, and because the phytoplankton gradually enter a dormant period, the density of phytoplankton begins to decrease, while the PP also begins to decrease [13,28]. The spatial distribution of PP values shows a low value in the center of the lake and a high value near the lakeshore and estuary, where more nutrients can be received from the lakeshore drainage or tributary streams for phytoplankton growth [29].

The interannual fluctuation trend in the PP value from 2002 to 2023 is consistent with the research results of other scholars on the primary productivity of Qinghai Lake. The reasons for the fluctuation are mainly the nutrient input and water temperature [33–35]. The results of this study show that the PP value shows a trend of periodic fluctuation and increase from 2002 to 2023. It is speculated that this may be related to the significant increase in the average temperature of Qinghai Lake in the past 50 years [36]. According to this study, the average temperature of Qinghai Lake shows interannual fluctuation with a gradual upward trend [37]. It is speculated that the fluctuating trend in water temperature affects the interannual fluctuation in PP. In addition, the increased precipitation in the Qinghai Lake basin caused the increase in runoff in the Buha River estuary [38], which resulted in more nutrients being imported into Qinghai Lake. Increased chlorophyll-a and PP will increase the food supply for fish resources in Qinghai Lake [32].

5. Conclusions

The model of phytoplankton primary productivity of Qinghai Lake established in this study can estimate the monthly mean value of phytoplankton primary productivity of Qinghai Lake during the non-freezing period (May to October), and can be used to analyze the interannual spatial and temporal distribution trends of phytoplankton primary productivity of a long time series. The spatial distribution of chlorophyll-a and PP generally showed a small value in the center of the lake, and a large value in the lakeshore and estuary area. Seasonal changes showed an increasing and then decreasing trend. Chlorophyll-a and PP were lowest in the early stages of thawing (May), increased with increasing water temperature, and decreased to a low value in October. The multiannual monthly mean values ranged from 0.24 to 0.40 μg/L and 40 to 369 mg $C/m^2/day$, characteristic of oligotrophic lakes. The interannual fluctuation trend in PP showed a fluctuating upward trend with a period of 2~4 years, and the range of its value was about 20~400 mg $C/m^2/day$. The results can be explained by the growth cycles of phytoplankton and naked carp. Compared with traditional measurement methods, the PP estimation method of Qinghai Lake based on satellite remote sensing established in this study not only has the advantages of wide monitoring coverage and strong monitoring timeliness, but also significantly reduces the cost of field sampling and avoids the risks of field work. It can also enable continuous systematic observation and historical inversion of phytoplankton productivity in Qinghai Lake. In the future, it could provide important support for the dynamic monitoring of the ecological environment, and the conservation and management of fish resources, by linking it with analysis of the trophic food chain in Qinghai Lake.

Author Contributions: Conceptualization, X.B. and Y.D.; methodology, X.B. and P.S.; software, P.S.; validation, X.B. and P.S.; formal analysis, X.B. and P.S.; investigation, H.Q. and Y.L.; resources, H.Q.; data curation, X.B. and P.S.; writing—original draft preparation, X.B.; writing—review and editing, X.B. and Y.Z.; visualization, X.B. and P.S.; supervision, F.X., Q.F. and Y.Z.; project administration, X.B.; funding acquisition, H.Q. and Y.D. All authors have read and agreed to the published version of the manuscript.

Funding: This research was funded by Hubei Key Laboratory of Three Gorges Project for Conservation of Fishes OF FUNDER, grant number 2022055-ZHX and Hubei Province International Cooperation project, grant number 2022EHB029 (Funder is X.B. and Y.D.).

Data Availability Statement: The satellite product data for temperature, chlorophyl-a, photosynthetically active radiation, and the diffuse attenuation coefficient at 490 nm were downloaded from the website https://oceancolor.gsfc.nasa.gov/ (accessed on 31 December 2023). The daily photoperiod data were sourced from the website http://data.cma.cn/ (accessed on 31 December 2023).

Acknowledgments: We thank all partners in Qinghai Lake Naked Carp Rescue Centre for their help in collecting field data in Qinghai Lake.

Conflicts of Interest: The authors declare no conflicts of interest.

References

1. Theng, V.; Sith, R.; Uk, S.; Yoshimura, C. Phytoplankton productivity in a tropical lake-floodplain system revealed by a process-based primary production model. *Ecol. Model.* **2023**, *479*, 110317. [CrossRef]
2. Jia, J.; Wang, Y.; Lu, Y.; Sun, K.; Lyu, S.; Gao, Y. Driving mechanisms of gross primary productivity geographical patterns for Qinghai–Tibet Plateau lake systems. *Sci. Total Environ.* **2021**, *791*, 148286. [CrossRef]
3. Li, Z.; Gao, Y.; Wang, S.; Lu, Y.; Sun, K.; Jia, J.; Wang, Y. Phytoplankton community response to nutrients along lake salinity and altitude gradients on the Qinghai-Tibet Plateau. *Ecol. Indic.* **2021**, *128*, 107848. [CrossRef]
4. Weng, C.; Xu, M.; Lei, F.; Rose, K.A. Management strategy of the naked carp (*Gymnocypris przewalskii*) in the Qinghai lake using matrix population modeling. *J. Environ. Manag.* **2023**, *336*, 117596. [CrossRef] [PubMed]
5. James, R.C.; Paul, D.F.; Mcdonald, G.; Justin, E.O.; Richard, G.K.; James, R.V.; Scott, C.D.; Benjamin, A.S.V.M. An autonomous, in situ light-dark bottle device for determining community respiration and net community production. *Limnol. Oceanogr.-Methods* **2018**, *16*, 323–338. [CrossRef]
6. Cai, Y.; Ke, C.Q.; Duan, Z. Monitoring ice variations in Qinghai Lake from 1979 to 2016 using passive microwave remote sensing data. *Sci. Total Environ.* **2017**, *607*, 120–131. [CrossRef] [PubMed]

7. Hou, P.; Weidman, R.P.; Liu, Q.; Li, H.; Duan, L.; Zhang, X.; Liu, F.; Gao, Y.; Xu, J.; Li, H.; et al. Recent water-level fluctuations, future trends and their eco-environmental impacts on Lake Qinghai. *J. Environ. Manag.* **2023**, *333*, 117461. [CrossRef] [PubMed]
8. Zhang, Y.; Yang, K.; Chen, H.; Dong, Y.; Li, W. Origin, composition, and accumulation of dissolved organic matter in a hypersaline lake of the Qinghai-Tibet Plateau. *Sci. Total Environ.* **2023**, *868*, 161612. [CrossRef] [PubMed]
9. Doernhoefer, K.; Oppelt, N. Remote sensing for lake research and monitoring—Recent advances. *Ecol. Indic.* **2016**, *64*, 105–122. [CrossRef]
10. Zhou, Y.; Hu, Z.; Geng, Q.; Ma, J.; Liu, J.; Wang, M.; Wang, Y. Monitoring and analysis of desertification surrounding Qinghai Lake (China) using remote sensing big data. *Environ. Sci. Pollut. Res.* **2022**, *30*, 17420–17436. [CrossRef]
11. Guo, Q.; Zhang, D.; Cao, L.; Zhan, J. A Remote Sensing Method to Inverse Chemical Oxygen Demand in Qinghai Lake. In Proceedings of the International Geoscience and Remote Sensing Symposium 2021, Brussels, Belgium, 11–16 July 2021. [CrossRef]
12. Mascarenhas, V.; Therese, K. *Marine Optics and Ocean Color Remote Sensing*; Springer eBooks; Springer: Berlin/Heidelberg, Germany, 2018. [CrossRef]
13. Behrenfeld, M.J.; Westberry, T.K.; Boss, E.S.; O'Malley, R.T.; Siegel, D.A.; Wiggert, J.D.; Franz, B.A.; McClain, C.R.; Feldman, G.C.; Doney, S.C.; et al. Satellite-detected fluorescence reveals global physiology of ocean phytoplankton. *Biogeosciences* **2009**, *6*, 779–794. [CrossRef]
14. Ding, Q.; Chen, W. Spatial-Temporal Variation of China's Offshore Net Primary Production Based on Vertically Generalized Production Model. *Ocean Dev. Manag.* **2016**, *33*, 31–35.
15. Behmel, S.; Damour, M.; Ludwig, R.; Rodriguez, M.J. Water quality monitoring strategies—A review and future perspectives. *Sci. Total Environ.* **2016**, *571*, 1312–1329. [CrossRef]
16. Mattei, F.; Michele, S. Collection and analysis of a global marine phytoplankton primary-production dataset. *Earth Syst. Sci. Data* **2021**, *13*, 4967–4985. [CrossRef]
17. Matthew, J.C.; John, J.C.; David, M.K. *Approaches to Measuring Marine Primary Production*; Elsevier eBooks; Elsevier: Amsterdam, The Netherlands, 2019. [CrossRef]
18. Zui, T.; Yan, W.; Ma, S.; Lv, T.; Zhou, X. A Phytoplankton Class-Specific Marine Primary Productivity Model Using MODIS Data. *IEEE J. Sel. Top. Appl. Earth Obs. Remote Sens.* **2017**, *10*, 5519–5528. [CrossRef]
19. Zhang, L.; Wen, X.; Qi, X. Climate change and its impact in the Qinghai Lake Basin. *Qinghai Sci. Technol.* **2019**, *26*, 84–91.
20. Liu, F.; Zeng, Y. Spatial-temporal change in vegetation Net Primary Productivity and its response to climate and human activities in Qinghai Plateau in the past 16 years. *J. Acta Ecol. Sin.* **2019**, *39*, 1528–1540.
21. Li, X.; Li, L.; Wang, G.; Yao, K.; He, Z.; Xiao, J. Water Level and Agriculture and Husbandry Characteristics in Qinghai Lake Basin: The Response to Climate Warming and Wetting. *Chin. Agric. Sci. Bull.* **2018**, *34*, 119–127.
22. Zhao, L.; Wang, S.-Y.S.; Meyer, J. Influence of Interdecadal climate Change on the water level of Qinghai Lake in the Tibetan Plateau. In Proceedings of the 35th Annual Meeting of the Chinese Meteorological Society 2018, Hefei, China, 24–26 October 2018.
23. Behrenfeld, M.J.; Falkowski, P.G. Photosynthetic Rates Derived from Satellite-Based Chlorophyll Concentration. *Limnol. Oceanogr.* **1997**, *42*, 1–20. [CrossRef]
24. Jiang, P.; Shi, J.; Zhang, Y.; Qi, H.; Sun, X. Microsatellite variation analysis of genetic diversity in six wild populations of naked common carp *Gymnocypris przewalski* (Kessler). *J. Acta Ecol. Sin.* **2009**, *29*, 939–945.
25. Wen, Z.; Song, K.; Fang, C.; Yang, Q.; Liu, G.; Shang, Y.; Wang, X. Estimation of K d (PAR) in inland waters across China in relation to the light absorption of optically active components. *Environ. Sci. Pollut. Res.* **2019**, *26*, 30098–30111. [CrossRef] [PubMed]
26. Bi, R. Characteristics and Changes of Water Quality Parameters of Qinghai Lake in 2015. *J. Water Resour. Res.* **2018**, *7*, 74–83. [CrossRef]
27. Shi, J. Artificial reproduction and fry breeding of naked carp from Qinghai Lake. *Fresh Water Fish.* **2000**, *30*, 3–6.
28. Lomas, M.W.; Moran, S.B.; Casey, J.R.; Bell, D.W.; Tihalo, M.; Whitefield, J.; Kelly, R.P.; Mathis, J.T.; Cokelet, E.D. Spatial and seasonal variability of primary production on the Eastern Bering Sea shelf. *Deep Sea Res. Part II Top. Stud. Oceanogr.* **2012**, *65–70*, 126–140. [CrossRef]
29. Siswanto, E.; Ishizaka, J.; Yokouchi, K. Optimal primary production model and parameterization in the eastern East China Sea. *J. Oceanogr.* **2006**, *62*, 361–372. [CrossRef]
30. Cai, L.; Zhu, G.; Li, X. Characteristic of phytoplankton primary productivity and influencing factors in littoral zone of Lake Taihu. *Acta Ecol. Sin.* **2013**, *33*, 7250–7258.
31. Hu, M.; Wang, J. Mixed-pixel Decomposition and Super-resolution Reconstruction of RS Image. *Prog. Geogr.* **2010**, *29*, 747–756.
32. Mesquita, M.C.B.; Prestes, A.C.C.; Gomes, A.M.A.; Marinho, M.M. Direct Effects of Temperature on Growth of Different Tropical Phytoplankton Species. *Microb. Ecol.* **2020**, *79*, 1–11. [CrossRef]
33. Zhu, J.; Zhou, H.; Han, B.; Li, T. Feature analysis of phytoplankton pigments in Qinghai Lake. In *OCEANS*; IEEE: Shanghai, China, 2016.
34. Yao, W.; Shi, J.; Qi, H.; Yang, J.; Jia, L.; Pu, J. Study on the phytoplankton in Qinghai Lake during summer of 2006–2010. *Freshw. Fish.* **2011**, *41*, 22–28.
35. Feng, L.; Liu, J.; Ali, T.A.; Li, J.; Li, J.; Kuang, X. Impacts of the decreased freeze-up period on primary production in Qinghai Lake. *Int. J. Appl. Earth Obs. Geoinf.* **2019**, *83*, 101915. [CrossRef]
36. Xiao, F.; Ling, F.; Du, Y.; Feng, Q.; Yan, Y.; Chen, H. Evaluation of spatial-temporal dynamics in surface water temperature of Qinghai Lake from 2001 to 2010 by using MODIS data. *J. Arid Land* **2013**, *5*, 452–464. [CrossRef]

37. Su, D.; Hu, X.; Wen, L.; Zhao, L.; Li, Z. Variation Characteristic of Surface Water Temperature and Its Response to Climate Change in Qinghai Lake. *J. Yellow River* **2018**, *40*, 25–29.
38. Wang, H.; Liu, J.; Xie, Z.; Ma, L. Trend and Attribution Analysis of Runoff in Qinghai Lake Basin. *J. Water Resour. Power* **2018**, *36*, 18–21+32.

water

MDPI

Article

The Use of Macrophytes for the Removal of Chlorpyrifos from the Aquatic Environment

Elżbieta Sobiecka [1,*], Milena Mroczkowska [1], Tomasz P. Olejnik [2] and Agnieszka Nowak [3]

1 Institute of Natural Products and Cosmetics, Faculty of Biotechnology and Food Sciences, Lodz University of Technology, ul. Stefanowskiego 2/22, 90-357 Lodz, Poland; milena.mroczkowska.500@guest.p.lodz.pl
2 Department of Sugar and Food Safety Management, Faculty of Biotechnology and Food Sciences, Lodz University of Technology, ul. Wolczanska 171/173, 90-530 Lodz, Poland; tomasz.olejnik@p.lodz.pl
3 Institute of Fermentation Technology and Microbiology, Lodz University of Technology, ul. Wolczanska 171/173, 90-530 Lodz, Poland; agnieszka.nowak@p.lodz.pl
* Correspondence: elzbieta.sobiecka@p.lodz.pl

Abstract: Phytoremediation is one of the effective technologies for removing pollutants from the aquatic environment. Toxic compounds such as chlorpyrifos can affect the physiological processes of aquatic plants, causing secondary oxidative stress in plant tissues. Macrophytes, like other organisms inhabiting the contaminated ecosystem, have developed a system of defense mechanisms, thanks to which plants can still exist in their natural ecosystem. Our research is a summary of the previously presented results of the effectiveness of purifying contaminated water with chlorpyrifos in the phytoremediation process and the second type of phytoremediation supported by microorganisms, which intensify the process of removing contaminants from the environment. This research concerned changes in nonenzymatic and enzymatic antioxidants in Canadian seaweed, needle spikerush and water mint caused by chlorpyrifos. The research determines changes in the total concentration of polyphenols, flavonoids and dyes (chlorophyll A, chlorophyll B, anthocyanins and carotenoids) as well as differences in the activity of guaiacol peroxidase and glutathione S-transferase. The analysis of the results showed an increase in the content of polyphenols and flavonoids. The reverse trend was observed in the case of the pigment content. The appearance of chlorpyrifos in the environment caused an increase in the activity of the examined enzymes. The process involving microorganisms that were obtained from places contaminated with pesticide proved to be more effective. This shows the cooperation of species living in an investigated ecosystem.

Keywords: antioxidant defense system; autochthonic microorganisms; chlorpyrifos; macrophytes; phytoremediation

1. Introduction

The intensive development of civilization caused the high negative effects on the environment. The increase in awareness and the desire to improve the quality of life, which initiated the dynamic development of industry with the appearance of synthetic compounds, have a negative impact on the aquatic ecosystems. This required the development of new and effective processes based on biological, chemical and thermal methods, whose utilitarian use will allow to get safety products [1–3]. One of the technologies based on biological methods used in the degradation of organic compounds is a phytoremediation that uses plants for reduction, degradation, assimilation and metabolizing environmental pollution such as heavy metals, hydrocarbons or pesticides [4,5]. Some water plants that appear naturally in the aquatic ecosystems in Poland are able to participate in biological processes to decrease or move the pollutants amounts [6]. The native plants are able to change their physiological functions to exist in a polluted environment; this is their response to abiotic stress. This phenomenon allows the plants to participate in water treatment as a main tool for phytoremediation. This process occurs naturally and can be intensified using microbial consortia to increase the effectiveness of pollution removal [7–9].

Plant protection products that have been used for years have a beneficial effect on inhibiting the development of microorganisms. However, intensive use of pesticides in agriculture may pose threats to the proper functioning of aquatic and terrestrial ecosystems, both fauna and flora [10–12]. Improperly selected doses of chemical compounds may remain in the environment and migrate within ecosystems. The presence of plant protection products was found in all types of flowing waters—rainwater, surface and groundwater, which may increase during agrochemical treatments. Organophosphate pesticides were very popular in the world and used intensely. These include, among others, chlorpyrifos which was distributed by the Dow Chemical Company since 1965 as a foliar pesticide [13]. These compounds were used to control pests in industrial crops, in vegetable and orchards [14]. Organophosphate pesticides are esters of phosphoric acid and its derivatives. Their activities relate to the electrophilic nature of the phosphorus ester bond [15]. The presence of organophosphorus compounds in the environment may cause unfavorable changes in the physiological processes of plants such as: photosynthesis, respiration, cell division, synthesis of growth regulators, water uptake, as well as changes in the structure of cell organelles and limiting plant productivity [16]. The presence of pesticides in the environment is one of the causes of secondary oxidative stress in plant tissues, characterized by intensive production of reactive oxygen species (ROS). An undesirable effect caused by toxic pesticide particles may be a change in the proper functioning of protein, lipid or nucleic acid molecules [17].

One of the groups found in aquatic ecosystems are aquatic plants called macrophytes. The presence of toxic chemical compounds such as pesticides resulted in the activation of defense mechanisms in plants tissues [18]. The ability to quickly adapt macrophytes to new environmental conditions enabled the initiation of research using them. The phytoremediation process was used to optimize the system of macrophytes antioxidant mechanisms, consisting of nonenzymatic and enzymatic antioxidants [19]. The system of enzymes that remove free radicals include guaiacol peroxidase (GPX) and glutathione S-transferase (GST) [20–22]. Biochemical reactions involving enzymes, e.g., GPX, contribute to detoxification processes that remove various types of toxic compounds [23]. This enzyme participates in various physiological processes, such as auxin catabolism, wound healing and defense mechanisms against infections caused by pathogens [24]. GPX may also participate in the capture of pollutants in the root sphere and the oxidative degradation of compounds found in the plant's environment [25]. The second enzyme that enters the system of enzymatic mechanisms is glutathione S-transferase (GST). In 1970, a reaction involving GST was first described, catalyzing the decomposition of atrazine by conjugation with glutathione tripeptide in Sorghum and corn plants [26]. The results initiated intensive research on the participation of GST in the detoxification processes of herbicides and other plant xenobiotics [27]. According to the literature, the GPX and GST activities can be induced in plants by abiotic factors including insecticides, heavy metals, infection caused by pathogens. Both the enzymes can be treated as stress markers [28].

The nonenzymatic antioxidant system of plants consists of low molecular weight compounds such as phenols, flavonoids, pigments—chlorophyll A and B, anthocyanins, carotenoids [29–31]. Phenolic compounds participate in reactions leading to the removal of reactive oxygen species and inhibition of the formation of free radicals [32]. Research on the properties of flavonoids has proven that they are responsible for the color of plants, the smell of flowers and the taste of fruit [33,34]. This feature qualifies them as natural repellents, and any action of toxic compounds, including pesticides, causes changes in their proper functioning [35].

In our research, we focused on the chosen macrophytes use in a pure phytoremediation process and a phytoremediation assisted by microorganisms to remove chlorpyrifos from an aquatic environment. We compared effectiveness both types of the processes. Because the enzymatic and nonenzymatic antioxidants defense system is responsible for water plants' stress tolerance we also investigated the polyphenols, flavonoids and pigments (chlorophyll a, chlorophyll b, anthocyanins and carotenoids) contents as well as the activities changes of guaiacol peroxidase (GPX) and glutathione S-transferase (GST) caused by chlorpyrifos presence during the phytoremediation process. There are many works in the literature describing research on the biological degradation of pesticides. However, there are many questions about specific compounds such as chlorpyrifos. Our research focused on the answers to such questions regarding a single compound of chlorpyrifos as an example of an organophosphorus pesticide, and, more precisely, on the use of phytoremediation appearing in natural conditions as a part of the processes taking place to clean aquatic ecosystems. In addition, we confirmed that in the natural environment, water treatment occurs as a complex process where all organisms that live in the habitat participate in it. It is also important that the plants and microorganisms conducting the water purification process are characteristic for a given climate zone—in our case, a moderate zone. The above shows the novelty of our applied research, which brings elements of innovation to the general knowledge.

2. Materials and Methods

2.1. Materials

2.1.1. Water Plants

Three species of macrophytes were used in our studies: Canadian waterweed (*Elodea canadensis* Michx.), needle spikerush (*Eleocharis acicularis* L.) and water mint (*Mentha aquatica* L.), which originated from organic farming "Ogrody Wodne", Miedzychod, Poland. The plants were chosen because of their natural location. They are characteristic for the moderate climate zone and are often found in areas of water reservoirs.

2.1.2. Microorganisms

In the process of supported phytoremediation the autochthonous microorganisms isolated from soils were used in the studies: *Bacillus cereus, Bacillus licheniformis, Oerskovia paurometabola.* The details concerning the species are presented in Table 2, Results.

2.1.3. Chlorpyrifos

In our research, we used chlorpyrifos, a commercial plant protection product. For the analytical part, we used the chlorpyrifos standard (Sigma-Aldrich Production GmbH, Buchs, Switzerland; CAS: 2921-88-2). Selected physicochemical properties of the chlorpyrifos are presented in Table 1.

Table 1. Chlorpyrifos chosen properties.

Chemical Formula	$C_9H_{11}Cl_3NO_3PS$
Molar mass	350.57 g/mol
Melting point	42–43.5 °C
Density	1.398 g/cm^3 in 43.5 °C
Solubility in water	2 mg/dm^3
Vapor pressure	1.87×10^{-5} mmHg at 25 °C
Octanol–Water Partition Coefficient (K_{OW})	4.70
Henry's constant	4.2×10^{-6} atm·m^3/mol at 25 °C
Soil Sorption Coefficient (K_{oc})	360 to 31 000 depending on soil type and environmental conditions

2.2. Experiments

2.2.1. Isolation of Microorganisms

Soil samples were taken from agricultural areas where the following crops were supported by pesticide spraying: A—corn (central Poland), B—celery (central Poland), C—strawberries (Germany), D—apple trees (Germany). The collected soil samples were sieved through sieves with a mesh size of 2 mm. The sieved fractions were used for further studies. An enrichment procedure was used to isolate bacteria that effectively degrade insecticides. The process was carried out in a mineral medium: $(NH_4)_2SO_4$ 2.0 g/dm^3 (Pol-Aura sp z o.o., Lodz, Poland), Na_2HPO_4 $12H_2O$ 1.5 g/dm^3 (Pol-Aura sp z o.o., Lodz, Poland), KH_2PO_4 1.5 g/dm^3 (Pol-Aura sp z o.o., Lodz, Poland), $MgSO_4 \cdot 7H_2O$ 0.01 g/dm^3 (Pol-Aura sp z o.o., Lodz, Poland), $FeSO_4 \cdot 7H_2O$ 0.01 g/dm^3 (Pol-Aura sp z o.o., Lodz, Poland), $CaCl_2$ $2H_2O$ 0.001 g/dm^3 (Pol-Aura sp z o.o., Lodz, Poland). Chlorpyrifos was added to the medium at a concentration of 50 mg/dm^3.

To 90 mL of the medium prepared in this way, 10 g of soil was added and culturing was carried out at 30 °C for 72 h. After this time, the culture of microorganisms was passaged on a medium for the determination of the total number of microorganisms, using Plate Count Agar (PCA) with the following composition: casein peptone 5.0 g/dm^3, yeast extract 2.5 g/dm^3 (Pol-Aura sp z o.o., Lodz, Poland), glucose 1.0 g/dm^3 (Pol-Aura sp z o.o., Lodz, Poland), agar 12 g/dm^3 (Pol-Aura sp z o.o., Lodz, Poland). The plates were incubated in an incubator at 37 °C for 48 h.

After this time, three strains of morphologically different bacteria were isolated from each soil sample. Pure bacterial cultures were stored in cryobanks at −30 °C. The strains from the pure culture were then inoculated into 100 mL mineral medium, which contained 50 mg/ dm^3 of chlorpyrifos. After 72 h of incubation at 30 °C, the concentration of chlorpyrifos was measured. The optical density of the culture (OD_{550}) was also measured using a DEN-1B densitometer (Biosan). Three strains with the highest biodegradation activity were selected for further research.

The bacteria were identified using molecular methods based on 16S rRNA gene analysis. Genomic DNA was extracted using the Genomic Mini kit (A&A Biotechnology) according to the methodology provided by the manufacturer. The reaction mixture was prepared in a volume of 25 µL containing 12 µL of polymerase (1.5 units) REDTaq™ ReadyMix™ (Sigma), 0.2 µL of each universal primer (27F and 1492R) and 11.6 µL of water and 1 µL of DNA. The 16S rRNA gene was amplified via PCR in the MJ Mini Gradient Thermal Cycler (Bio-Rad) in a cycle consisting of initial denaturation at 94 °C for 2 min, denaturation at 94 °C for 1 min, primer annealing at 50 °C for 1 min (34 repetitions), extension 72 °C for 3 min and final extension at 72 °C for 3 min.

The PCR reaction products were analyzed with horizontal electrophoresis in 1% (*w*/*v*) agarose gel in 0.5 × TBE buffer (Sigma-Aldrich Sp z o.o., Poznan, Poland). The amplified PCR products were purified employing the Clean-Up AX kit (A&A Biotechnology) and then subjected to a sequencing reaction. The obtained nucleotide sequences were compared with the BLAST 2.10.0+ program (the Basic Local Alignment Search Tool) with the sequences available in the NCBI database (the National Center of Biotechnology Information database).

2.2.2. Phytoremediation Process

All details concerning the cultivation process were described in the previously published paper of Sobiecka et al. [36]. The pure phytoremediation process (F1) was carried out in an aquatic environment polluted by different concentrations of chlorpyrifos: 50 µg/dm^3, 100 µg/dm^3, 150 µg/dm^3. Macrophytes were also cultivated in the medium without adding the tested pesticide, as a reference test. The second phytoremediation process was conducted with assistance of microbial consortia (F2). In this process, the plant cultivation environment was enriched with inoculum of three isolated microbial consortia. To prepare the inoculum, each strain was activated on PCA medium. After 72 h of culturing in 37 °C, suspensions of the tested strains were made in physiological saline. The density of each suspension was determined densitometrically so that the final number of cells was 1.0×10^6 cells/cm^3. Then, 0.650 cm^3 of each of the suspensions prepared in this way was added to 1 dm^3 of the medium in which the phytoremediation process was carried out. The initial abundance of each strain in the environment was approximately 2.0×10^3 cells/cm^3.

2.3. Determination of Chlorpyrifos

After the end of phytoremediation supported by microorganisms, the bacterial suspension had to be removed before performing the chromatographic analysis. For this purpose, a sample containing bacterial cells was extracted by Solid Phase Extraction (SPE). The extraction was carried out in extraction columns filled with a C-18 bed from Phenomenex on a 12-station SPE system from J.T. Baker. After conditioning the column by passing 5 mL of methylene chloride and then 10 mL of distilled water, the samples were applied to the column. A sample of the bacterial suspension was loaded onto the column under pressure 5mBar.

Methylene chloride (Pol-Aura sp z o.o., Lodz, Poland) was used to extract the chlorpyrifos from the column. Elution was carried out twice using 10 mL of reagent. Each portion of the solvent remained in contact with the stationary phase from 20 s to 1 min, because then the leaching effect is most effective.

Chlorpyrifos concentrations before and after both phytoremediation processes were determined with gas chromatography on a two-dimensional gas chromatograph coupled with a mass spectrometer, with a TOF ion time-of-flight detector (Pegasus 4D, LECO Corp., St. Joseph, MI, USA). An extremely important advantage of this electron capture detector is its high sensitivity and selectivity to impurities containing elements with high electronegativity.

After the extraction process, the obtained eluates were dosed to the chromatographic column in various temperature programs of the oven, which enabled the determination of the effect of the oven temperature on the chromatographic separation of chlorpyrifos. The retention time of the pesticide with a positive response was then recorded. The following temperature program was used: initial temperature 70 °C, then increase of 15 °C/min to 300 °C, at which chlorpyrifos in the tested samples gave a positive response with the correct peak shape.

2.4. Determination of Enzymatic Antioxidants

The enzymatic antioxidants system analytical details were described in a previous paper of Sobiecka et al. [37]. The information concerned the determination of a glutathione S-transferase (GST) and guaiacol peroxidase (GPX). The standards came from the Polish Office of Sigma-Aldrich Sp z o.o., Poznan, Poland.

2.5. Determination of Nonenzymatic Antioxidants

The nonenzymatic antioxidants system analytical details were described in a previous paper of Sobiecka et al. and concerned the determination of polyphenols, flavonoids and pigments has been repeated [36]. All the required standards were bought in the Polish Office of Sigma-Aldrich Sp z o.o., Poznan, Poland.

2.6. Statistical Analysis

The STATISTICA Version 10 (StatSoft, Cracow, Poland) was used for statistical calculations. The results present the average of three biological samples measurements. The single-factor analysis of ANOVA variance was used. The Duncan multiple-range post hoc test ($p < 0.05$) in order to show statistically significant differences between the tested samples was used to analysis.

3. Results

Our research on the purification of water contaminated with chlorpyrifos was preceded by the isolation of bacterial strains naturally occurring in temperate climate zones—

including Poland. The bacteria were isolated from places contaminated with pesticides. Next, they were cultivated in a solution polluted by chlorpyrifos to check their biodegradation effectiveness, as seen in Figure 1. The three most effective strains were identified, and these results are summarized in Table 2.

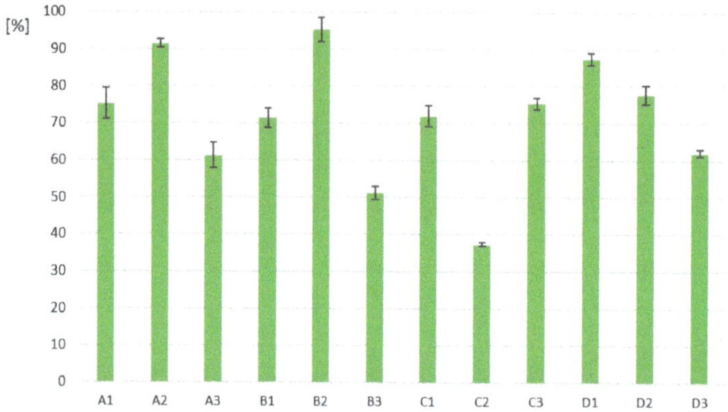

Figure 1. Biodegradation of chlorpyrifos by soil bacteria originating from A—a corn field, B—celery cultivation, C—a strawberry field, D—an apple orchard.

Table 2. Summary of the most efficient strains.

Identified Strain [Accession Number in GenBank]	Reference Strain in GenBank [Accession Number]	Similarity	Taxonomy
Bacillus cereus A2 [PP473786]	*Bacillus cereus* strain CBC-4 [MK285635]	100%	Domain *Bacteria*, Phylum *Bacillota*, Class *Bacilli*, Order *Caryophanales*, Family *Bacillaceae*, Genus *Bacillus*
Bacillus licheniformis B2 [PP473787]	*Bacillus licheniformis* strain HO-A7 [MT495615]	99.9%	Domain *Bacteria*, Phylum *Bacillota*, Class *Bacilli*, Order *Caryophanales*, Family *Bacillaceae*, Genus *Bacillus*
Oerskovia paurometabola D1 [PP473788]	*Oerskovia paurometabola* strain CSE_1 [KX027337]	99.9%	Domain *Bacteria*, Phylum *Actinomycetota*, Order *Micrococcales*, Family *Promicromonosporaceae*, Genus *Oerskovia*

In the main part of our studies we estimated the effectiveness of phytoremediation—one of the biological technologies for pollutant removal from aquatic environment. During the phytoremediation process (F1) and phytoremediation supported by microorganisms (F2) various concentrations of chlorpyrifos: 50 $\mu g/dm^3$, 100 $\mu g/dm^3$ and 150 $\mu g/dm^3$, in the aquatic environment were monitoring (Figure 2). The enrichment of the aquatic environment with the bacteria species (*Bacillus cereus, Bacillus licheniformis, Oerskovia paurometabola*) intensified the biological process, which resulted in an increase in the effectiveness of the phytoremediation process. The presence of macrophytes and microorganisms enlarged their scope to get rid of toxins from the environment. The investigated macrophytes reduced the concentration of chlorpyrifos to a large extent after the end of F1 and F2 processes according to the initial concentrations, Figure 2.

The used of a hybrid phytoremediation method in the removal of chlorpyrifos significantly decreased the pesticide amount from the aquatic environment. After the F2 process, the pesticide concentration was from 1.55% to 14.00% of its initial values, and after the F1 process from 5.35% to 34.00%. Analyzing the results, it was found that the phytoremediation process supported by microorganisms conducted by macrophytes is more effective. The plants selected for this research, Canadian waterweed, needle spikerush and water mint, showed increased activity of an enzymatic antioxidant systems in response to abiotic stress caused by the presence of chlorpyrifos in the environment. These results confirmed that biological methods, including commonly occurring aquatic plants, can be effective in cleaning an aquatic environment contaminated with chlorpyrifos.

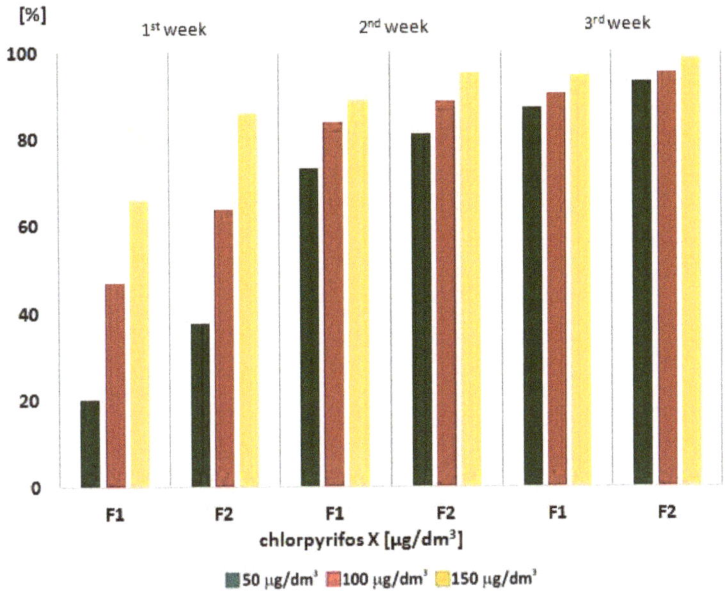

Figure 2. Chlorpyrifos removal from an aquatic environment in F1—phytoremediation; and F2—phytoremediation supported by microorganisms.

Water plants have developed the ability to accumulate pollutants in their tissues in order to grow in difficult environmental conditions in the presence of toxic substances. In order to protect against oxidative stress, macrophytes have developed mechanisms consisting of nonenzymatic and enzymatic antioxidants system.

For the second part of the presented results of our studies, we focused on the behavior of the chosen water plants (Canadian waterweed (A), needle spikerush (B) and water mint (C)) in the presence of three various concentrations of chlorpyrifos: 50 $\mu g/dm^3$, 100 $\mu g/dm^3$ and 150 $\mu g/dm^3$. One of the processes conducted in the aquatic environment was enriched by an inoculum of bacterial consortia. We analyzed a few nonenzymatic and enzymatic antioxidants that can be markers of the abiotic stress caused by various concentrations of chlorpyrifos.

The contents of polyphenols and flavonoids were measured in leaves of the chosen macrophytes (Table 3). The increased concentration of chlorpyrifos in the cultures intensified the concentration of both investigated compounds in the tissues of the tested plants. The increasing amounts of polyphenols and flavonoids correlated with the increasing chlorpyrifos concentrations in the water solution. All of the macrophytes were able to grow and develop in the presence of toxic substances.

Table 4 shows the average content of four pigments in the leaves of Canadian seaweed, needle spikerush and water mint. The presence of chlorpyrifos in the plant growth environment resulted in a reduction in the content of pigments in the tissues of the tested plants. This correlation was directly proportional and the concentration of chlorpyrifos 150 $\mu g/dm^3$ significantly inhibited the pigment content in plant tissues compared to the control sample.

Table 3. The content of polyphenols and flavonoids in chosen macrophytes after phytoremediation processes F1and F2.

Nonenzymatic Antioxidants	Phytoremediation	Plant	Chlorpyrifos Concentration [μg/dm^3] 0	50	100	150
Polyphenols [mg CEA/g f.m.]	F1	A	3.66 ± 0.02	8.80 ± 0.01	10.46 ± 0.01	18.57 ± 0.02
		B	4.45 ± 0.01	11.85 ± 0.03	14.61 ± 0.02	15.89 ± 0.02
		C	5.60 ± 0.02	14.59 ± 0.02	16.64 ± 0.01	19.51 ± 0.02
	F2	A	6.60 ± 0.03	13.86 ± 0.01	19.81 ± 0.02	23.73 ± 0.02
		B	6.85 ± 0.02	15.83 ± 0.03	19.59 ± 0.02	23.09 ± 0.01
		C	6.82 ± 0.02	14.52 ± 0.01	18.85 ± 0.01	24.10 ± 0.02
Flavonoids [mg QE/g f.m.]	F1	A	4.56 ± 0.03	7.24 ± 0.01	9.30 ± 0.02	13.29 ± 0.02
		B	3.81 ± 0.02	7.66 ± 0.01	10.56 ± 0.03	14.67 ± 0.01
		C	5.22 ± 0.01	6.28 ± 0.02	10.64 ± 0.01	16.82 ± 0.01
	F2	A	3.99 ± 0.02	8.22 ± 0.03	10.88 ± 0.01	15.06 ± 0.01
		B	4.50 ± 0.01	8.56 ± 0.01	12.79 ± 0.03	16.38 ± 0.01
		C	5.19 ± 0.02	7.85 ± 0.02	11.53 ± 0.02	19.21 ± 0.01

Notes: F1—phytoremediation; F2—phytoremediation assisted by microorganisms; A—Canadian waterweed (*Elodea canadensis* Michx.), B—needle spikerush (*Eleocharis acicularis*), C—water mint (*Mentha aquatica* L.).

Table 4. The content of nonenzymatic antioxidants in chosen macrophytes after phytoremediation processes.

Nonenzymatic Antioxidant [mg/g f.m.]	Phytoremediation	Plant	Chlorpyrifos Concentration [μg/dm^3] 0	50	100	150
Chlorophyll a	F1	A	12.15 ± 0.02	10.28 ± 0.02	7.25 ± 0.03	2.36 ± 0.01
		B	13.02 ± 0.02	12.18 ± 0.03	7.38 ± 0.03	5.12 ± 0.02
		C	13.71 ± 0.01	12.27 ± 0.01	8.12 ± 0.02	6.08 ± 0.01
	F2	A	12.10 ± 0.02	11.12 ± 0.03	7.97 ± 0.01	3.62 ± 0.03
		B	15.32 ± 0.03	13.78 ± 0.01	8.02 ± 0.01	5.42 ± 0.02
		C	13.08 ± 0.02	12.24 ± 0.03	7.35 ± 0.03	6.71 ± 0.03
Chlorophyll b	F1	A	13.52 ± 0.03	10.00 ± 0.02	7.24 ± 0.02	3.14 ± 0.03
		B	13.01 ± 0.02	12.15 ± 0.02	8.18 ± 0.03	7.02 ± 0.03
		C	12.97 ± 0.02	11.71 ± 0.03	7.93 ± 0.01	3.14 ± 0.02
	F2	A	12.05 ± 0.03	10.27 ± 0.02	8.19 ± 0.03	3.28 ± 0.02
		B	14.12 ± 0.02	12.32 ± 0.01	8.27 ± 0.02	5.64 ± 0.03
		C	13.09 ± 0.03	11.02 ± 0.01	7.37 ± 0.03	6.27 ± 0.03
Anthocyanins	F1	A	29.52 ± 0.03	26.36 ± 0.01	23.09 ± 0.03	18.38 ± 0.03
		B	35.12 ± 0.02	31.51 ± 0.02	27.11 ± 0.01	20.32 ± 0.03
		C	33.41 ± 0.01	31.81 ± 0.01	23.72 ± 0.03	17.42 ± 0.02
	F2	A	30.21 ± 0.01	27.21 ± 0.02	23.34 ± 0.01	18.34 ± 0.01
		B	35.32 ± 0.02	31.37 ± 0.01	27.81 ± 0.02	20.18 ± 0.01
		C	32.41 ± 0.01	29.46 ± 0.02	23.47 ± 0.03	17.24 ± 0.02
Carotenoids	F1	A	25.18 ± 0.03	19.21 ± 0.03	15.32 ± 0.01	8.71 ± 0.01
		B	24.28 ± 0.02	22.08 ± 0.02	14.54 ± 0.03	5.21 ± 0.02
		C	23.58 ± 0.01	20.36 ± 0.02	13.37 ± 0.01	7.83 ± 0.03
	F2	A	25.04 ± 0.01	18.54 ± 0.01	16.23 ± 0.01	9.14 ± 0.01
		B	24.52 ± 0.02	23.21 ± 0.01	14.75 ± 0.02	8.35 ± 0.01
		C	23.24 ± 0.01	21.87 ± 0.03	14.41 ± 0.01	7.08 ± 0.03

Notes: F1—phytoremediation; F2—phytoremediation assisted by microorganisms; A—Canadian waterweed (*Elodea canadensis* Michx.), B—needle spikerush (*Eleocharis acicularis*), C—water mint (*Mentha aquatica* L.).

Our research also analyzed changes in the activities of selected enzymatic oxidants, glutathione S-transferase (GST) and guaiacol peroxidase (GPX), in leaf and root tissues. The results are presented in Table 5. It was observed that the concentration of chlorpyrifos

150 μg/dm^3 intensifies the enzyme activity the most. Activity increased more than three times compared to the initial value in all plant species. For water mint the highest increase in GST activity was observed in leaves and roots. In a case of Canadian waterweed and needle spikerush the activity intensification was observed in leaves. Generally, the similar tendency was observed in activity changes of GXP. The highest concentration of pollutant caused the most intensive activity of the enzyme. We observed more intensified activities in leaves compared roots.

Table 5. The enzymes activities changes in chosen macrophytes after phytoremediation processes.

Enzymatic Antioxidant	Phytoremediation	Plant	Chlorpyrifos Concentration [μg/dm^3]			
			0	50	100	150
GST in leaves [nmol CDNB/g f.m.]	F1	A	4.40 ± 0.01	9.67 ± 0.02	12.68 ± 0.02	16.89 ± 0.01
		B	4.82 ± 0.01	10.61 ± 0.01	12.93 ± 0.01	20.24 ± 0.03
		C	5.88 ± 0.02	9.22 ± 0.03	14.44 ± 0.01	19.77 ± 0.01
GST in leaves [nmol CDNB/g f.m.]	F2	A	5.93 ± 0.03	12.62 ± 0.02	20.07 ± 0.03	22.18 ± 0.01
		B	6.73 ± 0.01	11.71 ± 0.02	18.83 ± 0.01	23.14 ± 0.03
		C	6.60 ± 0.03	11.99 ± 0.01	19.97 ± 0.03	20.07 ± 0.01
GST in roots [nmol CDNB/g f.m.]	F1	A	2.93 ± 0.03	5.29 ± 0.02	9.87 ± 0.01	12.34 ± 0.03
		B	2.33 ± 0.01	6.84 ± 0.03	11.33 ± 0.03	15.23 ± 0.01
		C	6.36 ± 0.02	12.38 ± 0.02	19.51 ± 0.01	22.61 ± 0.01
GST in roots [nmol CDNB/g f.m.]	F2	A	1.98 ± 0.01	3.24 ± 0.01	9.86 ± 0.02	13.19 ± 0.01
		B	2.67 ± 0.01	7.33 ± 0.02	14.80 ± 0.01	17.79 ± 0.02
		C	6.14 ± 0.03	12.59 ± 0.01	20.16 ± 0.03	23.29 ± 0.01
GPX in leaves [mmol TG/g f.m.]	F1	A	3.66 ± 0.02	8.80 ± 0.01	10.46 ± 0.03	18.57 ± 0.03
		B	4.45 ± 0.01	11.85 ± 0.01	14.61 ± 0.03	15.89 ± 0.01
		C	5.60 ± 0.01	14.59 ± 0.02	16.64 ± 0.01	19.51 ± 0.03
GPX in leaves [mmol TG/g f.m.]	F2	A	6.60 ± 0.01	13.86 ± 0.01	19.81 ± 0.03	23.75 ± 0.01
		B	6.86 ± 0.02	15.80 ± 0.01	19.60 ± 0.01	23.09 ± 0.03
		C	6.92 ± 0.03	14.53 ± 0.01	18.85 ± 0.03	24.11 ± 0.02
GPX in roots [mmol TG/g f.m.]	F1	A	4.18 ± 0.01	6.78 ± 0.03	9.83 ± 0.02	14.27 ± 0.03
		B	4.34 ± 0.03	12.70 ± 0.01	16.41 ± 0.03	16.10 ± 0.03
		C	6.36 ± 0.03	12.38 ± 0.03	19.51 ± 0.01	22.61 ± 0.01
GPX in roots [mmol TG/g f.m.]	F2	A	2.27 ± 0.03	9.18 ± 0.01	12.47 ± 0.02	17.36 ± 0.03
		B	5.09 ± 0.01	16.74 ± 0.01	18.64 ± 0.03	21.45 ± 0.01
		C	6.14 ± 0.03	12.59 ± 0.02	20.16 ± 0.01	23.29 ± 0.03

Notes: F1—phytoremediation; F2—phytoremediation assisted by microorganisms; A—Canadian waterweed (*Elodea canadensis* Michx.), B—needle spikerush (*Eleocharis acicularis*), C—water mint (*Mentha aquatica* L.).

4. Discussion

The results presented in this paper compare the changes of chosen nonenzymatic and enzymatic antioxidants defense system for pure phytoremediation process (F1) and the phytoremediation assisted by microorganisms (F2). Changes in the activities and concentrations of compounds that are part of the plant defense system enable the removal of pollutants from the environment. Living organisms, both plants and microorganisms, are exposed to toxic compounds which induce the response to stress. In the presence of chlorpyrifos, the plants activated the cell organelles that induce biochemical processes of transcriptional up-regulation of phenylpropanoid pathway [38]. The activation of biosynthetic enzymes and up-regulation of key genes of phenylpropanoid branch allowed to stimulate a phenolic biosynthesis [39,40].

The presence of chlorpyrifos resulted in a decrease in the content of chloroplast pigments in the studied macrophytes, caused by oxidative stress. The toxic compound caused an increased content of free radicals. The direct effect of these reactions was the inhibition of the activity of enzymes responsible for the process of chlorophyll synthesis [41].

Similar results were obtained by the Shixiang and Yadav groups [42,43]. Scientists have also observed that chlorophyll concentrations decrease as the concentration of pesticides in the environment increases. Higher concentrations of pollutants adversely affect the physiology of macrophytes.

Abiotic stress caused by the presence of pesticides in the environment also affects the enzymatic defense system of plants. Our studies have proven that the activities of both enzymes: glutathione S-transferase (GST) and guaiacol peroxidase (GPX) in plant tissues increase with increasing concentration of chlorpyrifos. Both enzymes are involved in the biochemical reactions of plants responsible for the detoxification process of xenobiotics such as pesticides. The plant defends itself against accumulating pollutants in its tissues [44]. Our study results regarding changes in GST activity were similar to those obtained in Tlidjen's study [45]. In the case of the activity of guaiacol peroxidase (GPX), which is responsible for the proper course of physiological processes in plant tissues, a correlation with the growth of chlorpyrifos in the environment was observed. The increased activity of GPX was also observed in research of Sharma group [46] and Bertrand, in which different concentrations of chlorpyrifos were tested in culture in small pondweed (*Potamogeton pusillus*) tissue [14]. The obtained results of the increase in activity confirm that guaiacol peroxidase is one of the enzymes responsible for the plant response to abiotic stress caused by chlorpyrifos. The analysis of the results also confirms that both enzymes—GPX and GST—are involved in the process of removing organic peroxides and hydrogen peroxide, which are products of aerobic metabolism [47,48]. These enzymes significantly reduce the oxidative stress effects on plant tissues.

The proper management of waste disposal is one of the elements of the economy that requires not only technological, but also economic and social activities, taking into account the needs and interests of man together with balancing adverse impact on the natural environment.

During the research, the hypothesis that the concentration of abiotic factors directly affects the activity of enzymes was also confirmed. The highest concentration of chlorpyrifos stimulated enzyme activity most intensely.

The investigated nonenzymatic antioxidative stress system compounds: polyphenols, flavonoids and pigments (chlorophyll a, chlorophyll b, anthocyanins and carotenoids) effected in amount changes which was correlated to the chlorpyrifos concentration. The aquatic environment polluted by the highest chlorpyrifos concentration (150 $\mu g/dm^3$) caused an increase in polyphenols and flavonoids in plants tissues while the content of pigments diminished.

The ability of the chosen macrophytes to adapt in the presence of toxic chlorpyrifos molecules allowed for the testing of a remediation process to clean a polluted ecosystem [49,50]. For the study we used three species of plants naturally occurring in the aquatic environment of our temperate climate in Poland.

5. Conclusions

To summarize our research, we confirm the proper choice of macrophytes used in our studies: Canadian waterweed, needle spikerush and water mint. The chosen water plants provided an efficient process of pollutant removal. The phytoremediation process was intensified in the presence of the following microbial consortia: *Bacillus cereus, Bacillus licheniformis, Oerskovia paurometabola*, selected from the polluted sites. The investigated method was a useful technology for cleansing the aquatic environment of chlorpyrifos, resulting in chlorpyrifos removals of 86% to 98% after the F2 process, compared to its initial values, and 66% to 94% after the F1 process.

Author Contributions: Conceptualization, E.S. and M.M.; methodology, E.S., M.M. and A.N.; validation, M.M. and T.P.O.; formal analysis, M.M. and A.N.; investigation, M.M. and A.N.; writing—original draft preparation, E.S. and T.P.O.; writing—review and editing, E.S., T.P.O. and A.N.; visualization, T.P.O.; supervision, E.S. All authors have read and agreed to the published version of the manuscript.

Funding: This research received no external funding.

Data Availability Statement: The data are contained within the article.

Conflicts of Interest: The authors declare no conflict of interest.

References

1. Rani, K.; Dhania, G. Bioremediation and biodegradation of pesticide from contaminated soil and water—A noval approach. *Int. J. Curr. Microbiol. App. Sci.* **2014**, *3*, 23–33.
2. Meng, X.; Guo, Y.; Wang, Y.; Fan, S.; Wang, K.; Han, W. A systematic review of photolysis and hydrolysis degradation modes, degradation mechanisms, and identification methods of pesticides. *J. Chem.* **2022**, *2022*, 9552466. [CrossRef]
3. Sobiecka, E.; Kołaciński, Z.; Rincón, J.M.; Szymański, Ł.; Olejnik, T.P. Coloured sintered glass-ceramics from hospital incineration fly ash. *Mat. Lett.* **2019**, *252*, 34–37. [CrossRef]
4. Susarla, S.; Medina, V.F.; McCutcheon, S.C. Phytoremediation: An ecological solution to organic chemical contamination. *Ecol. Eng.* **2002**, *18*, 647–658. [CrossRef]
5. Seridou, P.; Fyntrilakis, K.; Syranidou, E.; Kalogerakis, N. Hydroponic phytoremediation of antimony by *Tamarix smyrnensis* and *Nerium oleander*. *J. Chem. Technol. Biotechnol.* **2023**, *98*, 2214–2223. [CrossRef]
6. Futughe, A.E.; Purchase, D.; Jones, H. Phytoremediation using Native Plants. In *Phytoremediation: In-Situ Applications*, 1st ed.; Shmaefsky, B., Ed.; Concepts and Strategies in Plant Sciences; Springer Nature: Cham, Switzerland, 2020; pp. 285–327.
7. Shilev, S.; Babrikova, I.; Babrikov, T. Consortium of plant growth-promoting bacteria improves spinach (*Spinacea oleracea* L.) growth under heavy metal stress conditions. *J. Chem. Technol. Biotechnol.* **2020**, *95*, 932–939. [CrossRef]
8. Arora, N.K.; Fatima, T.; Mishra, J.; Mishra, I.; Verma, S.; Verma, R.; Verma, M.; Bhattacharya, A.; Verma, P.; Mishra, P. Ha-lo-tolerant plant growth promoting rhizobacteria for improving productivity and remediation of saline soils. *J. Adv. Res.* **2020**, *26*, 69–82. [CrossRef]
9. Tchuisseu Tchakounté, G.V.; Berger, B.; Patz, S.; Becker, M.; Turecková, V.; Novák, O.; Tarkowská, D.; Fankem, H.; Silke, R. The response of maize to inoculation with *Arthrobacter* sp. and *Bacillus* sp. in phosphorus-deficient, salinity-affected soil. *Microorganisms* **2020**, *8*, 1005. [CrossRef] [PubMed]
10. Tudi, M.; Daniel Ruan, H.; Wang, L.; Lyu, J.; Sadler, R.; Connell, D.; Chu, C.; Phung, D.T. Agriculture development, pesticide application and its impact on the environment. *Int. J. Environ. Res. Public Health* **2021**, *18*, 1112. [CrossRef] [PubMed]
11. De Souza, R.M.; Seibert, D.; Quesada, H.B.; de Jesus Bassetti, F.; Fagundes-Klen, M.R.; Bergamasco, R. Occurrence, impacts and general aspects of pesticides in surface water: A review. *Process Saf. Environ. Prot.* **2020**, *135*, 22–37. [CrossRef]
12. Rumschlag, S.L.; Mahon, M.B.; Hoverman, J.T.; Raffel, T.R.; Carrick, H.J.; Hudson, P.J.; Rohr, J.R. Consistent effects of pesticides on community structure and ecosystem function in freshwater systems. *Nat. Commun.* **2020**, *11*, 6333. [CrossRef] [PubMed]
13. Espinoza-Navarro, O.; Ponce-LaRosa, C.; Bustos-Obregón, E. Organophosphorous Pesticides: Their Effects on Biosentinel Species and Humans. Control and Application in Chile. *Int. J. Morphol.* **2017**, *35*, 1069–1074. [CrossRef]
14. Bertrand, L.; Marinob, D.J.; Monferránc, M.V.; Améa, M.V. Can a low concentration of an organophosphate insecticide cause negative effects on an aquatic macrophyte? Exposure of *Potamogeton pusillus* at environmentally relevant chlorpyrifos concentrations. *Environ. Exp. Bot.* **2017**, *198*, 139–147. [CrossRef]
15. Pehkonen, S.O.; Zhang, Q. The degradation of organophosphorus pesticides in natural waters: A critical review. *Crit. Rev. Environ. Sci. Technol.* **2002**, *32*, 17–72. [CrossRef]
16. Zhang, H.; Yuan, X.; Xiong, T.; Wang, H.; Jiang, L. Bioremediation of co-contaminated soil with heavy metals and pesticides: Influence factors, mechanisms and evaluation methods. *Chem. Eng. J.* **2020**, *398*, 125657. [CrossRef]
17. Elerŝek, T.; Filipič, M. Chapter 12: Organophosphorus Pesticides—Mechanism of Their Toxicity. In *The Impacts of Pesticide Exposure*; Stoytcheva, M., Ed.; IntechOpen Limited: London, UK, 2011; pp. 243–260.
18. Singh, A.; Kumar, A.; Yadav, S.; Singh, I.K. Reactive oxygen species-mediated signaling during abiotic stress. *Plant Gene* **2019**, *18*, 100173. [CrossRef]
19. Anee, T.I.; Nahar, K.; Rahman, A.; Mahmud, J.A.; Bhuiyan, T.F.; Alam, M.U.; Fujita, M.; Hasanuzzaman, M. Oxidative damage and antioxidant defense in *Sesamum indicum* after different waterlogging durations. *Plants* **2019**, *8*, 196. [CrossRef]
20. Nianiou-Obeidat, I.; Madesis, P.; Kissoudis, C.; Voulgari, G.; Chronopoulou, E.; Tsaftaris, A.; Labrou, N.E. Plant glutathione transferase-mediated stress tolerance: Functions and biotechnological applications. *Plant Cell Rep.* **2017**, *36*, 791–805. [CrossRef] [PubMed]
21. Lotfi, N.; Vahdati, K.; Hassani, D.; Kholdebarin, B.; Amiri, R. Peroxidase, guaiacol peroxidase and ascorbate peroxidase activity accumulation in leaves and roots of walnut trees in response to drought stress. *Acta Hortic.* **2010**, *861*, 309–316. [CrossRef]
22. Choudhury, F.K.; Rivero, R.M.; Blumwald, E.; Mittler, R. Reactive oxygen species, abiotic stress and stress combination. *Plant J.* **2017**, *90*, 856–867. [CrossRef] [PubMed]
23. Kawano, T. Roles of the reactive oxygen species-generating peroxidase reactions in plant defense and growth induction. *Plant Cell Rep.* **2003**, *21*, 829–837. [CrossRef]
24. Bela, K.; Horváth, E.; Gallé, Á.; Szabados, L.; Tari, I.; Csiszár, J. Plant glutathione peroxidases: Emerging role of the antioxidant enzymes in plant development and stress responses. *J. Plant Physiol.* **2015**, *176*, 192–201. [CrossRef]

25. Hemeda, H.M.; Klein, B.P. Effects of naturally occurring antioxidants on peroxidase activity of vegetable extracts. *J. Food Sci.* **1990**, *55*, 184–185. [CrossRef]
26. Varga, B.; Janda, T.; László, E.; Veisz, O. Influence of abiotic stresses on the antioxidant enzyme activity of cereals. *Acta Physiol. Plant.* **2012**, *34*, 849–858. [CrossRef]
27. Edwards, R.; Dixon, D.D. Metabolism of Natural and Xenobiotics Substrates by the Plant Glutathione S-Transferase Superfamily. In *Ecological Studies, Molecular Ecotoxicology of Plants*; Sandermann, H., Ed.; Springer-Verlag: Berlin/Heidelberg, Germany, 2004; Volume 170, pp. 17–50.
28. Hasanuzzaman, M.; Nahar, K.; Anee, T.I.; Fujita, M. Glutathione in plants: Biosynthesis and physiological role in environmental stress tolerance. *Physiol. Mol. Biol. Plants* **2017**, *23*, 249–268. [CrossRef]
29. Uarrota, V.G.; Moresco, R.; Schmidt, E.C.; Bouzon, Z.L.; da Costa Nunes, E.; de Oliveira Neubert, E.; Peruch, L.A.M.; Rocha, M.; Maraschin, M. The role of ascorbate peroxidase, guaiacol peroxidase, and polysaccharides in cassava (Manihot esculenta Crantz) roots under postharvest physiological deterioration. *Food Chem.* **2016**, *197*, 737–746. [CrossRef] [PubMed]
30. Yadav, N.R. Toxic Effect of Chlorpyrifos and Dimethoate on Protein and Chlorophyll-a Content of *Spirulina platensis*. *Int. J. Eng. Sci. Adv. Res.* **2015**, *1*, 24–26.
31. Dvořáková Březinová, T.; Vymazal, J. Phenolic compounds in wetland macrophytes. *Sci. Agric. Bohem.* **2018**, *49*, 1–8.
32. Ghasemzadeh, A.; Ghasemzadeh, N. Flavonoids and phenolic acids: Role and biological activity in plants and humans. *J. Med. Plants Res.* **2011**, *5*, 6697–6703.
33. Belščak-Cvitanović, A.; Durgo, K.; Huđek, A.; Bačun-Družina, V.; Komes, D. Overview of Polyphenols and Their Properties. In *Polyphenols: Properties, Recovery, and Applications*; Woodhead Publishing: New York, NY, USA, 2018; pp. 3–44.
34. Solovchenko, A.; Yahia, E.M.; Chen, C. Pigments. In *Postharvest Physiology and Biochemistry of Fruits and Vegetables*; Yahia, E.M., Ed.; Woodhead Publishing (Elsevier): Duxford, UK, 2019; pp. 225–252.
35. Panche, A.N.; Diwan, A.D.; Chandra, S.R. Flavonoids: An overview. *J. Nutr. Sci.* **2016**, *5*, 1–15. [CrossRef] [PubMed]
36. Dias, M.C.; Pinto, D.C.G.A.; Silva, A.M.S. Plant flavonoids: Chemical characteristics and biological activity. *Molecules* **2021**, *26*, 5377. [CrossRef] [PubMed]
37. Sobiecka, E.; Mroczkowska, M.; Olejnik, T.P. The Influence of Chlorpyrifos on the Nonenzymatic Antioxidants Content in Macrophytes Leaves. *Antioxidants* **2022**, *11*, 684. [CrossRef] [PubMed]
38. Sobiecka, E.; Mroczkowska, M.; Olejnik, T.P. The Enzymatic Antioxidants Activities Changes in Water Plants Tissues Exposed to Chlorpyrifos Stress. *Antioxidants* **2022**, *11*, 2104. [CrossRef] [PubMed]
39. Kumar, S.; Pandey, A.K. Chemistry and biological activities of flavonoids: An Overview. *Sci. World J.* **2013**, *2013*, 162750. [CrossRef] [PubMed]
40. Sharma, A.; Shahzad, B.; Rehman, A.; Bhardwaj, R.; Landi, M.; Zheng, B. Response of Phenylpropanoid Pathway and the Role of Polyphenols in Plants under Abiotic Stress. *Molecules* **2019**, *24*, 2452. [CrossRef]
41. Sharma, A.; Thakur, S.; Kumar, V.; Kanwar, M.K.; Kesavan, A.K.; Thukral, A.K.; Bhardwaj, R.; Alam, P.; Ahmad, P. Presowing Seed Treatment with 24-Epibrassinolide Ameliorates Pesticide Stress in *Brassica juncea* L. through the Modulation of Stress Markers. *Front. Plant Sci.* **2016**, *7*, 1569. [CrossRef]
42. Sharma, A.; Kumar, V.; Yuan, H.; Kanwar, M.K.; Bhardwaj, R.; Thukral, A.K.; Zheng, B. Jasmonic Acid Seed Treatment Stimulates Insecticide Detoxification in *Brassica juncea* L. *Front. Plant Sci.* **2018**, *9*, 1609. [CrossRef] [PubMed]
43. Shixiang, G.A.O.; Huiyun, P.; Xiaolu, L.I.; Xiaohua, X.U. Phytotoxicity of four herbicides on *Ceratophyllum demersum, Vallisneria natans* and *Elodea nuttallii*. *J. Environ. Sci.* **2009**, *21*, 307–312.
44. Komives, T.; Gullner, G. Phase I Xenobiotic Metabolic Systems in Plants. *Z. Nat.* **2005**, *60*, 179–185.
45. Tlidjen, S.; Meksem Amara, L.; Bouchlaghem, S.; Sbartai, H.; Djebar, M.R. Oxidative stress in *Elodea canadensis* and *Lemna minor* exposed to Calliofop 36EC. *Glob. J. Biodivers. Sci. Manag.* **2012**, *2*, 29–37.
46. Vighi, I.; Benitez, L.; Amaral, M.; Moraes, G.; Auler, P.; Rodrigues, G.; Deuner, S.; Maia, L.; Braga, E. Functional characterization of the antioxidant enzymes in rice plants exposed to salinity stress. *Biol. Plant.* **2017**, *61*, 540–550. [CrossRef]
47. Gill, S.S.; Tuteja, N. Reactive oxygen species and antioxidant machinery in abiotic stress tolerance in crop plants. *Plant Physiol. Biochem.* **2010**, *48*, 909–930. [CrossRef] [PubMed]
48. Del Rio, L.A.; Corpas, F.J.; Sandalio, L.M.; Palma, J.M.; Gomez, M.; Barroso, J.B. Reactive oxygen species, antioxidant systems and nitric oxide in peroxisomes. *J. Exp. Bot.* **2002**, *372*, 1255–1272. [CrossRef]
49. Lv, T.; Carvalho, P.N.; Zhang, L.; Zhang, Y.; Button, M.; Arias, C.A.; Weber, K.P.; Brix, H. Functionality of microbial communities in constructed wetlands used for pesticide remediation: Influence of system design and sampling strategy. *Water Res.* **2017**, *110*, 241–251. [CrossRef] [PubMed]
50. Yadav, M.; Shukla, A.K.; Srivastva, N.; Upadhyay, S.N.; Dubey, S.K. Utilization of microbial community potential for removal of chlorpyrifos: A review. *Crit. Rev. Biotechnol.* **2016**, *36*, 727–742. [CrossRef] [PubMed]

water

MDPI

Article

A Comprehensive Assessment of the Hydrological Evolution and Habitat Quality of the Xiangjiang River Basin

Fengtian Hong, Wenxian Guo * and Hongxiang Wang

College of Water Resources, North China University of Water Resources and Electric Power, Zhengzhou 450045, China; sci181818@163.com (F.H.); naib1106687@163.com (H.W.)
* Correspondence: guowenxian@ncwu.edu.cn; Tel.: +86-150-3712-1238

Abstract: Human disturbance and climatic factors alter the hydrological state of rivers in many ways and have a degree of negative impact on the quality of watershed habitats; quantifying the impact of both human disturbance and climatic factors on hydrological change can help improve the quality of watershed habitats. Therefore, in this research, an integrated watershed assessment framework is proposed to analyse the watershed from four perspectives: hydrological situation, environmental flows, drivers, and habitat quality. A meteorological streamflow model based on the Long Short-Term Memory (LSTM) model was employed to analyse the hydrological evolution and quantify the influence of the drivers from the perspective of hydrological and environmental flows. The Integrated Valuation of Ecosystem Services and Tradeoffs (InVEST) model was then used to evaluate the spatial and temporal evolution of habitat quality in the basin. And, finally, the grey correlation theory was used to reveal the response of habitat quality to hydrological changes. Studies have shown that annual flow and precipitation are increasing in the Xiangjiang River (XJR) basin, while its annual potential evapotranspiration is decreasing significantly. After 1991, the hydrological conditions of the XJR were highly variable, with the combined rate of change of the most Ecologically Relevant Hydrological Indicators, ERHIs-IHA and ERHIs-EFCs, reaching 26.21% and 121.23%, respectively. Climate change and human disturbance are the main drivers of change for both (with contributions of 60% and 71%, respectively). Between 1990 and 2020, the habitat quality in the basin declined over time (from 0.770 to 0.757), with areas of high habitat value located mainly in mountainous areas and habitat degradation being concentrated in urban areas in the middle and lower reaches, gradually evolving towards areas of high habitat value in the periphery. There is a strong correlation between watershed habitat quality and the ERHIs. The results of the study can provide a scientific basis for maintaining regional ecological security and rational allocation of water resources.

Keywords: ERHIs; LSTM model; InVEST model; habitat quality; comprehensive evaluation

1. Introduction

As the lifeblood of a watershed ecosystem, rivers perform an important ecological service [1]. Although watershed ecosystems have a degree of self-healing capacity, today's climate extremes and frequent human disturbances threaten to break the ceiling of this capacity. According to the Intergovernmental Panel on Climate Change (IPCC)'s Sixth Assessment Report, the global average temperature rose by ca. 1.09 °C between 2011 and 2020, relative to pre-industrial times [2]. Myhre et al. [3] noted that extreme precipitation events are expected to increase in frequency with global warming. Frequent human disturbances are mainly reflected in the changes in land use caused by the construction of water projects, industry, and agricultural development. It has been shown that only 37% of rivers over 1000 km that maintain free flow exist globally [4]. The above climatic extremes and human disturbance are the two main causes of changes in natural river flow patterns. Such unnatural changes can cause significant disruption to essential ecosystem services throughout the basin, with some weakening of basic ecosystem functions including material cycling and energy transfer, as well as negative impacts on basin habitat quality (i.e., the ability of ecosystems to provide conditions suitable for the continued development of individuals and populations) [5]. A prerequisite for ensuring the effective management of water resources in a changing environment is the scientifically sound dissection of a river's hydrological evolution and its drivers, as well as the revealing of the habitat quality's response to hydrological change.

The Indicators of Hydrologic Alteration (IHA) were proposed by Richter et al. [6]. They are recognised as the most comprehensive set of indicators currently available for assessing rivers' hydrologic conditions, not only for systematically characterising flow variability but also for establishing the ecosystem impacts associated with each indicator. Song et al. [7] used the IHA-based Range of Variability Approach (RVA) to analyse data from 32 stations in China, finding that the rivers' hydrological situation in China was moderately variable. And, Gao et al. [8] found a significant reduction in the early fry of "four major domestic fish" species in a study of fish stocks in the Hengyang and Changzhutan river sections of the XJR. Moreover, Richter et al. [9], who proposed the Environmental Flow Components (EFCs), suggested that maintaining adequate flows during dry periods is essential to maintaining suitable river habitats, and that extreme flow events play an important ecological function. Based on the IHA and the EFCs, Gunawardana et al. [10] found that hydropower development within the Srepok River Basin primarily affects decline rates and reversals. The IHA and the EFCs are now widely used globally, but in a review of 171 hydrological indicators, Olden et al. [11] found that there was a sinkhole of information between most indicators. Using the IHA, Smakhtin et al. [12] point out that there is a high autocorrelation between annual minimum multi-day flows with a difference of less than 6%, and that annual maximum multi-day flows exhibit the same characteristics. Principal Components Analysis (PCA) is considered to be an effective method for solving this problem. For the 32 IHA indicators, Cheng et al. [13] used PCA to successfully screen seven ERHIs for the estimation of environmental flow at the outlet of Dongting Lake and found that it could retain the valid information of the IHA well.

In addition, "observation–simulation" comparison is considered to be an important method for quantifying the effects of human disturbance and climatic factors on flow variability [14]. There are two types of models commonly used in streamflow simulation: one is the hydrological model, which contains mainly distributed hydrological models, and the other is the conceptual hydrological model. While the former's simulations are somewhat physical in nature, they are also problematic, with the large observational dataset required and the numerous model parameters raising the threshold for their application. The uncertainty in the model's parameters and the spatial and temporal transformation of the dataset are also problematic; furthermore, some of these models were developed for specific study areas, thus limiting their applicability [15]. The latter model, while having fewer model parameters, also limits its simulation time scales, with most conceptual hydrological simulations stopping at the monthly scale flow level [16]. In response to these problems, a large number of scholars have turned to methods for studying data characteristics, and, with the rapid development of computers in recent years, data-driven models such as artificial neural networks have been sought after by most scholars in the field of hydrology, with the Long Short-Term Memory (LSTM) model being widely used in hydrology. The LSTM models solve the problem of vanishing gradients in traditional machine learning and eliminate long- and short-term dependence in time-series. Therefore, it is more suitable for dealing with long time-series datasets, including simulating flow-variation processes under natural conditions [17]. For example, Fan et al. [18] constructed a meteorological streamflow model of the Poyang Lake basin based on the LSTM model and realised the process of simulating the daily flow under natural conditions. Cao et al. [19] also introduced this method and combined it with ecological flow indicators to quantify the effects of climate change and human activities in terms of ecological water demand.

Human activities and climate change have influenced hydrological evolution and disturbed watershed habitats' quality [20]. In the early days, researchers focused on habitat-specific and biodiversity-based field surveys; however, this method is time-consuming and expensive, suitable only for small-scale surveys, and is difficult to implement at the catchment scale. With the rapid development of geo-information technology, ecological models based on remote sensing techniques have been widely used to assess watershed habitat quality [21]. One of the most established and commonly used models is the Integrated Valuation of Ecosystem Services and Tradeoffs (InVEST) model, developed jointly by Stanford University, the University of Minnesota, the Nature Conservancy, and the Wildlife Fund Society. Its strength lies in establishing a link between suitability and threat for different land-use types and then assessing the distribution and degradation according to the sensitivity of each habitat to the sources of threat. Using this model, Zhang et al. [22] found that land-use changes in the Yangtze River Delta region between 1975 and 2010 led to a significantly reduced habitat quality.

In recent decades, the XJR basin has experienced rapid demographic, economic, industrial, and agricultural development, as well as rapid urban expansion. These changes have not only affected the alteration of river flow regimes in the basin but have also been accompanied by a degradation of habitat quality [23]. For example, urban expansion implies the expansion of urban population, industrial development, etc., which will increase the local water demand and affect the hydrological situation of rivers to some extent. Urban expansion also leads to the loss of surrounding habitats, habitat fragmentation, and habitat quality degradation, which seriously threatens biodiversity and human well-being [24]. Zeng et al. [25] conducted a study on the fish community of the Xiangjiang River. They discovered that human activities have had a significant impact on the fish habitat, resulting in a decline in fish biodiversity and abundance. In addition, the impact of human activities and climate change on hydrology is multifaceted; for instance, in a given month, although the average flow increases, the low flow may lower. Most previous studies have quantified the amount of change in river runoff due to climate change and human activities based on an annual scale, for instance, by applying the Budyko model [26]. And, few studies have quantified the contribution of human activities and climate change to hydrologic change in terms of hydrologic conditions and environmental flows, and few studies have linked them to habitat quality. Therefore, this study proposes an integrated watershed assessment framework to analyse watersheds from four perspectives: hydrological situation, environmental flows, drivers, and habitat quality. This study is divided into four main steps as follows: (1) The most Ecologically Relevant Indicators (ERHIs) were obtained by screening for the IHA and the EFCs, respectively, using Principal Components Analysis (PCA). (2) The reconstruction of flows in their natural state based on LSTM models and the quantification of the effects of climatic factors and human disturbances on the hydrological situation and environmental flows using a separation framework were carried out. (3) Land-use data was used to construct the InVEST model in order to evaluate the spatial and temporal evolutionary characteristics of the watershed habitat's quality. And (4), the relationship between hydrological change and watershed habitat quality through grey correlation theory was revealed. The results of this study may provide a new idea for the hydrological analysis of river basins and are expected to provide a scientific basis for the management of water resources in the XJR and promote the ecological protection of the river basin.

2. Study Area and Data

2.1. Study Area Overview

The XJR (110°50′–114°25′ E, 24°5′–28°25′ N) is located in the hilly region of southeast China and is the largest tributary of the Dongting Lake system, as well as a first-class tributary of the Yangtze River (Figure 1). The watershed has abundant precipitation and a dense network of rivers, with a total length of 856 km on the main stream and an asymmetrical feather pattern of tributaries on both sides of the river. The recharge source of flow is mainly rainfall, which is influenced by its spatial and temporal distribution, and, in its natural state, the flow in the XJR basin is very unevenly distributed within the year.

As a result, a large number of reservoirs have been built on its main tributaries. But, this has also led to significant changes in the natural hydrological situation of the XJR, which has largely affected the habitat and biological abundance of the wetlands of the XJR basin and Dongting Lake.

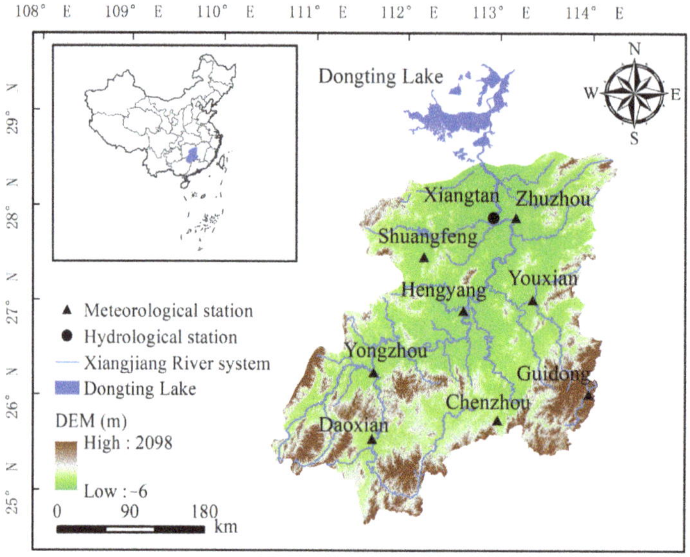

Figure 1. Location of the Xiangjiang River basin and the distribution of meteorological and hydrological stations.

2.2. Data Source and Processing

In this research, meteorological data from eight national meteorological stations in the XJR basin and flow data from the XJR basin's hydrological control station (Xiangtan Station) were selected (Table 1 records the basic information of the relevant stations). In particular, weather station data (including wind speed, relative humidity, temperature, sunshine hours, and precipitation) are provided by the China Meteorological Data Service Center (http://data.cma.cn/, accessed on 16 March 2022). The flow data from the Xiangtan station is obtained from the Yangtze River Basin Hydrological Yearbook. The potential evapotranspiration was calculated for each meteorological station using the Penman–Monteith formula [27]. The Thiessen polygon principle was also used to calculate the precipitation and potential evapotranspiration for the whole basin [28]. In addition, most neural network models require normalized pre-processing of the dataset before simulation. In this paper, the original dataset is pre-processed using a normalization formula to ensure fast and stable convergence of the model, while the normalized output is subjected to a corresponding denormalization operation; the specific steps can be found in [29]. The land-use data used in the InVEST model (1990, 1995, 2000, 2005, 2010, 2015, and 2020) were derived from the Data Center for Resources and Environmental Sciences of the Chinese Academy of Sciences (http://www.resdc.cn, accessed on 16 March 2022).

Table 1. Basic information on hydrological and meteorological stations.

Station	Station Type	Control Area (km^2)	Altitude (m)	Longitude (E)	Latitude (N)
Xiangtan	Hydrological station	81,600.00	63.80	112.93	27.87
Chenzhou		10,601.17	368.80	112.97	25.73
Shuangfeng		11,905.68	100.00	112.17	27.45
Yongzhou		12,826.50	172.60	111.62	26.23
Daoxian	Meteorological station	16,456.87	192.00	111.60	25.53
Zhuzhou		16,444.20	74.60	113.17	27.87
Guidong		6170.81	835.90	113.95	26.00
Hengyang		9593.82	104.90	112.60	26.88
Youxian		10,166.43	115.20	113.35	27.00

Note: The control area of the meteorological stations in the basin is calculated based on the Thiessen principle.

3. Methodology

3.1. Hydrological Variability Determination

The Mann–Kendall trend test determines whether there is a significant trend change in the time-series data. In this study, the Mann–Kendall trend test was used to calculate three series of flow, precipitation, and potential evapotranspiration in the Xiangjiang River basin, and the specific principles can be found in [30]. When the statistic $Z < 0$ indicates that the series shows a decreasing trend, $Z > 0$ indicates that the series shows an increasing trend, and, when $|Z| > 1.96$, it indicates that the series trend passes the 95% significance test. Meanwhile, to analyse the variation of flow series more intuitively, we determined the mutation years of the flow series using the Sliding *t*-test and the Cumulative Anomaly test [31].

3.1.1. Sliding *t*-Test

Sliding *t*-test is widely used in the analysis of hydrological time-series mutability. The method tests for mutation points by examining whether the difference between the means of the two sample groups is significant. For the time-series (x), there are a total of n sample sizes, with a certain moment as the reference point. The samples of the sequence x_1 and x_2 before and after the base point are n_1 and n_2, respectively, with mean $\overline{x_1}$ and $\overline{x_2}$ (in m^3/s) and variance ${s_1}^2$ and ${s_2}^2$ (in m^6/s^2). Then, the statistics are as follows:

$$t = \frac{\overline{x_1} - \overline{x_2}}{s \times \sqrt{\frac{1}{n_1} + \frac{1}{n_2}}} \tag{1}$$

of which

$$s = \sqrt{\frac{n_1 {s_1}^2 + n_2 {s_2}^2}{n_1 + n_2 - 2}} \tag{2}$$

3.1.2. Cumulative Anomaly Test

The principle of the Cumulative Anomaly test is to accumulate the difference between each data point and the mean of the series in order to determine the year of mutation. Due to its simple structure and easy implementation, this method has been widely used in the field of hydrology. For a flow series Q, the cumulative distance level x_t at any moment t is expressed as follows:

$$x_t = \sum_{i=1}^{t} (Q_i - \overline{Q}) \tag{3}$$

where Q_i is the value of the ith time-period of the flow series, $\overline{Q}$ is the mean value of the flow series, and the units of the parameters are m^3/s.

3.2. The Most Ecologically Relevant Hydrological Indicators

3.2.1. The Indicators of Hydrologic Alteration

To better evaluate the hydrological situation of a basin, Richter et al. [32] developed 33 Indicators of Hydrologic Alteration (IHA) in terms of flow, time, frequency, delay, and rate of change. Based on the IHA indicator, Richter proposed a method to quantify the extent of change in the IHA indicator following hydrological disturbance, the Range of Variability Approach (RVA). It is calculated as follows:

$$D_i = \left| \frac{N_{0,i} - N_f}{N_f} \right| \times 100\% \tag{4}$$

$$D_0 = \left(\frac{1}{32}\sum_{i=1}^{32} {D_i}^2\right)^{0.5} \tag{5}$$

where D_i indicates the degree of change in the ith hydrological indicator (when D_i is at 0–33%, it is considered a low degree of change; when D_i is at 33–67%, it is considered a moderate degree of change; and when D_i is at 67–100%, it is considered a high degree of change); $N_{0,i}$ indicates the number of years in which the ith hydrological indicator falls within the RVA target threshold after hydrological variation; N_f indicates the number of years that the IHA value is expected to be at the RVA target threshold after hydrological variation; and D_0 is the degree of change in the combined hydrological indicators.

3.2.2. The Environmental Flow Components

Extreme flows, including high-flow events and low-flow events, are considered necessary to maintain the health of river ecosystems. The Environmental Flow Components (EFCs) are based on this and consist of five flow processes: low flows, extreme low flows, high-flow pulses, small floods, and large floods, with a total of 34 hydrological indicators [33]. The EFCs indicators were evaluated by conducting calculations before and after the disturbances and using coefficients of variation (C). The C_i mainly reflects the degree of variation of the *i*th indicator from the mean value and is calculated as follows:

$$C_i = \left| \frac{S}{m} \right| \tag{6}$$

where S is the standard deviation before and after hydrological variation, and m is the mean value before and after hydrological variation. The overall change in the coefficients of variation of the EFCs indicator is calculated using the principle of calculating the overall degree of change using the weighted average of the IHA indicators above (Equation (2)).

3.2.3. Principal Components Analysis

Principal Components Analysis (PCA) is a statistical method for multivariate analysis, the basic principle of which is to reduce the dimensionality of a large number of relevant variables into a few uncorrelated variables using orthogonal transformations, and to retain as much information as possible [34]. The principles for determining the number of principal components n are the following: (1) a cumulative contribution of 70–90% and (2) an eigenvalue ≥ 1. The Indicators of Hydrologic Alteration and Environmental Flow Components indicators were screened separately using PCA. And, finally, the most Ecologically Relevant Hydrological Indicators-the Indicators of Hydrologic Alteration (ERHIs-IHA) and the most Ecologically Relevant Hydrological Indicators-the Environmental Flow Components (ERHIs-EFCs) were obtained. They are used to evaluate hydrological situation and environmental flows separately.

3.3. The Long Short-Term Memory Model

3.3.1. Model Structure

The LSTM model is a variant of the Artificial Neural Network (ANN) model, whose special design structure (cell state and "gate" structure) allows it to avoid the problem of long-term dependency in long-time sequence prediction. Figure 2 reflects a schematic representation of the operation and structure of the LSTM, the details of which can be found in [35]. The study refers to the basin meteorological flow model proposed by Gauch et al. [36]. In this research, natural flows were reconstructed using measured data (including precipitation, potential evapotranspiration, temperature, sunshine hours, wind speed, relative humidity, etc.) from meteorological stations in the basin as the input. The NSE, R^2, and $RMSE$ were used as evaluation indicators (Equations (4) and (5)).

$$NSE = 1 - \frac{\sum_{i=1}^{n}(Q_{obs,i} - Q_{sim,i})^2}{\sum_{i=1}^{n}\left(Q_{obs,i} - \overline{Q_{obs}}\right)^2};\ RMSE = \sqrt{\frac{\sum_{i=1}^{n}(Q_{obs,i} - Q_{sim,i})^2}{n}} \quad (7)$$

$$R^2 = \left[\frac{\left(\left(Q_{obs,i} - \overline{Q_{obs}}\right)\sum_{i=1}^{n}\left(Q_{sim,i} - \overline{Q_{sim}}\right)\right)^2}{\sum_{i=1}^{n}\left(Q_{obs,i} - \overline{Q_{obs}}\right)^2\sum_{i=1}^{n}\left(Q_{sim,i} - \overline{Q_{sim}}\right)^2}\right] \quad (8)$$

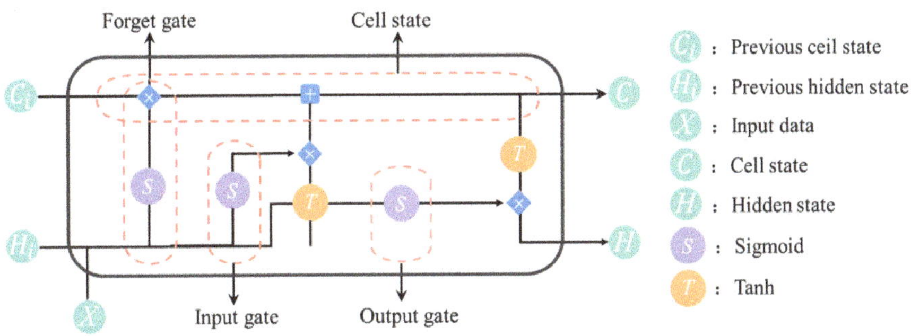

Figure 2. Structure of the Long Short-Term Memory model.

3.3.2. Model Parameters

To better train the LSTM model, the model parameters suitable for the watershed in question were determined. This research refers to the study of Yin et al. [37], which took the 1961–1969 sequence of the XJR basin as the base period and further divided it into a calibration period (1961–1966) and a validation period (1967–1969). The LSTM model contains many important hyperparameters, including hidden size, epoch, dropout rate, batch size, etc., that usually need to be optimised before the model can be learned. In this study, we first determined the hidden size in the hidden layer, and, through training, we found that the model worked best with 150 neurons. As for the epoch, this study first set a longer epoch (200 times), and, after several training sessions, the best fit was found to be after 150 times, thus setting the epoch to 150 times. And, to prevent overfitting problems during the simulation, a discard layer was set up in the study, with a discard probability of 0.4, which meant that there was a 60% probability that the hidden units in this layer would be retained. In addition, it has been shown that the batch size has an impact on the prediction accuracy of the model. Therefore, we set up eight groups of batch sizes (1 d, 5 d, 10 d, 20 d, 40 d, 60 d, 80 d, and 100 d) to train the XJR meteorological streamflow model 15 times, respectively, and used NSE, R^2, and $RMSE$ as the evaluation indicators.

The model's results for the calibration and validation periods are shown in Figure 3, and it can be observed that the simulation works best at a batch size of 10 d.

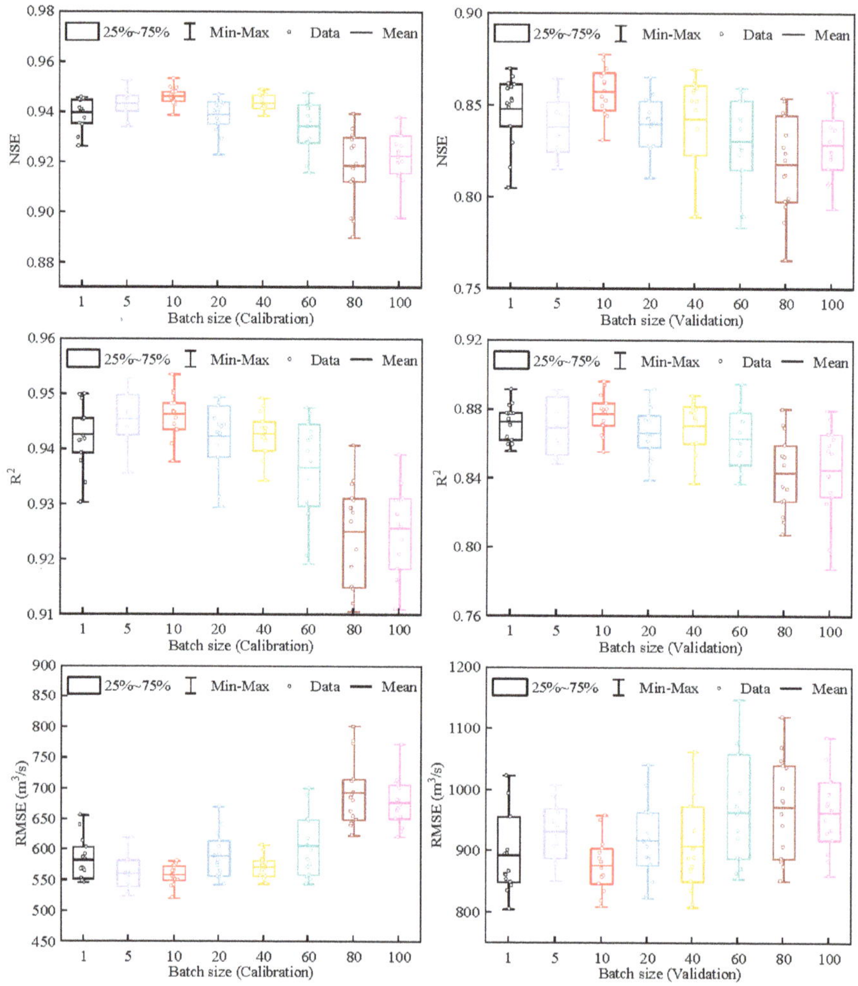

Figure 3. Model performance with different batch sizes (1, 5, 10, 20, 40, 60, 80, and 100).

Ultimately, the main model parameters were set as shown in Table 2. The results of the meteorological streamflow model's reconstructions for the XJR streamflow all follow these parameters. Figure 4 shows the overall effect of the streamflow reconstruction by the model. The observed and simulated streamflow series between 1961 and 2019 reached an R^2 of 0.85 and an *NSE* of 0.84; in the natural period (1961–1990) and in the variation period (1991–2019), the R^2 was 0.88 and 0.82, respectively, and the *NSE* was 0.86 and 0.82, respectively. These results indicate that the LSTM model performs well and can effectively capture the characteristics of streamflow variability.

Table 2. Model's parameter settings.

Type of Parameters	Parameter Name	Setting
Hyper parameters	Dropout rate	40
	Initial learning rate	0.02
	Epoch	150
	Batch size	10
	Layers	5
	Dropout period	40
	Hidden size	150
Common parameters	Training hardware	CPU
	Gradient threshold	1
	Network solving algorithm	adam

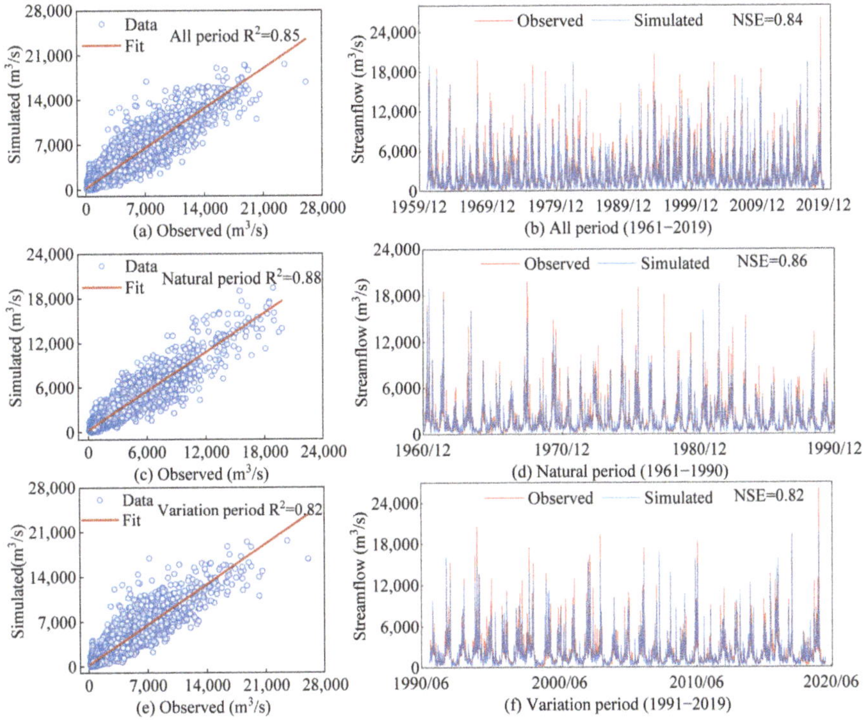

Figure 4. Comparison of observed and simulated flows in the Xiangjiang River basin. The left three panels (**a**,**c**,**e**) show the fitted plots of the measured and simulated flows. The three plots on the right (**b**,**d**,**f**) show the variations of the measured and simulated flow rates.

3.3.3. Separation Framework

Based on the streamflow series simulated by the meteorological streamflow model, an "observation–simulation" comparative analysis was used to quantify the effects of human disturbances and climatic factors on hydrological changes [38]. First, the degree of change (D) for the ERHIs-IHA indicator and the change in the coefficients of variation (C_V) for the ERHIs-EFCs indicator were calculated for the measured and simulated series, respectively. Assuming that human disturbance and climatic effects are independent of each other, the degrees of change (D_{obs}) of the observed flow's hydrological situation and the change in the coefficients of variation (C_{Vobs}) of environmental flows are as follows:

$$D_{obs} = D_h + D_c \tag{9}$$
$$C_{Vobs} = C_{Vc} + C_{Vh} \tag{10}$$

The simulated streamflow is streamflow under the influence of only climate change; thus, the degree of change (D_{sim}) of the simulated streamflow's hydrological situation and the change in the coefficients of variation (C_{Vsim}) of the environmental flow are as follows:

$$D_{sim} = D_c \tag{11}$$
$$C_{Vsim} = C_{Vc} \tag{12}$$

The difference between the D values of the observed and simulated sequences and the difference between the observed and simulated C_V values are as follows:

$$D_h = D_{obs} - D_{sim} \tag{13}$$
$$C_{Vh} = C_{Vobs} - C_{Vsim} \tag{14}$$

The contribution of human disturbances and climatic factors to the hydrological situation and environmental flows, respectively, are the following:

$$\eta_h = \frac{|D_h|}{|D_c| + |D_h|} \times 100\%; \ \eta_c = \frac{|D_c|}{|D_c| + |D_h|} \times 100\% \tag{15}$$

$$\eta_{Vh} = \frac{|C_{Vh}|}{|C_{Vc}| + |C_{Vh}|} \times 100\%; \ \eta_{Vc} = \frac{|C_{Vc}|}{|C_{Vc}| + |C_{Vh}|} \times 100\% \tag{16}$$

In the above equations, D_{obs} and C_{Vobs} indicate the degree of variation in the observed ERHIs-IHA indicators and the change in the coefficient of variation of the ERHIs-EFCs indicators, respectively. D_{sim} and C_{Vsim} indicate the degree of variation in the simulated ERHIs-IHA indicators and the change in the coefficient of variation of the ERHIs-EFCs indicators, respectively. D_h and D_c indicate changes in ecohydrological situation due to human disturbance and climatic factors, respectively. C_{Vh} and C_{Vc} indicate changes in environmental flows due to human disturbance and climatic factors, respectively. η_h and η_c indicate the contribution of human disturbances and climatic factors to changes in hydrological situation, respectively. η_{Vh} and η_{Vc} indicate the contribution of human disturbances and climatic factors to changes in environmental flows, respectively.

3.4. The Integrated Valuation of Ecosystem Services and Tradeoffs Model

The "Habitat Quality" module of the Integrated Valuation of Ecosystem Services and Tradeoffs (InVEST) model was used to assess habitat quality in the XJR basin. This module uses land-use data to reflect the impact of human activity on the environment: the higher the intensity of human activity, the greater the threat to the habitat and the lower the quality of the habitat. Therefore, the relevant parameters (Tables 3 and 4) were set with reference to relevant studies [39].

Table 3. Threat factors and their stress intensity.

Habitat Threat Factors	Maximum Impact Distance (km)	Weight	Recession Correlation
Agricultural land	4	0.6	Linear
Rural land	5	0.6	Exponential
Urban land	10	1.0	Exponential
Industrial mining	12	1.0	Exponential
Reservoir/Pond	6	0.6	Exponential

Table 4. Sensitivity of land-use type to habitat threat factors.

Land-Use Type	Habitat Suitability	Sensitivity				
		Agricultural Land	Rural Land	Urban Land	Industrial Mining	Reservoir/Pond
Agricultural land	0.3	0.0	0.6	0.8	0.8	0.6
Forest land	1.0	0.6	0.4	0.6	0.7	0.5
Grass land	1.0	0.8	0.5	0.4	0.6	0.6
Water body	0.7	0.5	0.3	0.7	0.5	0.7
Built-up land	0.0	0.0	0.0	0.0	0.0	0.4
Unused land	0.4	0.3	0.1	0.1	0.3	0.4

3.5. Grey Correlation Theory

Grey correlation analysis is a method of multi-factor statistical analysis that is mainly used to study the degree of correlation between series. The basic idea is to determine whether sequence curves are closely related according to their geometric similarity. The higher the geometric similarity between the sequence curves, the greater the correlation between the corresponding data sequences, thus achieving a quantitative description of the operation process and evolution of the system [40]. The method is based on uncertain information and can effectively measure the degree of association in order to grasp the main characteristics of things to use; it is widely used in the field of hydrology. This study used a grey correlation model to reveal the correlation between the hydrological changes (hydrological situation and environmental flows) and the habitat quality in the XJR basin.

3.6. Shannon Index

Hydrologic changes often impact aquatic organisms, and the Shannon Index (SI) is often used to reflect the evaluation of watershed biodiversity. Yang et al. (2008) [41] established the best-fit relationship between the IHA metrics and the Shannon Index, which has been widely used for evaluating rivers' biodiversity. In this paper, due to the lack of data on the number of riverine biomes and species in the Xiangjiang River Basin, it is not possible to directly calculate the SI indicators; therefore, by using the relationship equation constructed by Formula (17) for SI and hydrological indicators, it is then possible to initially, roughly assess the biodiversity of the river.

$$SI = \frac{D_{\min}/Q_{\min7} + D_{\min}}{Q_3 + Q_5 + Q_{\min3} + 2 \times Q_{\max3}} + R_{rate} \quad (17)$$

where D_{min} is the Julian date of the annual minimum daily flow; Q_3 and Q_5 are the average monthly flows in March and May, respectively; $Q_{\min3}$ and $Q_{\min7}$ are the annual minimum 3-day flow and the annual minimum 7-day flow; $Q_{\max3}$ is the annual maximum 3-day flow; and R_{rate} is the overflow rate.

4. Results

4.1. Trend and Mutation Analysis

The study conducted a Mann–Kendall trend test for annual flow, annual precipitation, and annual potential evapotranspiration in the XJR basin (Table 5). The statistics (Z) for the annual flow and precipitation were 1.11 and 1.23, respectively, both of which failed the 95% significance level test, also indicating that the annual flow and the annual precipitation in the XJR are on an upward, but not significant, trend. The annual flow's and annual precipitation's rate of rise were 1334 (m^3/s)/Year and 1.58 mm/Year, respectively (Figure 5). In contrast, the annual potential evapotranspiration statistic (Z) was −2.41, which passed the 95% significance level test, indicating a significant decrease in the annual potential evapotranspiration in the XJR basin; its rate of decrease was −1.26 mm/Year. Based on the

Sliding *t*-test and Cumulative Anomaly test used to analyse the mutation years of the XJR flow (Figure 6), the mutation point detected by both tests, together, was 1991. This research thus takes 1991 as the year of sudden hydrological change in the XJR and divides the flow sequence (1961–2019) into a natural period (1961–1990) and a variation period (1991–2019).

Table 5. Results of the test for trends in flow, precipitation, and potential evapotranspiration in the XJR basin.

Study Area	Flow		Precipitation		Potential Evapotranspiration	
	Z	Trend	Z	Trend	Z	Trend
Xiangjiang River basin	1.11	Rise	1.23	Rise	−2.41 *	Decline

Note: * is passing the 95% significance level test.

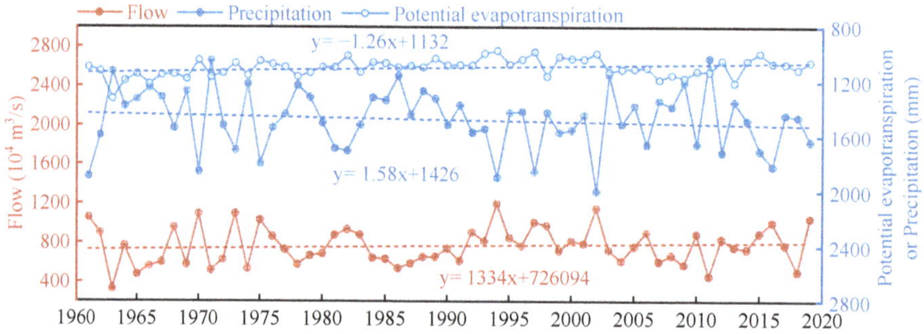

Figure 5. Changes in the inter-annual flow, precipitation, and potential evapotranspiration in the Xiangjiang River basin.

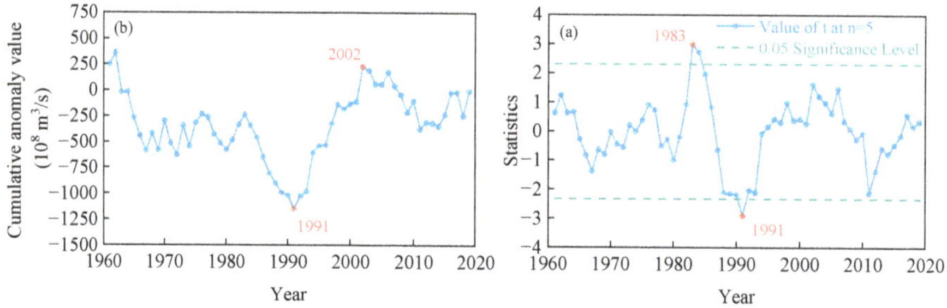

Figure 6. Mutation test for the Xiangjiang River flow (**a**) shows the results of the Sliding *t*-test, and (**b**) shows the results of the Cumulative Anomaly test.

4.2. The Most Ecologically Relevant Hydrological Indicators

4.2.1. Correlation Analysis of Indicators and Selection of ERHIs

The correlation between the 32 IHAs and between the 34 EFCs was analysed based on the Pearson correlation coefficients. And, Figure 7 reflects the high sink residual and strong correlation between both indicators, like between the annual minimum flows (1, 3, 7, 30, and 90 day), between the time of occurrence and the number of major floods, etc. Thus, the ERHIs in this study were selected from the IHAs and the EFCs using PCA. Figure 8a,b show the eigenvalues and cumulative contributions of the IHA and the EFCs at Xiangtan Station, respectively. From Figure 8a, it can be observed that the eigenvalues of the first seven principal components of the IHA indicators for the XJR basin are all greater than one and have a cumulative contribution of about 80%. From Figure 8b, it can be found that the eigenvalues of the first nine principal components of the XJR basin's EFCs indicators are all greater than one, and the cumulative contribution is about 80%. Based on the principle of principal component extraction, the PC1–PC7 in Figure 8a were selected as the main components of the required IHA indicators. And the PC1–PC9 in Figure 8b were selected as the main components of the required EFCs indicators in this study. In addition, a factor loading matrix of the principal components was calculated to further screen the ERHIs, with the criterion that the indicator with the highest or higher absolute value of the loading was used as the ERHIs. A month's flow may be correlated to the previous month's flow, but, as the 12-month average and low flows reflect the intra-year course of human and ecosystem water availability in a given month, we added these indicators to the ERHIs. The final screening of the ERHIs can be seen in Table 6.

Table 6. Changes before and after mutations in the ERHIs-IHA and ERHIs-EFCs in the Xiangjiang River basin.

ERHIs-IHA (label)	Measured average values		Measured thresholds		Degree of change (%)	
	1961–1990	1991–2019	Low	High	Obs	Sim
Mean flow in January (1)	829	1298	297	1361	24.4	5.55
Mean flow in February (2)	1357	1559	680	2034	1.25	3.45
Mean flow in March (3)	2048	2527	964	3131	8.37	6.40
Mean flow in April (4)	3855	3247	2261	5449	3.45	5.96
Mean flow in May (5)	4312	3867	2435	6189	23.15	3.45
Mean flow in June (6)	3788	4323	1940	5636	13.3	24.14
Mean flow in July (7)	2128	2866	9401	3710	3.45	2.30
Mean flow in August (8)	1451	2083	759	2143	11.33	34.17
Mean flow in September (9)	1219	1333	231	2207	11.72	3.44
Mean flow in October (10)	984.5	1066	460	1509	18.97	15.36
Mean flow in November (11)	1113	1223	439	1788	3.448	2.29
Mean flow in December (12)	805.2	1099	403	1379	19.78	9.48
Base flow index (13)	0.17	0.22	0.13	0.21	60.59	8.37
Date of maximum (14)	156.30	182.90	121.20	191.40	6.90	13.79
Low pulse count (15)	5.20	5.03	3.25	7.16	45.55	26.72
High pulse count (16)	6.30	8.24	4.30	8.30	58.62	31.03
Rise rate (17)	363.50	409.60	251.00	476.00	3.45	5.55
Overall degree of change (18)	——	——	——	——	26.21	15.71
ERHIs-IHA (label)	Measured average values		Coefficient of variation		Degree of variation (%)	
	1961–1990	1991–2019	Pre-1991	Post-1991	Obs	Sim
January Low Flow (19)	846	1104	0.47	0.41	11.67	12.58
February Low Flow (20)	1136	1186	0.30	0.33	12.46	21.03
March Low Flow (21)	1416	1641	0.28	0.27	5.15	9.95
April Low Flow (22)	1807	1836	0.20	0.20	0.66	15.76
May Low Flow (23)	1870	1946	0.14	0.14	3.91	0.87
June Low Flow (24)	1675	1942	0.22	0.14	38.03	31.59
July Low Flow (25)	1206	1500	0.31	0.26	14.61	3.968
August Low Flow (26)	1096	1392	0.24	0.34	38.19	48.94
September Low Flow (27)	1054	1152	0.39	0.32	18.45	16.35
October Low Flow (28)	924	926	0.34	0.42	21.48	23.06
November Low Flow (29)	1004	1026	0.45	0.40	10.54	1.70
December Low Flow (30)	808	906	0.43	0.44	3.216	17.91
High-flow peak (31)	4923	4711	0.17	0.24	44.41	13.53
High flow rise rate (32)	760	781	0.29	0.38	31.12	44.88
Small Flood duration (33)	31	39	0.60	0.74	22.63	13.06
Small Flood timing (34)	150	181.6	0.09	0.13	40.15	70.19
Large flood peak (35)	19,230	21,310	0.03	0.15	490.30	90.19
Overall variability (36)	——	——	——	——	121.23	35.07

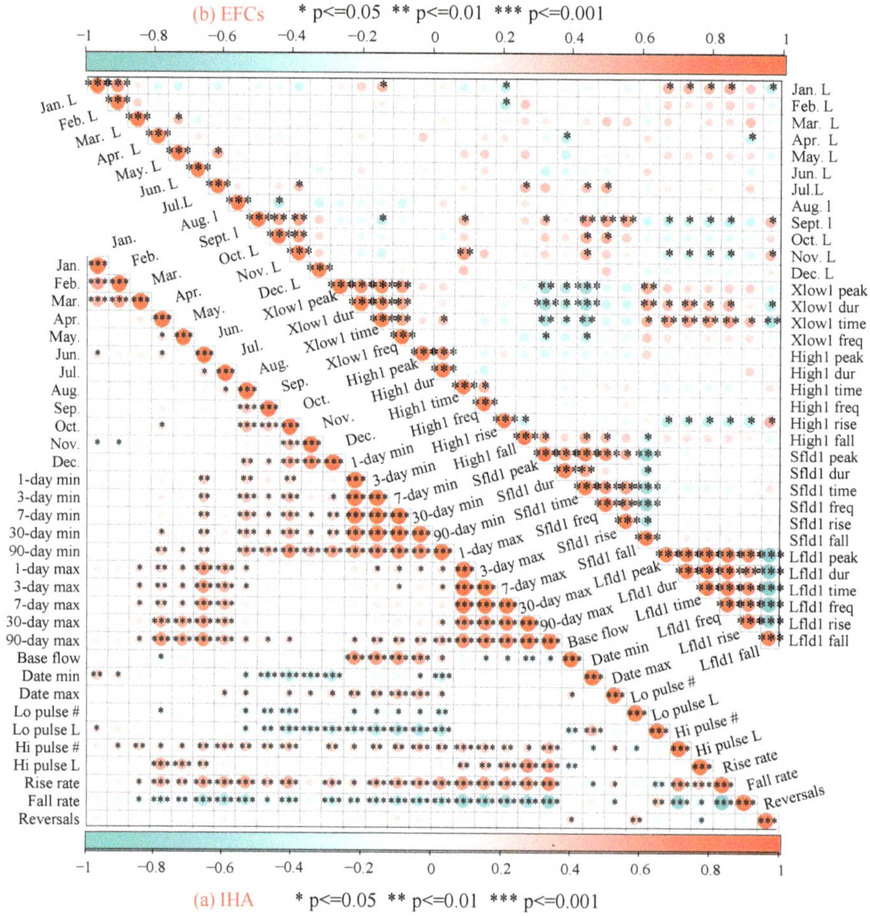

Figure 7. Correlation plots: (**a**) plot represents the correlation between the 32 IHA indicators and (**b**) represents the correlation between the 34 EFCs indicators; the specific values can be found in Appendix A. *p* represents the significant level.

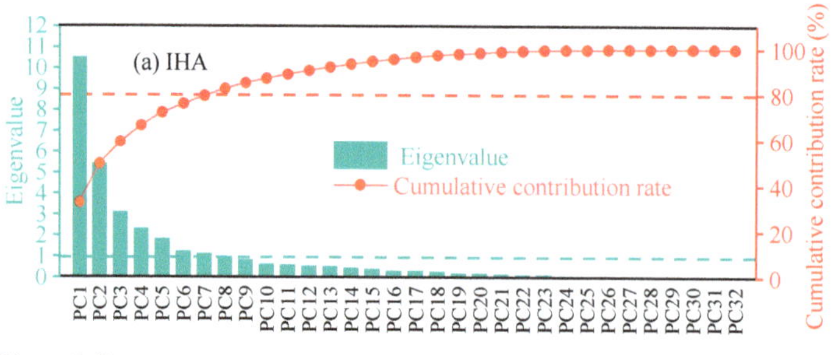

Figure 8. *Cont.*

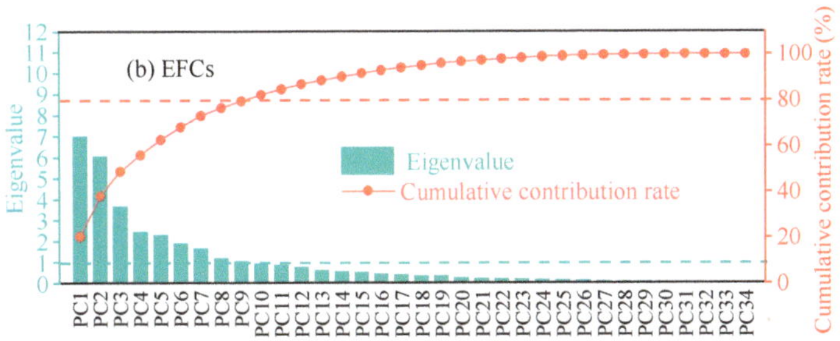

Figure 8. Changes in the characteristic values and cumulative contribution of the Xiangjiang River basin.

4.2.2. Inter-Annual Variation in ERHIs Indicators

The inter-annual variability of the screened ERHIs indicators was studied based on the observed flow series (Figure 9). In terms of the temporal characteristics of the ERHIs-IHAs, only April, May, and October saw a decrease in the mean flow (rates of 20.86 (m^3/s)/year, 11.30 (m^3/s)/year, and 0.20 (m^3/s)/year, respectively). The remaining monthly average flows showed an upward trend, with more pronounced increases in January, March, and July (rates of 11.07 (m^3/s)/year, 12.90 (m^3/s)/year, and 17.17 (m^3/s)/year, respectively). The base flow index showed a slight increase (0.0015); the date of maximum had a delay; and the flow rate's rise rate was increasing at a rate of 0.69 m^3/s/d per year. In addition, the low pulse count and high pulse count showed an increasing trend. For the ERHIs-EFCs, the monthly low flows only showed a decreasing trend in October (at a rate of 1.63 (m^3/s)/year), while remaining monthly low flows showed an increasing trend. The high flow's rise rate was increasing at a rate of 1.41 m^3/s/d per year. The small flood duration was extended and the small flood timing was delayed. In addition, the large flood peak increased significantly at a rate of 69.20 m^3/s per year, but the high-flow peak decreased at a rate of 7.65 m^3/s per year. These two indicator changes indicate that flows in the XJR basin have increased in many ways.

4.3. Ecohydrological Situation, Environmental Flow Evolution, and Quantitative Attribution

A meteorological streamflow model was used to reconstruct the natural flow of the XJR under the influence of climatic factors only and was combined with measured flows to calculate the ERHIs-IHA and the ERHIs-EFCs. Table 6 reflects the changes in the observed flow before and after hydrological variation while giving the degree of change (ERHIs-IHA) and the change in the coefficients of variation (ERHIs-EFCs) calculated based on the observed (obs) and simulated (sim) flow series. From this, it could be seen that the overall degree of change for the ERHIs-IHA and the overall change in the coefficients of variation for the ERHIs-EFCs obtained based on the measured flow were 26.21% and 121.23%, respectively. The overall degree of change for the ERHIs-IHA and the overall change in the coefficients of variation for the ERHIs-EFCs obtained based on the simulated flow were 15.71% and 35.07%, respectively. We attributed changes in the ERHIs in the XJR based on an "observation–simulation" comparison (Figure 10). Climatic factors contributed more to changes in the ecohydrological situation (60%) and human disturbance was the main driver of changes in the environmental flows (71%).

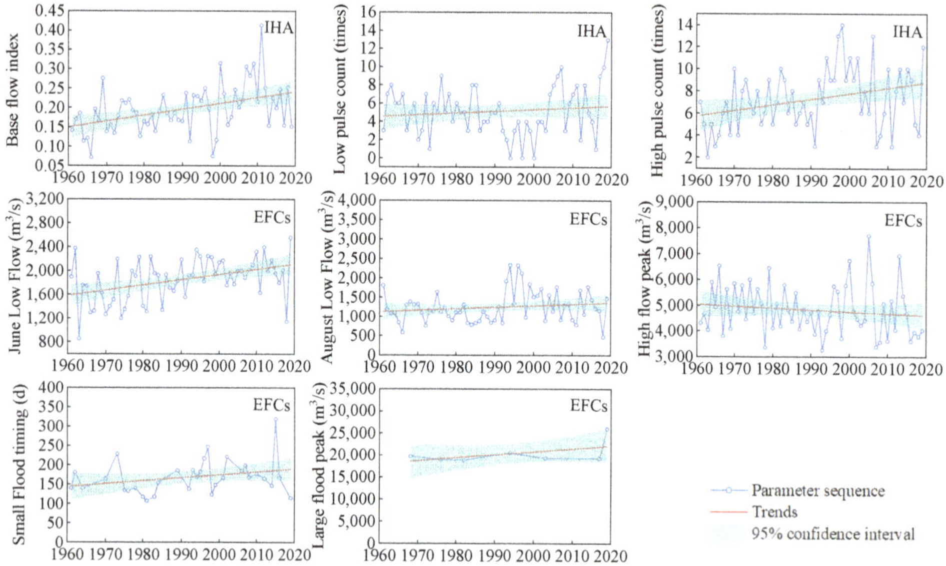

Figure 9. Inter-annual variation in the selected ERHIs (the extent of variation in these ERHIs was greater than 33%).

For the ERHIs-IHA, the base flow index, low pulse count, and high pulse count based on the measured flow all achieved a moderate change, while the three indicators based on the simulated flow showed a low change. Most of the simulated flow's variability was reduced compared to the measured monthly average flow's variability. In addition, the measured changes in the date of maximum and the rise rate were 6.90% and 3.45%, respectively, while the simulated flow's changes increased to 13.79% and 5.55%, respectively. For the ERHIs-EFCs, the variability of monthly dry flows in May–September (except August) and November was lower in the modelled results than in the measured flow's results. The measured variances for the high flow rise rate and the small flood timing were 31.12% and 40.15%, respectively, while the simulated flow's variances were even higher (44.88% and 70.19%). From the attribution results, it can also be seen (Figure 10c) that human disturbances contribute more than 50% to the mean low in January, mean flow in May, mean flow in September, mean flow in December, and base flow index (77%, 85%, 71%, 52%, and 86%, respectively), and that climate factors have a greater impact on the remaining ERHIs-IHA (53–81%). For the ERHIs-EFCs (Figure 10d), human disturbance had a greater impact on the changes in the four indicators of May low flow, July low flow, November low flow, high-flow peak, and large flood peak than on climatic factors (78–87%), while, for the other ERHIs-EFCs indicators, the contribution of climate change was more pronounced (51–94%).

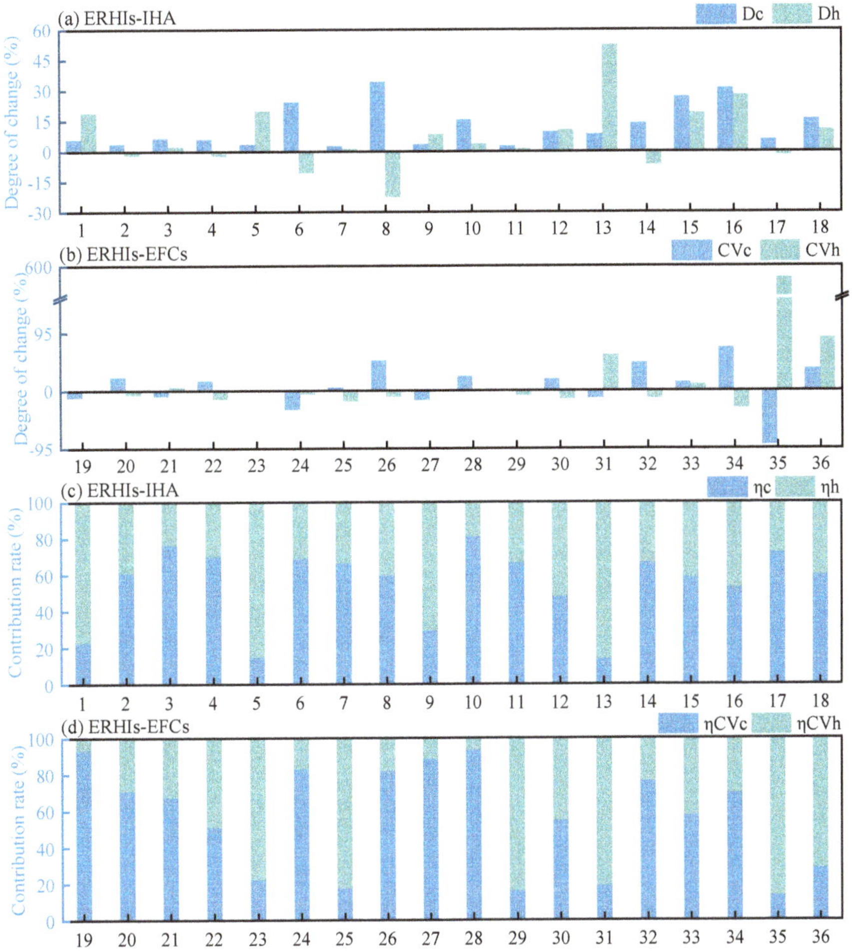

Figure 10. Impact of human disturbance and climatic factors on the ERHIs: (**a**,**b**) represent the degree of changes in the hydrological parameters (parameters 1–36, can refer to Table 6) driven by human activities and climate change; and (**c**,**d**) represent the contribution rates of human activities and climate change to changes in the hydrological parameters.

4.4. Habitat Quality Assessment and Its Response to Hydrological Change

The results of habitat degradation and habitat quality distribution in the XJR basin were obtained based on the InVEST model. Figure 11 reflects the spatial distribution of the habitat's degeneration degree in the XJR basin between 1990 and 2020. The high-degradation areas were concentrated in the middle and lower reaches of the basin, and the high-degradation areas spread to the surrounding areas over time, especially in the cities of Changsha, Xiangtan, and Zhuzhou, during which the average degradation of the basin increased from 0.0159 to 0.0181 (an increase of 13.84%). The maximum degradation increased from 0.1137 to 0.1190 (an increase of 4.66%); the low values of degradation were concentrated in the mountainous areas of the upper part of the basin. The spatial distribution of habitat quality in the XJR basin from 1990 to 2020 (Figure 12) shows that the areas with low habitat quality are also located in the urban areas in the middle and lower reaches of the basin, especially in the axis around Changsha and between Xiangtan and Zhuzhou, and in the Hengyang area in the middle reaches of the basin, while the habitat quality is higher in the Luoxiao Mountains, the Baku Lian Jiu Mountains, the Yangming Mountains, and the Nanling Mountains. In terms of temporal changes, the habitat quality in the XJR basin was on a declining trend between 1990 and 2020, with the average level of habitat quality in the basin decreasing from 0.7698 to 0.7569 (a decrease of 1.68%) during this period. The habitat quality decreased by 0.0004 (0.05%), 0.0007 (0.09%), 0.0015 (0.20%), 0.0009 (0.12%), 0.0041 (0.54%), and 0.0053 (0.70%) in the six periods between 1990 and 2020, respectively, while the area of low habitat quality in the basin gradually expanded to the periphery.

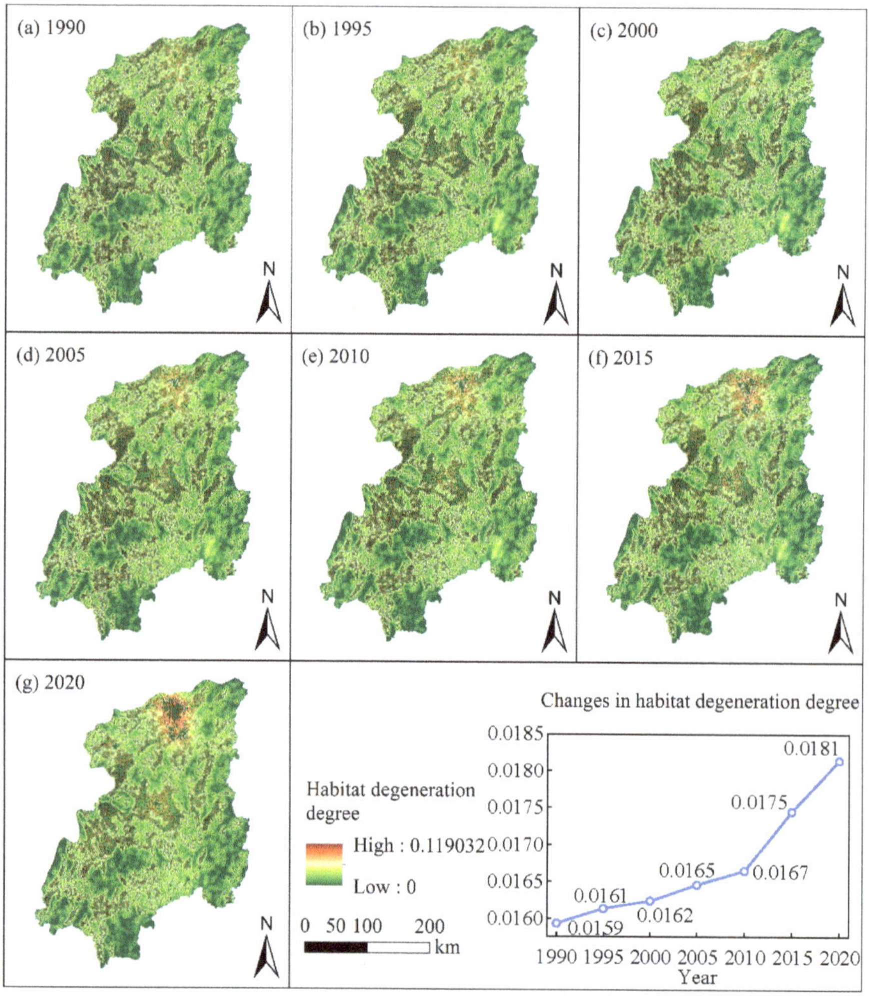

Figure 11. Distribution of habitat degeneration degree in the Xiangjiang River basin, 1990–2020.

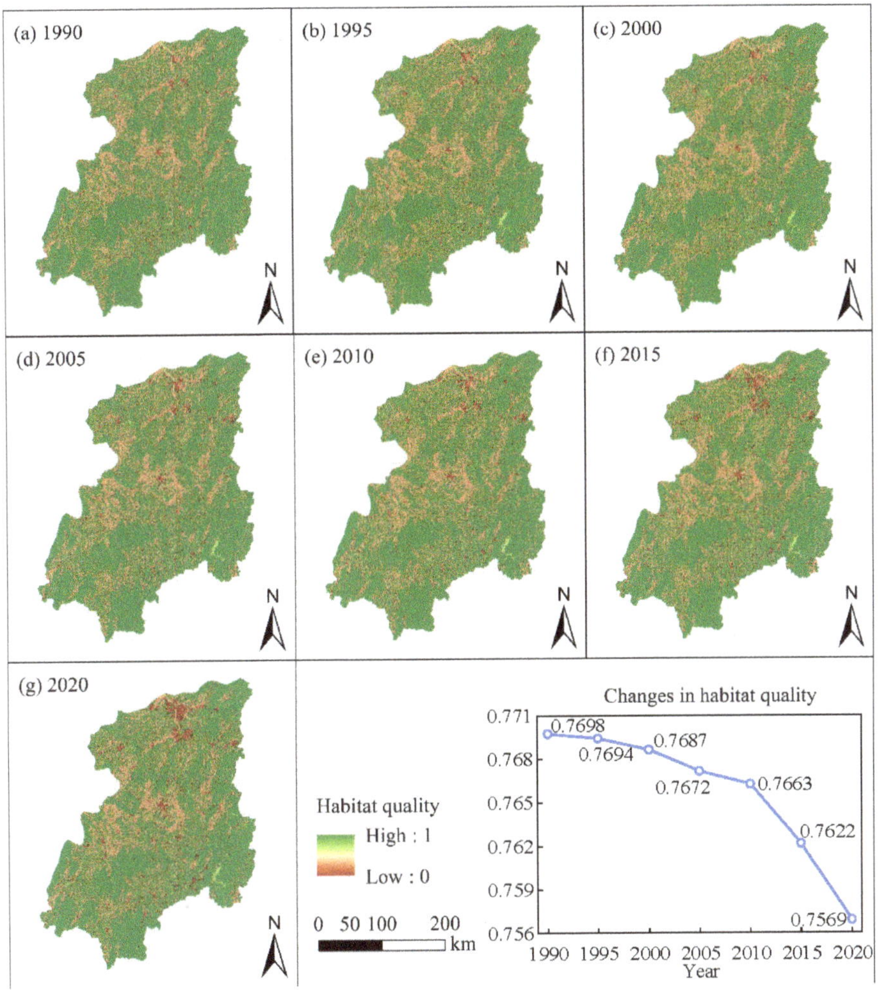

Figure 12. Spatial distribution of habitat quality in the Xiangjiang River basin, 1990–2020.

A grey correlation model was constructed from the time-series of habitat degeneration degree, habitat quality, and ERHIs in the XJR basin to quantitatively reveal the impact of hydrological changes on habitat quality (Figure 13). The habitat quality index had a strong correlation with each of the ERHIs (all the correlations were greater than 0.6). Of these, the ERHIs-EFCs were more highly correlated with the habitat's quality (habitat degeneration degree) than the ERHIs-IHA, overall. For both the habitat quality and habitat degeneration degree, the ERHIs with correlations greater than 0.90 were mean flow in August, March low flow, May low flow, August low flow, and large flood peak. It is also clear from this that habitat quality (habitat degeneration degree) responds more strongly to changes in environmental flows.

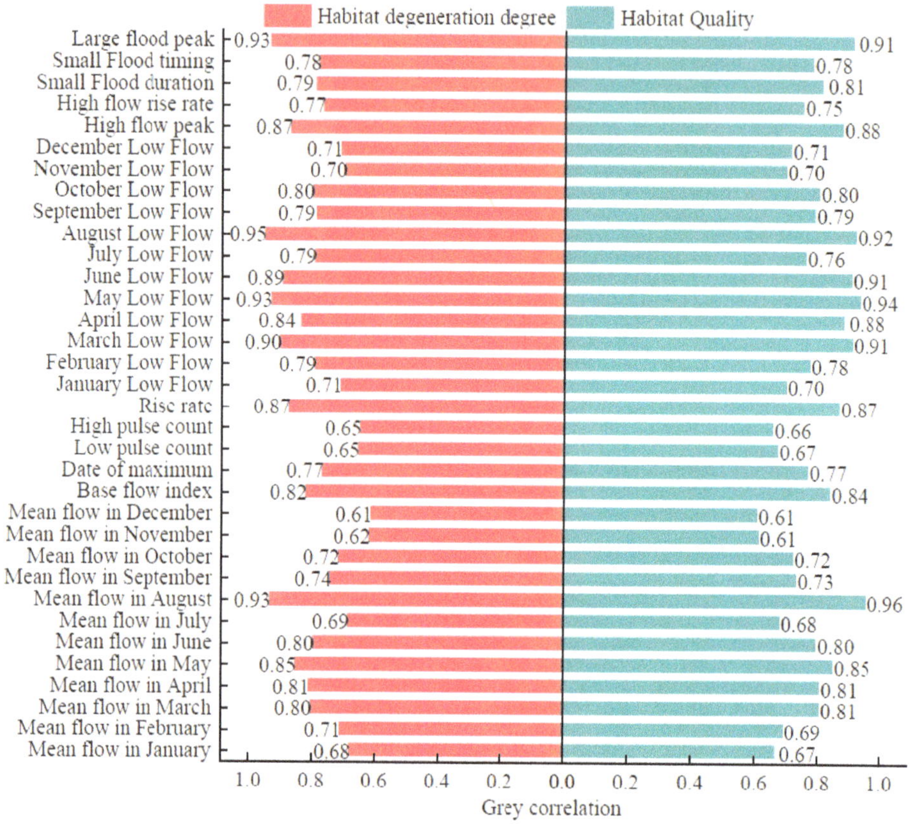

Figure 13. Grey correlation between ERHIs and habitat quality (habitat degeneration degree) in the Xiangjiang River.

4.5. Riverine Biological Conditions

The study adopted widely used SI indicators to evaluate the river biodiversity of the XJR (Figure 14a) and collected the number of XJR's four major domestic fish after hydrological variability (Figure 14b) [42,43]. It can be found that, before the hydrological mutation, the decline rate of the SI index in Xiangjiang River was −1.05/year; however, after the mutation, the decline rate of the SI index reached −1.45/year. At the same time, we found that the number of the four major domestic fish in the XJR basin also showed a decreasing trend after the hydrological variability, with a decreasing rate of 2703 fish/year. In addition, from the correlation between the catch of tetras and the ERHIs, we found that several ERHIs metrics showed a strong correlation (greater than 0.65) with them. These indicators were large flood peak, small flood duration, October low flow, June low flow, low pulse count, and mean flow in October.

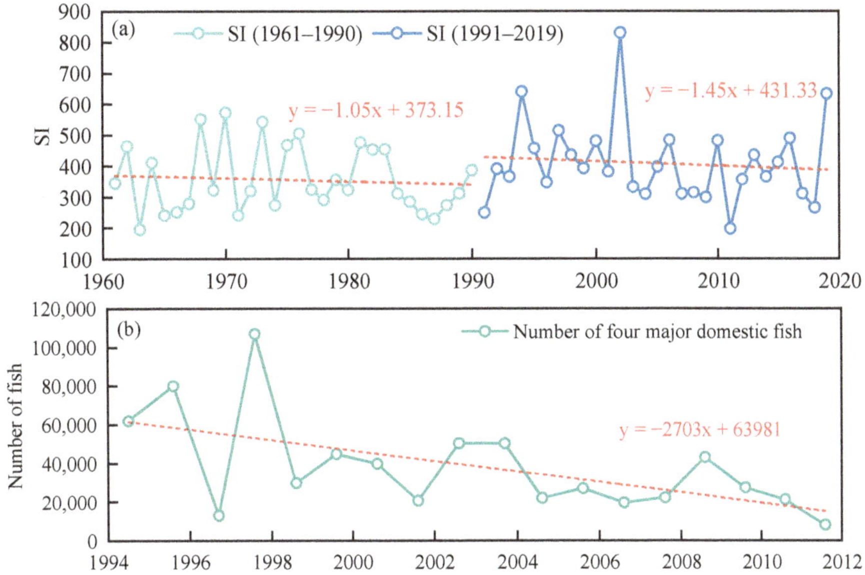

Figure 14. (**a**) Changes in the Shannon Index and (**b**) Situation of four major domestic fish in the Xiangjiang River.

5. Discussion

Rivers are among the most complex natural ecosystems, and fluctuations in flow in their natural state are considered critical for maintaining the health of watershed ecosystems [44]. However, research found that global environmental change (including climate and land-use change) has profoundly altered the flow patterns of most rivers worldwide. River ecosystems are sensitive to changes in flow [45]. Palmer and Ruhi [46] found that, when flow changes occur, they not only reduce primary productivity but also affect material cycling, ultimately leading to ecological degradation. The XJR, as the mother river of the Hunan Province as well as the largest river in the Dongting Lake system, not only feeds 60% of the provincial population but is also the habitat of many rare plants and animals [47]. This study found that the annual potential evapotranspiration in the XJR basin declined significantly, the precipitation and the flow were on an upward trend in terms of inter-annual variability, and the annual flow changed abruptly in 1991. Based on the measured daily flow of the XJR, the study used PCA to screen Indicators of Hydrologic Alteration and Environmental Flow Components, effectively avoiding information sink between indicators, and finally obtained 34 ERHIs.

It was found that the vast majority of the ERHIs-IHA showed an increasing trend after hydrological variation. After 1991, the average monthly flow increased more markedly in winter and decreased most markedly in spring (except March), which is inextricably linked to the human regulation of river water allocation. The XJR flows significantly less during the flood season, which represents the onset of flow homogenisation, with a reduced flow during the high-flow months and increased in flow during the low-flow months. Tonkin et al. [48] found that this phenomenon can lead to a reduction in organic matter in downstream floodplains, while potentially altering the adaptive range of vegetation habitats and reducing habitat quality. The degree of change in the base flow index, the low pulse count, and the high pulse count were 60.59%, 45.55%, and 58.62, respectively, all reaching a moderate change. And, the increase in the base flow index (from 0.17 to 0.22) had a negative impact on drifting eggs [49]. High and low pulses play an important role in maintaining the compatibility of organisms' life cycles, and their variation can directly or indirectly affect the population dynamics and structure of aquatic communities in river and floodplain systems [50]. Khatun and Pal [51] found that dam construction in the Tangon River basin resulted in reduced flow velocities in the lower river and a significant reduction in suitable fish habitats. Only two of the ERHIs-EFCs showed a decreasing trend (October low flow and high flow peak), with the high flow peak variation reaching 44.41%, the main reason for this phenomenon being the storage of water in the middle and upper reaches of the reservoir complex during the flood season. After the mutation, most of the monthly low flows were replenished, which effectively restored a certain level of dissolved oxygen and water temperature in the river, which could then provide basic living conditions for organisms [52]. After 1991, the small flood duration was extended from 30.54 d to 38.76 d. The small flood timing was also delayed from late spring to early summer. And, in a study of flooding by Bailly et al. [53], they found that prolonged small flood events may be beneficial in preventing encroachment of riparian vegetation on the river and that their duration is important for the reproduction and survival of fish. These findings were corroborated by our results, where the decline in the SI indicator intensified after hydrological variability, with the rate of decline decreasing from -1.05/year to a rate that would be -1.45/year. Four major domestic fish also showed a downward trend in their abundance (2703 fish/year). In conclusion, the operation of the upper and middle reaches of the XJR reservoir complex changed the hydrological situation and environmental flow composition in the basin, and the spatial and temporal evolution of hydrological processes in their natural state was disrupted. These changes negatively impacted the XJR's four major domestic fish. Therefore, this study developed a quantitative attribution study for both.

In the study, the human contribution to the change in mean flow in January was found to be 77%, but, for the change in January low flow, the human contribution was only 7%. This reasonably reflects the role of human activities aimed at recharging the dry season, as well as the fact that climatic elements are the primary cause of January low flow. This conclusion is supported by the findings of Guo et al. [54] that the role of human activities in the XJR flow is mainly reflected in the recharge of the downstream flow during the dry season and the release of water during the flood season. For changes in the other ERHIs, it is mostly climatic factors that contribute more. Most annual maximum floods in the XJR occur in April–July, when floods are primarily formed by cyclonic frontal storms, and their spatial and temporal variation characteristics are similar to those of heavy rainfall. Flood regulation efforts in the middle and upper reservoirs have led to a reduction in high flows. Best [55] also found that dam operation reduces high flows and affects fish diversity and floodplain area. The attribution results for the high-flow peak explain this phenomenon well (human interference contributing 81%). However, an increasing trend was observed for the large flood peak, which increased from 19,230 m^3/s before hydrological variation to 21,310 m^3/s after hydrological variation, with a variation of 490.30%. The increase in the frequency of extreme weather occurrences following the sudden change is the main reason for the dramatic increase in the magnitude of major floods in the XJR. Especially in July 2019, when a mega-flood exceeding the 1-in-50-year event occurred in the lower reaches of the XJR, from Hengshan to Xiangtan, including the 1-in-200-year event at Xiangtan station, with a peak flood flow of 26,060 m^3/s, far exceeding the historical measured maximum flow (20,600 m^3/s on 18 June 1994).

Human activities affect not only the hydrological changes in rivers, but also the quality of watershed habitats. The habitat quality in the XJR basin was on a declining trend between 1990 and 2020, but was overall at a high level (above 0.75). This is because the XJR basin is mainly composed of forest land (62.55–63.29% of the whole basin), with only (2.89–5.23%) of both building land and water bodies. As a result, areas of high value for habitat quality in the catchment dominate. However, the area of water bodies in the basin has increased from 1312 km^2 to 1528 km^2 (an expansion of 16.46%), and built-up land has expanded from 1411 km^2 to 3400 km^2 (an expansion of 140.96%), both of which have had a negative impact on the quality of habitats in the basin, where the expansion of water bodies also reflects the increased construction and operation of reservoirs in the basin. The construction and operation of reservoirs in the middle and upper reaches not only influences changes in river hydrology, but also continuously erodes habitats and causes the continuous fragmentation of habitats. In a study of Amazonian forests, Benchimol and Peres [56] found that man-made dams can reduce local habitat quality and adversely affect biodiversity. The expansion of construction land highlights the rapid urbanisation of the middle and lower reaches of the XJR, leading to the gradual expansion of the areas of lower habitat quality around the built-up areas of the city to the periphery, swallowing up the surrounding areas of higher habitat quality. Berta et al. [57], in their study of the Winike Watershed, found that urbanization poses a threat to and continues to degrade habitat quality. Urbanisation has been accompanied by population build-up, industrial and agricultural development, and a dramatic increase in water consumption in all sectors, putting enormous pressure on the local water-transfer sector, as well as compressing ecological water use and having a greater impact on local habitat quality. The correlation between habitat quality (habitat degeneration degree) and the ERHIs corroborates this finding, and the effect of extreme flow events on habitat quality is greater.

There are also some limitations and potential uncertainties in this study. The LSTM model on which the separation framework in the research is based is a black-box model, and its simulation results are a mapping of independent variables to dependent variables, which has some mechanistic resolution but is weak compared to traditional hydrological models [58]. Despite this, its excellent simulation performance and lightweight computation are still loved by many hydrological researchers. The results of Fan et al. (2020) [59] also demonstrate the reliability of the LSTM model in river flow simulations. Meanwhile, in the past two years, researchers have used LSTM models to construct meteorological flow models to reconstruct natural daily flows, and quantitative attribution studies have been conducted with relevant ecohydrological indicators [60]. This all indicates that the output of the model can be applied to hydrological analysis. In addition, LSTM models are subject to uncertainties, e.g., the model structure, parameter values, and input data can lead to model uncertainty and can affect the simulated flow; model uncertainties are inevitable and we have tried our best to optimize our model. For example, the XJR flow mainly originates from precipitation, so a wider range of influencing factors (including precipitation, temperature, wind speed, sunshine hours, etc.) are adopted in the paper to simulate the XJR flow in order to reduce the impact of model uncertainties on the output results. Also, we introduced relevant evaluation metrics (NSE, $RMSE$, and R^2) to evaluate the simulation effectiveness of the model, and the results also indicate that the constructed meteorological flow model is useful in this basin and can be used for the analysis of river ecohydrology. In addition, the final results are basically consistent with the results of previous studies based on hydrological models [61], which indicates that, although there are some uncertainties and limitations in the models, they have little influence on the results of this article, and the conclusions of this article are scientific and reliable [62].

In summary, the XJR basin is undergoing major ecohydrological changes and the habitat's quality is deteriorating, which poses a huge challenge to local water resources regulation and ecological conservation. The core of the problem is how to restore the natural flow regime of the river as far as possible while ensuring normal water use for human production and living. This study has analysed the hydrological situation and environmental flows, and the quantitative attribution results, habitat quality, and response to ERHIs can provide a scientific basis for managerial decisions. In addition, the LSTM model is more applicable because of its excellent simulation performance, smaller computational effort, and less data required; meanwhile, the data of the InVEST model are easily available. So, the comprehensive evaluation framework constructed in this study can be easily used to evaluate other watersheds, comprehensively analyse changes in watershed hydrology and habitat quality, and provide a scientific basis for the rational allocation of water resources and ecological restoration in the watershed.

6. Conclusions

This study presents a comprehensive assessment of the XJR basin in terms of four aspects: hydrological situation, environmental flows, drivers, and habitat quality. The annual precipitation and annual flow in the XJR basin showed a non-significant upward trend, but the annual potential evapotranspiration showed a significant downward trend. It was found that human disturbances and frequent climate extremes are inevitably altering the natural flow. Most of the ERHIs showed varying degrees of upward trend after hydrological variation (1991). The overall degree of change of the ERHIs-IHA and the overall change in the coefficients of variation of the ERHIs-EFCs obtained from the measured data reached 26.21% and 121.23%, respectively. And, it was found that the ecohydrological situation of the XJR basin is deteriorating, which has had adverse effects on river organisms. Quantitative attribution results indicated that climatic factors are the main factor influencing the ecohydrological evolution of the XJR, but the role of human disturbance cannot be ignored. The habitat quality in the basin decreased over the period 1990–2020, with high values mainly in mountainous areas and low values mostly in urban areas in the middle and lower reaches, gradually expanding to the periphery. The habitat quality (habitat degeneration degree) in the XJR basin has a strong correlation with each ERHIs and responds more strongly to changes in environmental flows.

The impact of human disturbance and climatic factors on watersheds is a complex issue, and the integrated evaluation framework in this paper may provide new insights for other researchers on watershed analysis. The results of the study show that the ecohydrological situation of the XJR has been significantly disturbed and that the quality of the watershed habitat has been damaged. In order to better maintain the natural hydrological evolution of the XJR and improve the quality of the river basin habitats, we propose the following recommendations. Policy makers need to enhance the basin water allocation capacity, which can be improved by constructing a basin-wide scientific water resource regulation and control system, at the same time accelerating the construction of a water-saving society and improving the efficiency of water resources' utilization. The relevant water-related departments can also strengthen their joint efforts and properly simulate natural flow events through artificial regulation, especially for environmental flow events.

Author Contributions: Conceptualization, F.H.; data curation, F.H.; formal analysis, F.H.; methodology, F.H.; supervision, W.G., F.H. and H.W.; validation, F.H.; writing original draft, F.H.; writing, review, and editing, F.H. All authors have read and agreed to the published version of the manuscript.

Funding: This work was supported by the National Nature Science Foundation of China (Grant No. 51779094); the Wisdom Introduction Project of Henan Province (GH2019032); the 2023 Special Fundamental Research Project Plan for Higher Education Institutions in the Henan Province (23ZX012); and the North China University of Water Resources and Electric Power for the Master's Innovation Capacity Enhancement Project (NCWUYC-2023025).

Data Availability Statement: Data will be made available on request.

Conflicts of Interest: The authors declare no conflict of interest.

Appendix A. Correlation of Hydrological Parameters

Label	Jan.	Feb.	Mar.	Apr.	May.	Jun.	Jul.	Aug.	Sep.	Oct.	Nov.	Dec.	1d-Min	3d-Min	7d-Min	30d-Min	90d-Min	1d-Max	3d-Max	7d-Max	30d-Max	90d-Max	Base index	Date min	Date max	Low-C	Low-D	High-C	High-D	Rise rate	Fall rate	Reversals
Jan.	1.00	0.63	0.52	0.14	0.11	0.30	0.01	-0.12	-0.17	-0.10	-0.29	-0.23	0.10	0.10	0.12	0.16	-0.03	0.22	0.19	0.15	0.11	0.22	-0.08	0.39	0.02	-0.11	0.26	0.24	0.14	0.16	-0.23	0.19
Feb.	0.63	1.00	0.64	0.15	0.05	0.24	0.00	-0.10	-0.06	-0.09	-0.28	-0.15	0.05	0.06	0.04	0.07	0.02	0.22	0.18	0.14	0.15	0.21	-0.18	0.26	-0.09	-0.13	0.24	0.31	0.15	0.21	-0.23	0.04
Mar.	0.52	0.64	1.00	0.25	0.04	0.24	0.08	-0.04	0.04	-0.10	-0.23	-0.16	0.10	0.11	0.12	0.15	0.03	0.32	0.31	0.30	0.23	0.30	-0.14	0.19	-0.10	-0.12	0.14	0.35	0.12	0.27	-0.32	0.09
Apr.	0.14	0.15	0.25	1.00	0.30	0.30	0.12	0.09	0.25	0.28	-0.04	0.02	0.21	0.25	0.23	0.31	0.36	0.34	0.33	0.40	0.55	0.67	-0.27	-0.09	-0.09	-0.28	-0.14	0.29	0.57	0.49	-0.44	-0.16
May.	0.11	0.05	0.04	0.30	1.00	0.15	0.00	0.12	0.07	0.18	0.08	0.17	0.09	0.19	0.18	0.17	0.27	0.32	0.32	0.33	0.54	0.62	-0.22	-0.17	-0.10	-0.19	-0.07	0.26	0.46	0.38	-0.34	-0.08
Jun.	0.30	0.24	0.24	0.30	0.15	1.00	0.32	0.20	0.22	0.09	-0.04	0.08	0.41	0.40	0.39	0.36	0.34	0.68	0.68	0.68	0.62	0.70	-0.11	0.00	0.16	-0.10	-0.10	0.41	0.42	0.59	-0.62	-0.01
Jul.	0.01	0.00	0.08	0.12	0.00	0.32	1.00	0.32	0.04	0.19	0.04	0.14	0.08	0.16	0.22	0.24	0.15	0.52	0.52	0.52	0.47	0.45	-0.16	-0.07	0.32	0.06	-0.16	0.42	0.19	0.56	-0.60	0.21
Aug.	-0.12	-0.10	-0.04	0.09	0.12	0.20	0.32	1.00	0.44	0.53	0.20	0.38	0.36	0.43	0.48	0.44	0.51	0.28	0.27	0.21	0.21	0.32	0.08	-0.33	0.27	-0.32	-0.32	0.42	0.13	0.61	-0.54	-0.04
Sep.	-0.17	-0.06	0.04	0.25	0.07	0.22	0.04	0.44	1.00	0.42	0.18	0.31	0.24	0.27	0.26	0.29	0.46	0.15	0.16	0.15	0.15	0.27	-0.05	-0.46	0.07	-0.37	-0.35	0.35	0.19	0.40	-0.33	-0.14
Oct.	-0.10	-0.09	-0.10	0.28	0.18	0.09	0.19	0.53	0.42	1.00	0.47	0.57	0.36	0.45	0.50	0.53	0.74	0.10	0.10	0.07	0.14	0.28	0.12	-0.56	0.31	-0.44	-0.50	0.42	0.12	0.58	-0.45	-0.12
Nov.	-0.29	-0.28	-0.23	-0.04	0.08	-0.04	0.04	0.20	0.18	0.47	1.00	0.80	0.19	0.23	0.26	0.28	0.48	-0.02	-0.01	0.01	0.06	0.02	0.07	-0.44	0.28	-0.14	-0.42	0.15	-0.03	0.25	-0.24	-0.02
Dec.	-0.23	-0.15	-0.16	0.02	0.17	0.08	0.14	0.38	0.31	0.57	0.80	1.00	0.22	0.32	0.35	0.38	0.51	0.14	0.14	0.13	0.14	0.17	0.02	-0.56	0.33	-0.23	-0.49	0.39	-0.05	0.43	-0.47	-0.02
1d-Min	0.10	0.05	0.10	0.21	0.09	0.41	0.08	0.36	0.24	0.36	0.19	0.22	1.00	0.97	0.92	0.86	0.66	0.23	0.20	0.18	0.14	0.26	0.67	-0.17	0.41	-0.29	-0.47	0.30	0.06	0.33	-0.35	0.13
3d-Min	0.10	0.06	0.11	0.25	0.19	0.40	0.16	0.43	0.27	0.45	0.23	0.32	0.97	1.00	0.98	0.91	0.74	0.27	0.24	0.22	0.22	0.36	0.66	-0.23	0.40	-0.31	-0.54	0.35	0.11	0.45	-0.46	0.15
7d-Min	0.12	0.04	0.12	0.23	0.18	0.39	0.22	0.48	0.26	0.50	0.26	0.35	0.92	0.98	1.00	0.96	0.76	0.29	0.26	0.23	0.23	0.39	0.66	-0.26	0.44	-0.28	-0.59	0.38	0.10	0.51	-0.52	0.22
30d-Min	0.16	0.07	0.15	0.31	0.17	0.36	0.24	0.44	0.29	0.53	0.28	0.38	0.86	0.91	0.96	1.00	0.79	0.27	0.23	0.21	0.24	0.39	0.59	-0.27	0.46	-0.30	-0.61	0.43	0.10	0.52	-0.53	0.22
90d-Min	-0.03	0.02	0.03	0.36	0.27	0.34	0.15	0.51	0.46	0.74	0.48	0.51	0.66	0.74	0.76	0.79	1.00	0.26	0.26	0.25	0.29	0.43	0.30	-0.49	0.32	-0.41	-0.58	0.41	0.21	0.61	-0.53	-0.07
1d-Max	0.22	0.22	0.32	0.34	0.32	0.68	0.52	0.28	0.15	0.10	-0.02	0.14	0.23	0.27	0.29	0.27	0.26	1.00	0.99	0.95	0.76	0.74	-0.23	-0.09	0.11	0.10	-0.15	0.44	0.39	0.66	-0.70	0.09
3d-Max	0.19	0.18	0.31	0.33	0.32	0.68	0.52	0.27	0.16	0.10	-0.01	0.14	0.20	0.24	0.26	0.23	0.26	0.99	1.00	0.97	0.78	0.74	-0.27	-0.11	0.09	0.10	-0.14	0.42	0.40	0.65	-0.68	0.05
7d-Max	0.15	0.14	0.30	0.40	0.33	0.68	0.52	0.21	0.15	0.07	0.01	0.13	0.18	0.22	0.23	0.21	0.25	0.95	0.97	1.00	0.83	0.76	-0.29	-0.14	0.05	0.13	-0.15	0.35	0.47	0.61	-0.64	0.05
30d-Max	0.11	0.15	0.23	0.55	0.54	0.62	0.47	0.21	0.15	0.14	0.06	0.14	0.14	0.22	0.23	0.24	0.29	0.76	0.78	0.83	1.00	0.89	-0.36	-0.12	0.03	-0.02	-0.12	0.33	0.72	0.66	-0.64	-0.06
90d-Max	0.22	0.21	0.30	0.67	0.62	0.70	0.45	0.32	0.27	0.28	0.02	0.17	0.26	0.36	0.39	0.39	0.43	0.74	0.74	0.76	0.89	1.00	-0.32	-0.18	0.06	-0.18	-0.20	0.52	0.69	0.81	-0.79	-0.03
Base index	-0.08	-0.18	-0.14	-0.27	-0.22	-0.11	-0.16	0.08	-0.05	0.12	0.07	0.02	0.67	0.66	0.66	0.59	0.30	-0.23	-0.27	-0.29	-0.36	-0.32	1.00	-0.05	0.29	0.05	-0.39	-0.18	-0.36	-0.19	0.15	0.30
Date min	0.39	0.26	0.19	-0.09	-0.17	0.00	-0.07	-0.33	-0.46	-0.56	-0.44	-0.56	-0.17	-0.23	-0.26	-0.27	-0.49	-0.09	-0.11	-0.14	-0.12	-0.18	-0.05	1.00	-0.07	0.19	0.48	-0.28	0.03	-0.31	0.23	0.15
Date max	0.02	-0.09	-0.10	-0.09	-0.10	0.16	0.32	0.27	0.07	0.31	0.28	0.33	0.41	0.40	0.44	0.46	0.32	0.11	0.09	0.05	0.03	0.06	0.29	-0.07	1.00	-0.08	-0.24	0.27	0.05	0.30	-0.31	0.08
Low-C	-0.11	-0.13	-0.12	-0.28	-0.19	-0.10	0.06	-0.32	-0.37	-0.44	-0.14	-0.23	-0.29	-0.31	-0.28	-0.30	-0.41	0.10	0.10	0.13	-0.02	-0.18	0.05	0.19	-0.08	1.00	-0.12	-0.29	-0.16	-0.23	0.14	0.40
Low-D	0.26	0.24	0.14	-0.14	-0.07	-0.10	-0.18	-0.32	-0.35	-0.50	-0.42	-0.49	-0.47	-0.54	-0.59	-0.61	-0.58	-0.15	-0.14	-0.15	-0.12	-0.20	-0.39	0.48	-0.24	-0.12	1.00	-0.22	0.02	-0.36	0.38	-0.22
High-C	0.24	0.31	0.35	0.29	0.26	0.41	0.42	0.42	0.35	0.42	0.15	0.39	0.30	0.35	0.38	0.43	0.41	0.44	0.42	0.35	0.33	0.52	-0.18	-0.28	0.27	-0.29	-0.22	1.00	-0.06	0.70	-0.73	0.12
High-D	0.14	0.15	0.12	0.57	0.46	0.42	0.19	0.13	0.19	0.12	-0.03	-0.05	0.06	0.11	0.10	0.10	0.21	0.39	0.40	0.47	0.72	0.69	-0.36	0.03	0.05	-0.16	0.02	-0.06	1.00	0.42	-0.33	-0.29
Rise rate	0.16	0.21	0.27	0.49	0.38	0.59	0.56	0.61	0.40	0.58	0.25	0.43	0.33	0.45	0.51	0.52	0.61	0.66	0.65	0.61	0.66	0.81	-0.19	-0.31	0.30	-0.23	-0.36	0.70	0.42	1.00	-0.94	0.03
Fall rate	-0.23	-0.23	-0.32	-0.44	-0.34	-0.62	-0.60	-0.54	-0.33	-0.45	-0.24	-0.47	-0.35	-0.46	-0.52	-0.53	-0.53	-0.70	-0.68	-0.64	-0.64	-0.79	0.15	0.23	-0.31	0.14	0.38	-0.73	-0.33	-0.94	1.00	-0.24
Reversals	0.19	0.04	0.09	-0.16	-0.08	-0.01	0.21	-0.04	-0.14	-0.12	-0.02	-0.02	0.13	0.15	0.22	0.22	-0.07	0.09	0.05	0.05	-0.06	-0.03	0.30	0.15	0.08	0.40	-0.22	0.12	-0.29	0.03	-0.24	1.00

Label	Jan. I	Feb. I	Mar. I	Apr. I	May. I	June. I	July I	Aug I	Sept. I	Oct. I	Nov. I	Dec. I	Xlow1 p	Xlow1 d	Xlow1 t	Xlow1 f	High1 p	High1 d	High1 t	High1 f	High1 r	High1 f	Sfd1 p	Sfd1 d	Sfd1 t	Sfd1 f	Sfd1 r	Sfd1 f	Lfd1 p	Lfd1 d	Lfd1 t	Lfd1 f	Lfd1 r	Lfd1 f
Jan. I	1.00	0.76	0.25	-0.38	-0.41	-0.02	-0.37	-0.05	-0.29	-0.29	-0.44	-0.05	0.22	0.37	0.53	0.13	-0.02	0.22	-0.38	-0.29	-0.52	0.13	-0.12	0.20	-0.25	-0.20	-0.28	-0.02	0.53	0.50	0.51	0.55	0.42	-0.52
Feb. I	0.76	1.00	0.35	-0.22	-0.14	0.01	-0.17	-0.12	-0.02	-0.11	-0.11	0.11	0.09	0.13	0.24	-0.02	-0.20	-0.02	-0.09	0.10	-0.54	0.43	0.12	0.20	0.02	0.07	-0.03	-0.24	0.34	0.27	0.29	0.37	0.30	-0.30
Mar. I	0.25	0.35	1.00	-0.15	0.47	-0.33	0.18	0.22	0.07	-0.07	-0.09	-0.04	-0.10	-0.19	0.09	-0.07	0.00	0.23	0.14	0.07	-0.25	0.33	0.26	-0.08	0.28	0.42	0.38	-0.30	0.34	0.18	0.28	0.30	0.44	-0.25
Apr. I	-0.38	-0.22	-0.15	1.00	0.18	0.27	-0.19	0.07	0.20	-0.25	-0.04	0.03	-0.05	0.25	-0.09	-0.14	0.39	0.06	0.13	0.27	0.27	-0.02	-0.30	-0.50	-0.19	-0.18	0.03	0.28	-0.26	-0.06	-0.22	-0.16	-0.49	0.16
May. I	-0.41	-0.14	0.47	0.18	1.00	-0.03	0.58	0.21	-0.05	-0.05	0.18	-0.14	0.13	0.05	0.07	0.09	-0.09	0.00	0.36	0.39	-0.08	0.38	-0.04	-0.46	0.07	0.22	0.22	0.13	0.32	0.28	0.31	0.31	0.30	-0.30
June. I	-0.02	0.01	-0.33	0.27	-0.03	1.00	-0.21	-0.21	0.05	0.02	0.01	-0.36	0.12	0.27	0.07	-0.04	0.20	-0.11	0.08	0.30	0.05	-0.06	-0.45	-0.36	-0.25	-0.30	0.06	0.33	-0.15	0.01	-0.11	-0.07	-0.34	0.07
July I	-0.37	-0.17	0.18	-0.19	0.58	-0.21	1.00	0.26	0.15	0.28	0.47	0.03	-0.15	-0.24	-0.22	-0.12	-0.12	0.14	0.24	0.14	-0.13	0.53	0.46	0.19	0.47	0.55	0.11	-0.27	0.22	0.11	0.21	0.15	0.36	-0.17
Aug I	-0.05	-0.12	0.22	0.07	0.21	-0.21	0.26	1.00	-0.16	-0.57	-0.08	-0.43	-0.35	-0.29	-0.27	-0.43	0.16	0.28	-0.07	0.03	0.07	0.27	0.14	0.16	0.12	-0.23	-0.15	0.02	0.05	0.01	0.05	0.02	0.11	-0.03
Sept. I	-0.29	-0.02	0.07	0.20	-0.06	0.05	0.15	-0.16	1.00	0.60	0.64	0.37	-0.31	-0.29	-0.50	-0.30	0.06	-0.03	0.52	0.20	-0.06	0.34	0.47	0.21	0.63	0.61	0.60	-0.43	-0.52	-0.49	-0.51	-0.52	-0.43	0.51
Oct. I	-0.29	-0.11	-0.07	-0.25	-0.05	0.02	0.28	-0.57	0.60	1.00	0.70	0.33	-0.14	-0.28	-0.29	-0.01	-0.16	-0.32	0.36	0.09	0.11	0.06	0.38	0.29	0.55	0.57	0.38	-0.29	-0.33	-0.33	-0.33	-0.34	-0.25	0.34
Nov. I	-0.44	-0.11	-0.09	-0.04	0.18	0.01	0.47	-0.08	0.64	0.70	1.00	0.35	-0.23	-0.31	-0.42	-0.04	-0.04	-0.12	0.70	0.40	0.05	0.22	0.35	0.21	0.66	0.38	0.24	-0.19	-0.48	-0.47	-0.47	-0.49	-0.36	0.48
Dec. I	-0.05	0.11	-0.04	0.03	-0.14	-0.36	0.03	-0.43	0.37	0.33	0.35	1.00	-0.06	-0.02	-0.15	0.19	-0.14	-0.08	0.43	0.08	0.00	-0.07	0.20	0.11	0.05	0.40	-0.05	-0.31	-0.26	-0.23	-0.25	-0.26	-0.23	0.24
Xlow1 p	0.22	0.09	-0.10	-0.05	0.13	0.12	-0.15	-0.35	-0.31	-0.14	-0.23	-0.06	1.00	0.78	0.81	0.70	0.31	0.33	-0.06	-0.16	-0.16	-0.30	-0.70	-0.57	-0.71	-0.36	-0.44	0.60	0.38	0.31	0.35	0.37	0.37	-0.34
Xlow1 d	0.37	0.13	-0.19	0.25	0.05	0.27	-0.24	-0.29	-0.29	-0.28	-0.31	-0.02	0.78	1.00	0.78	0.65	0.19	0.25	-0.10	-0.01	-0.30	-0.15	-0.74	-0.57	-0.71	-0.45	-0.44	0.62	0.51	0.60	0.53	0.57	0.25	-0.57
Xlow1 t	0.53	0.24	0.09	-0.09	0.07	0.07	-0.22	-0.27	-0.50	-0.29	-0.42	-0.15	0.81	0.78	1.00	0.62	0.30	0.47	-0.21	-0.36	-0.35	-0.21	-0.63	-0.49	-0.67	-0.40	-0.44	0.56	0.65	0.62	0.64	0.66	0.53	-0.64
Xlow1 f	0.13	-0.02	-0.07	-0.14	0.09	-0.04	-0.12	-0.43	-0.30	-0.01	-0.04	0.19	0.70	0.65	0.62	1.00	-0.04	0.12	0.10	0.07	-0.09	-0.44	-0.49	-0.41	-0.50	-0.24	-0.31	0.42	0.29	0.20	0.26	0.28	0.31	-0.24
High1 p	-0.02	-0.20	0.00	0.39	-0.09	0.20	-0.12	0.16	0.06	-0.16	-0.04	-0.14	0.31	0.19	0.30	-0.04	1.00	0.78	0.01	-0.33	0.28	-0.32	-0.39	-0.34	-0.33	0.24	-0.31	0.38	-0.23	-0.23	-0.23	-0.23	-0.18	0.23
High1 d	0.22	-0.02	0.23	0.06	0.00	-0.11	0.14	0.28	-0.03	-0.32	-0.12	-0.08	0.33	0.25	0.47	0.12	0.78	1.00	0.04	-0.43	-0.23	-0.08	-0.17	-0.23	-0.22	-0.03	-0.24	0.18	0.21	0.08	0.19	0.14	0.36	-0.14
High1 t	-0.38	-0.09	0.14	0.13	0.36	0.08	0.24	-0.07	0.52	0.36	0.70	0.43	-0.06	-0.10	-0.21	0.10	0.01	0.04	1.00	0.57	-0.08	0.00	-0.01	-0.28	0.18	0.25	0.28	0.06	-0.30	-0.32	-0.30	-0.32	-0.19	0.31
High1 f	-0.29	0.10	0.07	0.27	0.39	0.30	0.14	0.03	0.20	0.09	0.40	0.08	-0.16	-0.01	-0.36	0.07	-0.33	-0.43	0.57	1.00	-0.05	0.12	-0.09	-0.25	0.10	0.03	0.25	0.04	-0.17	-0.13	-0.18	-0.12	-0.22	0.16
High1 r	-0.52	-0.54	-0.25	0.27	-0.08	0.05	-0.13	0.07	-0.06	0.11	0.05	0.00	-0.16	-0.30	-0.35	-0.09	0.28	-0.23	-0.08	-0.05	1.00	-0.49	-0.13	0.02	-0.11	-0.23	-0.27	0.25	-0.58	-0.52	-0.58	-0.55	-0.55	0.56
High1 f	0.13	0.43	0.33	-0.02	0.38	-0.06	0.53	0.27	0.34	0.06	0.22	-0.07	-0.30	-0.15	-0.21	-0.44	-0.32	-0.08	0.00	0.12	-0.49	1.00	0.58	0.35	0.58	0.48	0.37	-0.47	0.31	0.32	0.32	0.32	0.23	-0.32
Sfd1 p	-0.12	0.12	0.26	-0.30	-0.04	-0.45	0.46	0.14	0.47	0.38	0.35	0.20	-0.70	-0.74	-0.63	-0.49	-0.39	-0.17	-0.01	-0.09	-0.13	0.58	1.00	0.78	0.92	0.76	0.54	-0.86	-0.16	-0.27	-0.19	-0.22	0.03	0.22
Sfd1 d	0.20	0.20	-0.06	-0.50	-0.46	-0.36	0.19	0.16	0.21	0.29	0.21	0.11	-0.57	-0.57	-0.49	-0.41	-0.34	-0.23	-0.28	-0.25	0.02	0.35	0.78	1.00	0.64	0.30	0.05	-0.56	-0.18	-0.22	-0.18	-0.20	-0.04	0.20
Sfd1 t	-0.25	0.02	0.28	-0.19	0.07	-0.25	0.47	0.12	0.63	0.55	0.66	0.05	-0.71	-0.71	-0.67	-0.50	-0.33	-0.22	0.18	0.10	-0.11	0.58	0.92	0.64	1.00	0.72	0.68	-0.72	-0.26	-0.31	-0.27	-0.29	-0.10	0.29
Sfd1 f	-0.20	0.07	0.42	-0.18	0.22	-0.30	0.55	-0.23	0.61	0.57	0.38	0.40	-0.36	-0.45	-0.40	-0.24	-0.24	-0.03	0.25	0.03	-0.23	0.48	0.76	0.30	0.72	1.00	0.68	-0.78	-0.03	-0.15	-0.06	-0.09	0.14	0.10
Sfd1 r	-0.28	-0.03	0.38	0.03	0.22	0.06	0.11	-0.15	0.60	0.38	0.24	-0.05	-0.44	-0.44	-0.44	-0.31	-0.31	-0.24	0.26	0.25	-0.27	0.37	0.54	0.05	0.68	0.66	1.00	-0.59	-0.18	-0.22	-0.19	-0.20	-0.09	0.20
Sfd1 f	-0.02	-0.24	-0.30	0.28	0.13	0.33	-0.27	0.02	-0.43	-0.29	-0.19	-0.31	0.60	0.62	0.56	0.42	0.38	0.18	0.06	0.04	0.25	-0.47	-0.86	-0.56	-0.72	-0.78	-0.59	1.00	0.17	0.25	0.19	0.22	0.02	-0.22
Lfd1 p	0.53	0.34	0.34	-0.26	0.32	-0.15	0.22	0.05	-0.52	-0.33	-0.48	-0.26	0.38	0.51	0.65	0.29	-0.23	0.21	-0.30	-0.17	-0.58	0.31	-0.16	-0.18	-0.25	-0.03	-0.18	0.17	1.00	0.92	0.99	0.99	0.87	-0.98
Lfd1 d	0.50	0.27	0.18	-0.06	0.28	0.01	0.11	0.01	-0.49	-0.33	-0.47	-0.23	0.31	0.60	0.62	0.20	-0.23	0.08	-0.32	-0.13	-0.52	0.32	-0.27	-0.22	-0.31	-0.15	-0.22	0.25	0.92	1.00	0.96	0.96	0.83	-0.99
Lfd1 t	0.51	0.29	0.28	-0.22	0.31	-0.11	0.21	0.05	-0.51	-0.33	-0.47	-0.25	0.35	0.53	0.64	0.26	-0.23	0.19	-0.30	-0.18	-0.58	0.32	-0.19	-0.18	-0.27	-0.06	-0.19	0.19	0.99	0.96	1.00	0.99	0.81	-0.99
Lfd1 f	0.55	0.37	0.30	-0.16	0.31	-0.07	0.15	0.02	-0.52	-0.34	-0.49	-0.26	0.37	0.57	0.66	0.28	-0.23	0.14	-0.32	-0.12	-0.55	0.32	-0.22	-0.20	-0.29	-0.09	-0.20	0.22	0.99	0.96	0.99	1.00	0.79	-0.99
Lfd1 r	0.42	0.30	0.44	-0.49	0.30	-0.34	0.36	0.11	-0.43	-0.25	-0.36	-0.23	0.37	0.25	0.53	0.31	-0.18	0.36	-0.19	-0.22	-0.55	0.23	0.03	-0.04	-0.10	0.14	-0.09	0.02	0.87	0.83	0.81	0.79	1.00	-0.75
Lfd1 f	-0.52	-0.30	-0.25	0.16	-0.30	0.07	-0.17	-0.03	0.51	0.34	0.48	0.24	-0.34	-0.57	-0.64	-0.24	0.23	-0.14	0.31	0.16	0.56	-0.32	0.22	0.20	0.29	0.10	0.20	-0.22	-0.98	-0.99	-0.99	-0.99	-0.75	1.00

References

1. Karr, J.R. Defining and measuring river health. *Freshw. Biol.* **1999**, *41*, 221–234. [CrossRef]
2. Zhou, T. New physical science behind climate change: What does IPCC AR6 tell us? *The Innovation* **2021**, *2*, 100173. [CrossRef] [PubMed]
3. Myhre, G.; Alterskjær, K.; Stjern, C.W.; Hodnebrog, Ø.; Marelle, L.; Samset, B.H.; Sillmann, J.; Schaller, N.; Fischer, E.; Schulz, M.; et al. Frequency of extreme precipitation increases extensively with event rareness under global warming. *Sci. Rep.* **2019**, *9*, 16063. [CrossRef] [PubMed]
4. Grill, G.; Lehner, B.; Thieme, M.; Geenen, B.; Tickner, D.; Antonelli, F.; Babu, S.; Borrelli, P.; Cheng, L.; Crochetiere, H.; et al. Mapping the world's free-flowing rivers. *Nature* **2019**, *569*, 215–221. [CrossRef]
5. Zhao, H.; He, J.; Liu, D.; Han, Y.; Zhou, Z.; Niu, J. Incorporating ecological connectivity into ecological functional zoning: A case study in the middle reaches of Yangtze River urban agglomeration. *Ecol. Inform.* **2023**, *75*, 102098. [CrossRef]
6. Richter, B.D.; Baumgartner, J.V.; Powell, J.; Braun, D.P. A method for assessing hydrologic alteration within ecosystems. *Conserv. Biol.* **1996**, *10*, 1163–1174. [CrossRef]
7. Song, X.; Zhuang, Y.; Wang, X.; Li, E.; Zhang, Y.; Lu, X.; Yang, J.; Liu, X. Analysis of hydrologic regime changes caused by dams in China. *J. Hydrol. Eng.* **2020**, *25*, 05020003. [CrossRef]
8. Gao, W.; Hu, K.; Gu, Q.; Ling, X.; Ruan, M.; Ren, R. Survey of Fish Resources in Hengyang and Changzhutan Sections of the Main Stream of Xiangjiang River and Measures for Conservation. *Low Carbon World* **2019**, *9*, 14–16.
9. Richter, B.D.; Warner, A.T.; Meyer, J.L.; Lutz, K. A collaborative and adaptive process for developing environmental flow recommendations. *River Res. Appl.* **2006**, *22*, 297–318. [CrossRef]
10. Gunawardana, S.K.; Shrestha, S.; Mohanasundaram, S.; Salin, K.S.; Piman, T. Multiple drivers of hydrological alteration in the transboundary Srepok River Basin of the Lower Mekong Region. *J. Environ. Manag.* **2021**, *278*, 111524. [CrossRef]
11. Olden, J.D.; Poff, N.L. Redundancy and the choice of hydrologic indices for characterizing streamflow regimes. *River Res. Appl.* **2003**, *19*, 101–121. [CrossRef]
12. Smakhtin, V.U.; Shilpakar, R.L.; Hughes, D.A. Hydrology-based assessment of environmental flows: An example from Nepal. *Hydrol. Sci. J.* **2006**, *51*, 207–222. [CrossRef]
13. Cheng, J.; Xu, L.; Jiang, J. Optimal selection of the most ecologically relevant hydrologic indicators and its application for environmental flow calculation in Lake Dongting. *J. Lake Sci.* **2018**, *30*, 1235–1245.
14. Jiang, S.; Zhou, L.; Ren, L.; Wang, M.; Xu, C.; Yuan, F.; Liu, Y.; Yang, X.; Ding, Y. Development of a comprehensive framework for quantifying the impacts of climate change and human activities on river hydrological health variation. *J. Hydrol.* **2021**, *600*, 126566. [CrossRef]
15. Arsenault, R.; Martel, J.L.; Brunet, F.; Brissette, F.; Mai, J. Continuous streamflow prediction in ungauged basins: Long Short-Term Memory Neural Networks clearly outperform hydrological models. *Hydrol. Earth Syst. Sci. Discuss.* **2022**, *2022*, 1–29. [CrossRef]
16. Hu, C.; Wu, Q.; Li, H.; Jiang, S.; Li, N.; Lou, Z. Deep learning with a long short-term memory networks approach for rainfall-runoff simulation. *Water* **2018**, *10*, 1543. [CrossRef]

17. Bai, P.; Liu, X.; Xie, J. Simulating runoff under changing climatic conditions: A comparison of the long short-term memory network with two conceptual hydrologic models. *J. Hydrol.* **2021**, *592*, 125779. [CrossRef]
18. Fan, H.; He, H.; Xu, L.; Zhang, M.; Jiang, J. Simulation and attribution analysis based on the long-short-term-memory network for detecting the dominant cause of runoff variation in the Lake Poyang Basin. *J. Lake Sci.* **2021**, *33*, 866–878.
19. Cao, Y.; Xu, L.; Fan, H.; Mao, Z.; Cheng, J.; Wang, D.; Wu, Y. Impact of climate and human activities on the changes of ecological flow indicators in the Lake Poyang Basin since 1960s. *J. Lake Sci.* **2022**, *34*, 232–246.
20. Sharma, A.; Baruah, A.; Mangukiya, N.; Hinge, G.; Bharali, B. Evaluation of Gangetic dolphin habitat suitability under hydroclimatic changes using a coupled hydrological-hydrodynamic approach. *Ecol. Inform.* **2022**, *69*, 101639. [CrossRef]
21. Nagendra, H.; Lucas, R.; Honrado, J.P.; Jongman, R.H.; Tarantino, C.; Adamo, M.; Mairota, P. Remote sensing for conservation monitoring: Assessing protected areas, habitat extent, habitat condition, species diversity, and threats. *Ecol. Indic.* **2013**, *33*, 45–59. [CrossRef]
22. Zhang, X.; Zhou, J.; Li, M. Analysis on spatial and temporal changes of regional habitat quality based on the spatial pattern reconstruction of land use. *Acta Geogr. Sin.* **2020**, *75*, 160–178.
23. Zhang, H.; Li, S.; Liu, Y.; Xu, M. Assessment of the Habitat Quality of Offshore Area in Tongzhou Bay, China: Using Benthic Habitat Suitability and the InVEST Model. *Water* **2022**, *14*, 1574. [CrossRef]
24. de Mello, K.; Taniwaki, R.H.; de Paula, F.R.; Valente, R.A.; Randhir, T.O.; Macedo, D.R.; Leal, C.G.; Rodrigues, C.B.; Hughes, R.M. Multiscale land use impacts on water quality: Assessment, planning, and future perspectives in Brazil. *J. Environ. Manag.* **2020**, *270*, 110879. [CrossRef]
25. Zeng, C.; Wen, Y.; Liu, X.; Yu, J.; Jin, B.; Li, D. Impact of anthropogenic activities on changes of ichthyofauna in the middle and lower Xiang River. *Aquac. Fish.* **2022**, *7*, 693–702. [CrossRef]
26. Wang, H.; Ma, Y.; Hong, F.; Yang, H.; Huang, L.; Jiao, X.; Guo, W. Evolution of Water–Sediment Situation and Attribution Analysis in the Upper Yangtze River, China. *Water* **2023**, *15*, 574. [CrossRef]
27. Djaman, K.; O'Neill, M.; Diop, L.; Bodian, A.; Allen, S.; Koudahe, K.; Lombard, K. Evaluation of the Penman-Monteith and other 34 reference evapotranspiration equations under limited data in a semiarid dry climate. *Theor. Appl. Climatol.* **2019**, *137*, 729–743. [CrossRef]
28. Faisal, N.; Gaffar, A. Development of Pakistan's new area weighted rainfall using Thiessen polygon method. *Pak. J. Meteorol.* **2012**, *9*, 1–10.
29. Hou, L.; Zhu, J.; Kwok, J.; Gao, F.; Qin, T.; Liu, T.Y. Normalization helps training of quantized lstm. *Adv. Neural Inf. Process. Syst.* **2019**, *32*, 1–11.
30. Hamed, K.H.; Rao, A.R. A modified Mann-Kendall trend test for autocorrelated data. *J. Hydrol.* **1998**, *204*, 182–196. [CrossRef]
31. Hu, J.; Ma, J.; Nie, C.; Xue, L.; Zhang, Y.; Ni, F.; Deng, Y.; Liu, J.; Zhou, D.; Li, L.; et al. Attribution Analysis of Runoff change in Min-tuo River Basin based on SWAT model simulations, china. *Sci. Rep.* **2020**, *10*, 2900. [CrossRef] [PubMed]
32. Richter, B.; Baumgartner, J.; Wigington, R.; Braun, D. How much water does a river need? *Freshw. Biol.* **1997**, *37*, 231–249. [CrossRef]
33. Wang, H.; Huang, L.; Guo, W.; Zhu, Y.; Yang, H.; Jiao, X.; Zhou, H. Evaluation of ecohydrological regime and its driving forces in the Dongting Lake, China. *J. Hydrol. Reg. Stud.* **2022**, *41*, 101067. [CrossRef]
34. Olsen, R.L.; Chappell, R.W.; Loftis, J.C. Water quality sample collection, data treatment and results presentation for principal components analysis–literature review and Illinois River watershed case study. *Water Res.* **2012**, *46*, 3110–3122. [CrossRef]
35. Kühnert, C.; Gonuguntla, N.M.; Krieg, H.; Nowak, D.; Thomas, J.A. Application of LSTM networks for water demand prediction in optimal pump control. *Water* **2021**, *13*, 644. [CrossRef]
36. Gauch, M.; Kratzert, F.; Klotz, D.; Nearing, G.; Lin, J.; Hochreiter, S. Rainfall–runoff prediction at multiple timescales with a single Long Short-Term Memory network. *Hydrol. Earth Syst. Sci.* **2021**, *25*, 2045–2062. [CrossRef]
37. Yin, H.; Wang, F.; Zhang, X.; Zhang, Y.; Chen, J.; Xia, R.; Jin, J. Rainfall-runoff modeling using long short-term memory based step-sequence framework. *J. Hydrol.* **2022**, *610*, 127901. [CrossRef]
38. Ahn, K.H.; Merwade, V. Quantifying the relative impact of climate and human activities on streamflow. *J. Hydrol.* **2014**, *515*, 257–266. [CrossRef]
39. Xu, H.; Dong, B.; Gao, X.; Xu, Z.; Ren, C.; Fang, L.; Wei, Z.; Liu, X.; Lu, Z. Habitat quality assessment of wintering migratory birds in Poyang Lake National Nature Reserve based on InVEST model. *Environ. Sci. Pollut. Res.* **2023**, *30*, 28847–28862. [CrossRef]
40. Gao, C.L.; Li, S.C.; Wang, J.; Li, L.P.; Lin, P. The risk assessment of tunnels based on grey correlation and entropy weight method. *Geotech. Geol. Eng.* **2018**, *36*, 1621–1631. [CrossRef]
41. Yang, Y.C.E.; Cai, X.; Herricks, E.E. Identification of hydrologic indicators related to fish diversity and abundance: A data mining approach for fish community analysis. *Water Resour. Res.* **2008**, *44*, 4. [CrossRef]
42. Yang, H.; Shen, L.; He, Y.; Tian, H.; Gao, L.; Wu, J.; Mei, Z.; Wei, N.; Wang, L.; Zhu, T.; et al. Status of aquatic organisms resources and their environments in Yangtze river system (2017–2021). *Aquac. Fish.* **2023**. [CrossRef]
43. Jia, Y.; Wang, L.; Li, S.; Cao, J.; Yang, Z. Species-specific bioaccumulation and correlated health risk of arsenic compounds in freshwater fish from a typical mine-impacted river. *Sci. Total Environ.* **2018**, *625*, 600–607. [CrossRef] [PubMed]
44. Sofi, M.S.; Bhat, S.U.; Rashid, I.; Kuniyal, J.C. The natural flow regime: A master variable for maintaining river ecosystem health. *Ecohydrology* **2020**, *13*, e2247. [CrossRef]

45. Comte, L.; Olden, J.D.; Tedesco, P.A.; Ruhi, A.; Giam, X. Climate and land-use changes interact to drive long-term reorganization of riverine fish communities globally. *Proc. Natl. Acad. Sci. USA* **2021**, *118*, e2011639118. [CrossRef]
46. Palmer, M.; Ruhi, A. Linkages between flow regime, biota, and ecosystem processes: Implications for river restoration. *Science* **2019**, *365*, eaaw2087. [CrossRef]
47. Hu, G.; Mao, D.; Li, Z.; Xu, Y. Evolution Processes and Characteristics of Annual Runoff and Sediment in Xiangjiang River from 1951 to 2011. *Bull. Soil Water Conserv.* **2014**, *34*, 166–172.
48. Tonkin, J.D.; Merritt, D.; Olden, J.D.; Reynolds, L.V.; Lytle, D.A. Flow regime alteration degrades ecological networks in riparian ecosystems. *Nat. Ecol. Evol.* **2018**, *2*, 86–93. [CrossRef]
49. Richter, B.D.; Baumgartner, J.V.; Braun, D.P.; Powell, J. A spatial assessment of hydrologic alteration within a river network. *Regul. Rivers Res. Manag. Int. J. Devoted River Res. Manag.* **1998**, *14*, 329–340. [CrossRef]
50. Petsch, D.K. Causes and consequences of biotic homogenization in freshwater ecosystems. *Int. Rev. Hydrobiol.* **2016**, *101*, 113–122. [CrossRef]
51. Khatun, R.; Pal, S. Effects of hydrological modification on fish habitability in riparian flood plain river basin. *Ecol. Inform.* **2021**, *65*, 101398. [CrossRef]
52. Bordalo, A.A.; Nilsumranchit, W.; Chalermwat, K. Water quality and uses of the Bangpakong River (Eastern Thailand). *Water Res.* **2001**, *35*, 3635–3642. [CrossRef]
53. Bailly, D.; Agostinho, A.A.; Suzuki, H.I. Influence of the flood regime on the reproduction of fish species with different reproductive strategies in the Cuiabá River, Upper Pantanal, Brazil. *River Res. Appl.* **2008**, *24*, 1218–1229. [CrossRef]
54. Guo, W.; Hong, F.; Yang, H.; Huang, L.; Ma, Y.; Zhou, H.; Wang, H. Quantitative evaluation of runoff variation and its driving forces based on multi-scale separation framework. *J. Hydrol. Reg. Stud.* **2022**, *43*, 101183. [CrossRef]
55. Best, J. Anthropogenic stresses on the world's big rivers. *Nat. Geosci.* **2019**, *12*, 7–21. [CrossRef]
56. Benchimol, M.; Peres, C.A. Widespread forest vertebrate extinctions induced by a mega hydroelectric dam in lowland Amazonia. *PLoS ONE* **2015**, *10*, e0129818. [CrossRef]
57. Berta Aneseyee, A.; Noszczyk, T.; Soromessa, T.; Elias, E. The InVEST habitat quality model associated with land use/cover changes: A qualitative case study of the Winike Watershed in the Omo-Gibe Basin, Southwest Ethiopia. *Remote Sens.* **2020**, *12*, 1103. [CrossRef]
58. Lees, T.; Buechel, M.; Anderson, B.; Slater, L.; Reece, S.; Coxon, G.; Dadson, S.J. Benchmarking data-driven rainfall–runoff models in Great Britain: A comparison of long short-term memory (LSTM)-based models with four lumped conceptual models. *Hydrol. Earth Syst. Sci.* **2021**, *25*, 5517–5534. [CrossRef]
59. Fan, H.; Jiang, M.; Xu, L.; Zhu, H.; Cheng, J.; Jiang, J. Comparison of long short term memory networks and the hydrological model in runoff simulation. *Water* **2020**, *12*, 175. [CrossRef]
60. Anshuka, A.; Chandra, R.; Buzacott, A.J.; Sanderson, D.; van Ogtrop, F.F. Spatio temporal hydrological extreme forecasting framework using LSTM deep learning model. *Stoch. Environ. Res. Risk Assess.* **2022**, *36*, 3467–3485. [CrossRef]
61. Guo, W.; Hong, F.; Ma, Y.; Huang, L.; Yang, H.; Hu, J.; Zhou, H.; Wang, H. Comprehensive evaluation of the ecohydrological response of watersheds under changing environments. *Ecol. Inform.* **2023**, *74*, 101985. [CrossRef]
62. Valentukevičienė, M.; Bagdžiūnaitė-Litvinaitienė, L.; Chadyšas, V.; Litvinaitis, A. Evaluating the impacts of integrated pollution on water quality of the trans-boundary neris (viliya) river. *Sustainability* **2018**, *10*, 4239. [CrossRef]

water

MDPI

Article

What Drives the Morphological Traits of Stress-Tolerant Plant *Cynodon dactylon* in a Riparian Zone of the Three Gorges Reservoir, China

Xiaolong Li [1,2], Shanze Li [3], Yawei Xie [4], Zehui Wei [1] and Zilong Li [1,*]

1 Ocean College, Zhejiang University, Zhoushan 316021, China; lixiaolong@mwr.gov.cn (X.L.); 22134120@zju.edu.cn (Z.W.)
2 Ministry of Water Resources of the People's Republic of China, Beijing 100054, China
3 State Key Laboratory of Simulation and Regulation of Water Cycle in River Basin, China Institute of Water Resources and Hydropower Research, Beijing 100038, China; lishanze@126.com
4 Nanjiang Hydrogeology Brigade of Chongqing Geological Survey Bureau, Chongqing 401121, China; xieyawei-2@163.com
* Correspondence: zilongli@zju.edu.cn; Tel.: +86-135-8873-6851

Abstract: The cyclical process of water storage and recession in the regular operation of the Three Gorges Reservoir creates a unique habitat stress that alters the structural and functional attributes of vegetation ecology within the riparian zone. The stress-tolerant plant *Cynodon dactylon* (L.) Pers is the dominant plant species in the riparian zone of the Three Gorges Reservoir. In this study, the riparian zone of the Daning River, a tributary located in the center of the Three Gorges Reservoir, was selected as our study area. To identify the drivers of the morphological traits of *C. dactylon* in the riparian zone of Daning River, we examined plant biomass and plant characteristics across different elevation gradients, with reference to abiotic factors to determine the distribution patterns of plant morphological traits. Results indicated that in the two main soil types of the riparian zone, plant biomass showed a consistent trend along the elevation gradient, with a "middle-height expansion" pattern; biomass increased and then decreased with rising water levels. Plant biomass positively correlated with soil total nitrogen and negatively correlated with soil pH, electrical conductivity, and total phosphorus. *C. dactylon* adapted to prolonged flooding in the riparian zone by having a significant negative correlation between plant height and erect stem length with soil moisture content to facilitate root respiration.

Keywords: Three Gorges Reservoir; riparian zone; soil; plant morphological traits; *Cynodon dactylon*

1. Introduction

The riparian zone of a reservoir, also known as the drawdown zone or disturbance zone, is a transitional ecosystem that alternately experiences submergence and exposure due to periodic fluctuations in water level [1]. These fluctuations can result from natural hydrological changes, such as seasonal water level fluctuations or anthropogenic manipulations, primarily due to cyclical water storage and discharge operations. Additionally, specific climatic events such as droughts can cause drawdown in reservoirs.

The Three Gorges Project is key for national infrastructure and a critical node in the water network of China. The Three Gorges Reservoir serves as an important strategic freshwater resource reserve for the nation and is a crucial ecological barrier in the upper reaches of the Yangtze River [2]. In 2010, the project achieved its experimental water storage target of 175 m for the first time. Since then, to maximize flood control benefits, the Three Gorges Reservoir has adopted a "winter storage and summer discharge" regulatory approach. The reservoir operates at a lower water level from April to October and at a higher water level from October to the following April. This unique operational scheduling forms a unique fluctuating environment in the riparian zone with a drop range of 30 m, between 145 m and 175 m [1].

The cyclic water impoundment and release during normal operation of the Three Gorges Reservoir creates a unique form of habitat stress, altering the structure and function of the ecological pattern of vegetation in the riparian zone. In the cross-section of the riparian zone at different elevations, the ecological pattern of the vegetation exhibits notable spatial heterogeneity [3]. The biodiversity in the riparian zone is significantly affected by hydrological disturbances. Large-scale, unnatural water level fluctuations severely disrupt the original distribution of plant communities, resulting in a reverse succession in the riparian zone [4–6]. After the water impoundment operation of the Three Gorges Reservoir in 2003, the riparian zone gradually formed a vegetation succession pattern dominated by a few species including the plant *Cynodon dactylon* (L.) Pers, accompanied by *Atractylodes macrocephala* Koidz. and *Homalomena aromatica* Gagnep [1,7]. Since *C. dactylon* is the most dominant plant species, covering almost all the riparian zones of the Three Gorges Reservoir, we selected it as our study object. The plant *C. dactylon* shows significant resistance to flooding and nutrient-poor conditions [8,9]. This might have significant effects on the material conversion and ecological effects of element cycling in the riparian zone [10]. During the flooding period, the submerged plants will release several nutrients into the water, resulting in water eutrophication [11]. During the water-level drawdown period, plants are in their growing season.

The growth behavior and life history strategies of plants at each stage of their growth and development are closely related to the morphological traits of plants and can determine the distribution pattern and population behavior of plant species in the habitat to a certain extent. The adaptability of species is developed in the process of species evolution, which is manifested in morphology. Different morphological growth strategies are adopted by plants to maximize the fitness of species at certain stages. Plant morphological traits play an important role in studying biogeography in many ecosystems such as forestry [12] and agricultural [13] ecosystems. Under drought stress, plants can obtain the water content required for normal growth and development through the plasticity of morphological and structural characteristics, and adapt to the stressed environment by changing their morphogenesis. However, the morphological traits of plants in the riparian zone of reservoirs have received little attention. Differences in abiotic factors caused by climate change and geographical location differences will lead to different morphological characteristics of vegetation [14,15].

To maintain the *C. dactylon* community, it is important to address the morphological traits of this plant across a gradient of elevations, as well as its relationships with regional abiotic factors in the riparian zones. Hence, in this study, we tested the hypotheses that (1) *C. dactylon* morphological traits varied across elevations between two types of soil (yellow loam and purple soil) and (2) *C. dactylon* morphological traits varied with abiotic drivers.

2. Methods

2.1. Study Area

Our study area was located in a riparian zone (31.2813° N; 109.8179° E) of the Daning River (Figure 1). The Daning River is a tributary on the left bank of the Three Gorges Reservoir area. The region has a subtropical monsoon climate. The average annual temperature is 19.8 °C, with an average annual precipitation over 1000 mm. The Three Gorges Reservoir has a water level drop of 30 m. The types of riparian zones mainly include rocky bank slopes and gentle soil slopes. Our study was focused on the soil bank. Purple soil and yellow loam were the main soil types in the riparian zone of the Three Gorges Reservoir [16].

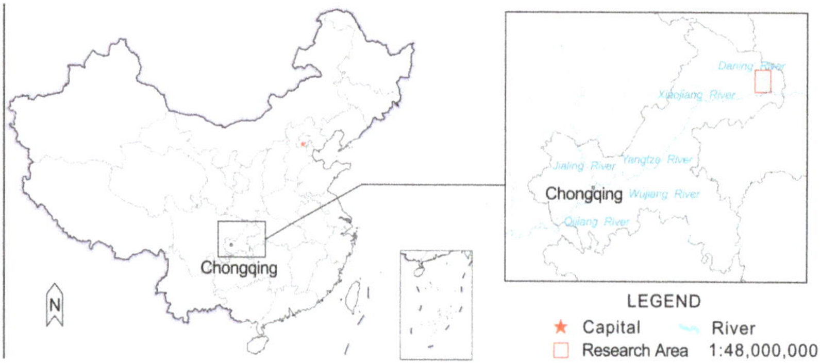

Figure 1. Study area.

2.2. Sample Collection and Testing

To give a comprehensive comparison of plant morphological traits between two types of soil, yellow loam and purple soil, across different elevations in a riparian zone of the Three Gorges Reservoir, China, we located three study sites in the riparian zones that are constituted of yellow loam and three study sites in the riparian zones that are constituted of purple soil. Thus, six riparian zones were selected. In each of the riparian zones, seven transects were selected from 145 m to 175 m (Figure 2). In each transect, we randomly set up three sampling quadrats as three replications of the plant community. At each sampling quadrat (1 m × 1 m), we recorded plant species and collected all the plant biomass. Plant biomass was dried in an oven (60 °C, 72 h), and weighed. We randomly selected six individuals of *C. dactylon* in each quadrat and recorded their erect stem length, number of ramets, average erect stem internode length, shoot height, first creeping stem, and root length.

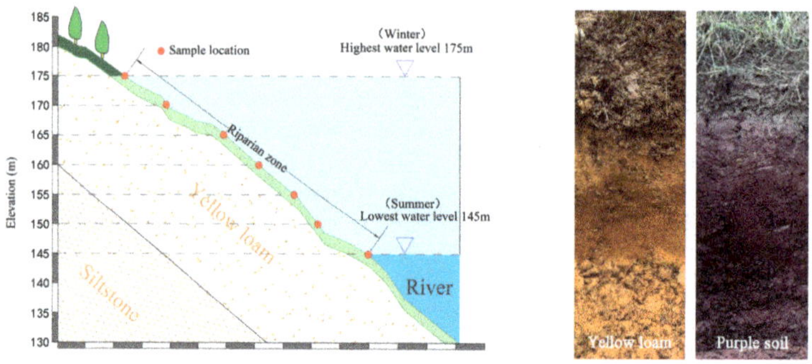

Figure 2. Schematic diagram of sampling locations in the Three Gorges Reservoir area (**Left**); photos of yellow loam and purple soil profiles (**Right**).

We examined environmental factors that might affect *C. dactylon* morphology, including soil types (yellow loam and purple soil), pH, electrical conductivity, total phosphorus, total nitrogen, moisture content, and temperature, and the duration of flooding. Three topsoil cores with a depth of 5 cm and a diameter of 5.05 cm were collected at each sampling point to obtain soil abiotic factors. Soil moisture content was determined by weighing wet soil cores and re-weighing them after drying for 48 h at 60 °C. Soil pH was analyzed by measuring the resulting supernatant of dry soil with deionized water at 1:5 w/v. Soil total nitrogen and total phosphorus were measured with a continuous flow analysis instrument [17]. The water level data were obtained from the Yangtze River Hydrology Bureau.

2.3. Data Analysis and Processing

We correlated yellow loam and purple soil with plant *C. dactylon* morphological traits (erect stem length, number of ramets, average erect stem internode length, shoot height, first creeping stem, and root length) using non-linear regression. The comparison between plant biomass in two soil types of riparian zone was carried out by using analysis of variance (ANOVA). We used Spearman's rank correlation analysis to examine potential relationships between plant morphology and soil abiotic factors (pH, electrical conductivity, total phosphorus, total nitrogen, moisture content, temperature, and duration of flooding). All statistical analyses were performed using SPSS 22 software (SPSS Inc., Chicago, IL, USA), where $p < 0.05$ was considered statistically significant [18].

3. Results

3.1. Dynamic Changes in Water Level in the Three Gorges Reservoir

The Three Gorges Reservoir has adopted a "winter storage and summer discharge" regulation mode. The water level slowly rises from 145 m to 175 m beginning at the end of September each year. The rising process generally takes about 60 days, during which time the riparian zone and its vegetation are gradually submerged. The high water level of 170~175 m then persists for approximately 60 days. Water discharge begins in April of the following year, and the water level gradually drops from 175 m to 145 m, a process that typically takes 200 days. During this time, the riparian zone and its vegetation gradually emerge from the water. A low water level of 145~150 m persists for approximately 90 days. The variations in water level in the Three Gorges Reservoir riparian zone over the past three years are shown in Figure 3.

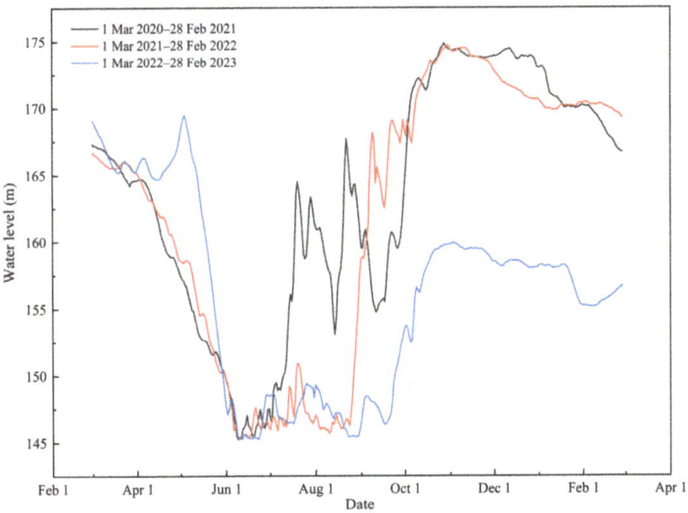

Figure 3. Variations of water level in the Three Gorges Reservoir.

3.2. Characteristics of Soil Type and Plant Biomass Distribution in the Riparian Zone

Through our survey, we found that the purple soil and yellow loam were the main soil types in the Daning River riparian zones. The proportions of clay and silt in yellow loam and purple soil reached as high as 66.44% and 56.4%, respectively, showing fine texture. As the drying time increases, the total nitrogen and phosphorus content in the soil gradually decreases.

The plant biomass at different elevations in the riparian zone under the two soil types is shown in Figure 4. This indicated significant differences in the plant biomass at different elevations in the yellow loam and purple soil riparian zones ($F = 6.159$, $p < 0.001$), but the trend of plant biomass changes with water level elevation is consistent. That is, from an elevation of 145 m to 160 m, plant biomass increased with elevation. The plant biomass in yellow loam peaked at an elevation of 160 m to approximately 1395.40 g/cm^2, whereas the biomass in purple soil peaked at an elevation of 155 m to approximately 1291.17 g/cm^2. From an elevation of 165 m to 180 m, plant biomass showed a trend of decreasing.

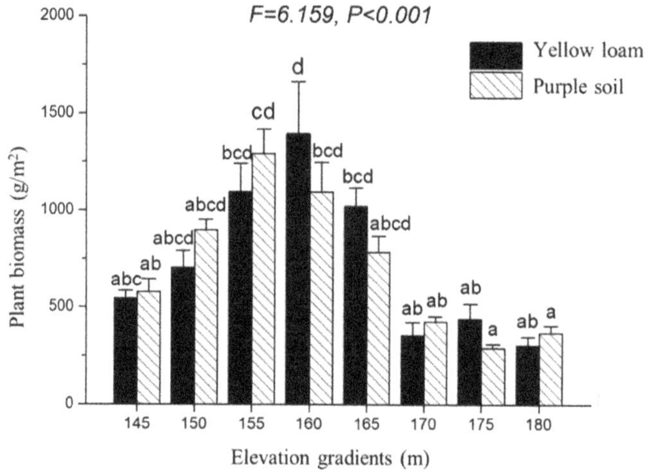

Figure 4. Distribution of plant biomass along different elevation gradients in the riparian zone of two soil types. Data are shown as means ± SE (N = 54). All ANOVA tests were significant ($p < 0.05$ in each case). The letter above each bar represents the results of post hoc Tukey's HSD test; bars sharing the same letter are not significantly different from one another.

3.3. Plant Community and Morphological Traits of C. dactylon under Different Soil Types and Elevations

The investigation results revealed that the dominant plant species within the 145 m to 165 m elevation range in the riparian zone of the Daning River were primarily *C. dactylon*, occasionally accompanied by *A. macrocephala* and *H. aromatica*. At elevations of 170 m and above, higher plant diversity was observed, including species such as *Daucus carota* L., *Ambrosia artemisiifolia* L., *Conyza canadensis* (L.) Cronq., *Digitaria sanguinalis* (L.) Scop., *Bidens frondosa* L., *Vitex negundo* L., *Melilotus officinalis* (L.) Desr., *Beckmannia syzigachne* (Steud.) Fern., and *Setaria viridis* (L.) Beauv. These species are mainly annual and perennial herbaceous plants. At higher elevations, shrubs, trees, and farmland were also present.

Within the elevation range of 145 m to 165 m in the riparian zone, the predominance of *C. dactylon* as the main plant species was related to its inherent plant characteristics. In the lower elevation riparian zone, where the duration of flooding was longer and species diversity was lower, *C. dactylon* exhibited strong tolerance to flooding. Even when the riparian zone soil remained inundated for more than 200 days during winter, *C. dactylon* could still grow during the following summer when water was released from the reservoir. However, when the elevation exceeded 165 m, the duration of soil inundation became shorter, and the number of viable species increased, leading to greater interspecific competition and the loss of dominance by *C. dactylon*. *Hemarthria compressa*, *S. viridis*, and *Euphorbia humifusa* became the dominant plant species at elevations of 170 m or above.

From 145 m to 180 m elevations, the length of the upright stems of *C. dactylon* increased with elevation (yellow loam riparian zone: $R^2 = 0.494$, $p < 0.05$; purple soil riparian zone: $R^2 = 0.674$, $p < 0.05$). The number of tillers of *C. dactylon* varied significantly across different elevations. In the yellow loam riparian zone, the number of tillers initially decreased and then increased with elevation ($R^2 = 0.664$, $p < 0.05$). In the purple soil riparian zone, the number of tillers significantly decreased with increasing elevation ($R^2 = 0.944$, $p < 0.001$). The average internode length showed a trend of increasing with elevation (yellow loam riparian zone: $R^2 = 0.262$, $p = 0.137$; purple soil riparian zone: $R^2 = 0.460$, $p = 0.084$). From the 145 m to 175 m elevations, the height of *C. dactylon* increased with elevation (yellow loam riparian zone: $R^2 = 0.314$, $p = 0.111$; purple soil riparian zone: $R^2 = 0.891$, $p < 0.01$). The primary stolon length of *C. dactylon* exhibited a trend of decreasing followed by increasing with elevation (yellow loam riparian zone: $R^2 = 0.463$, $p < 0.05$; purple soil riparian zone: $R^2 = 0.714$, $p < 0.05$). The root length of *C. dactylon* showed a trend of increasing with elevation in both the yellow loam riparian zone ($R^2 = 0.370$, $p = 0.087$) and the purple soil riparian zone ($R^2 = 0.648$, $p < 0.05$). Figure 5 illustrates the variations in the morphological characteristics of *C. dactylon* under different soil types and elevations of water level.

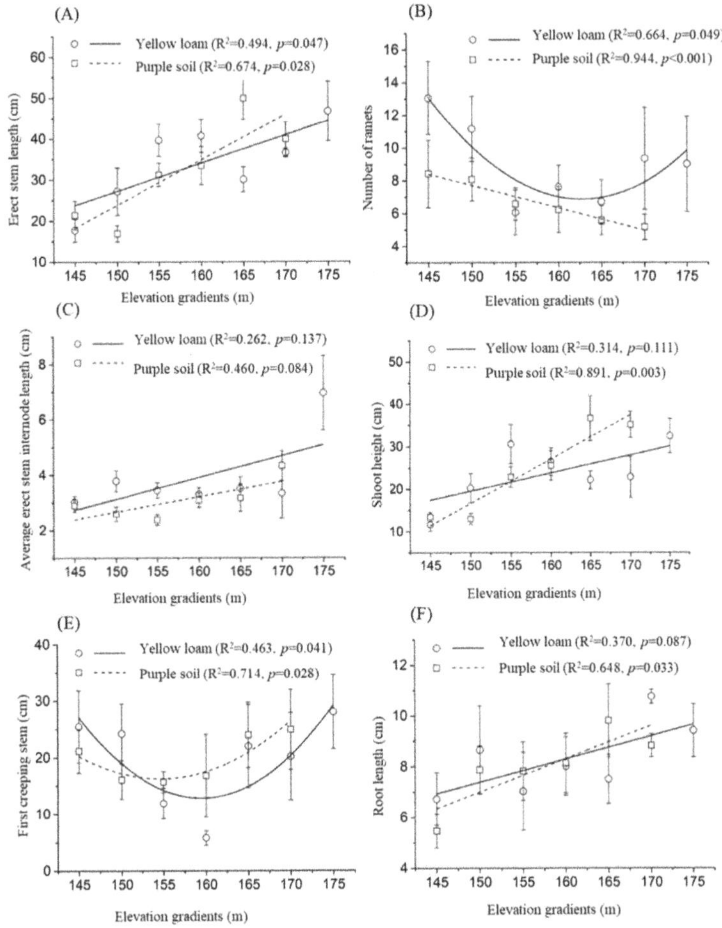

Figure 5. Variation patterns of *C. dactylon* morphology at different elevations of water level. Data are shown as means ± SE (N = 54). (**A**) Erect stem length, (**B**) number of ramets, (**C**) average erect stem internode length, (**D**) shoot height, (**E**) first creeping stem, and (**F**) root length of *C. dactylon* at different elevations of water level.

3.4. Relationship between Riparian Plant Biomass and C. dactylon Morphological Traits with Environmental Factors

3.4.1. Relationship between Riparian Plant Biomass and Environmental Factors

The relationships between plant biomass and environmental factors are shown in Figure 6. The biomass of plant *C. dactylon* in the Daning River was significantly positively correlated with soil total nitrogen content (R = 0.44, $p < 0.05$) and flood duration (R = 0.4, $p < 0.05$). Plant biomass also showed a positive correlation with soil moisture content and temperature. The biomass of plant *C. dactylon* was negatively correlated with soil pH, electrical conductivity, and total phosphorus.

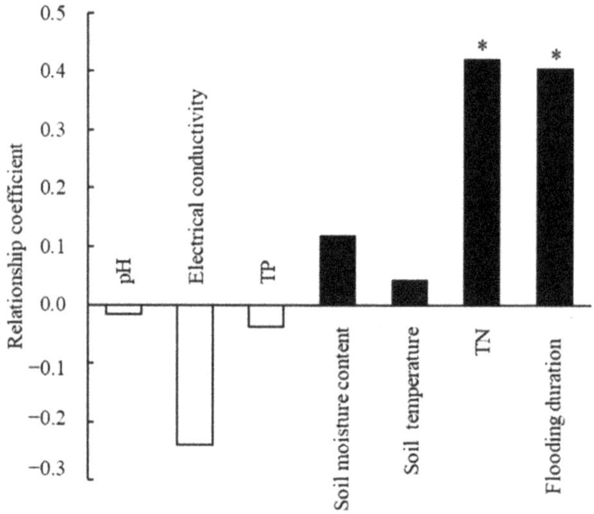

Figure 6. Relationship between plant biomass distribution and environmental factors in the riparian zone of the Daning River. Data are shown as means ± SE ($N = 36$). * $p < 0.05$.

3.4.2. Relationship between C. dactylon Morphological Traits and Environmental Factors

The Spearman's rank correlation between the morphological characteristics of *C. dactylon* and soil environmental factors is shown in Figure 7. Results indicated a correlation between the morphological traits of *C. dactylon* and environmental factors. The plant height of *C. dactylon* was highly positively correlated with its upright stem length (R = 0.95, $p < 0.001$) and significantly negatively correlated with its first creeping stem length (R = −0.44, $p < 0.05$). The upright stem length (R = −0.48, $p < 0.05$) and plant height (R = −0.47, $p < 0.05$) of *C. dactylon* were significantly negatively correlated with soil moisture content.

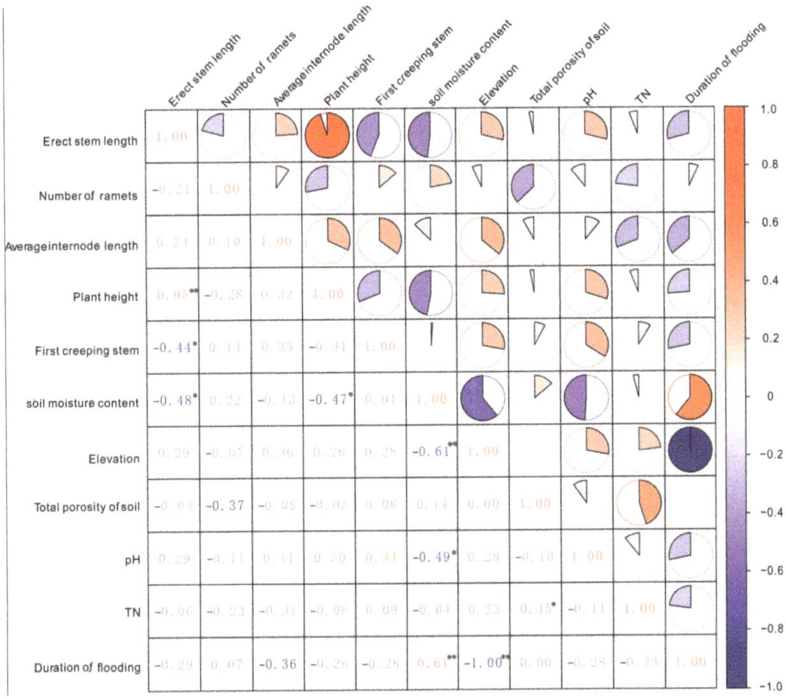

Figure 7. Spearman's rank correlation between *C. dactylon* morphological traits and soil environmental factors. $N = 108$, ** $p < 0.01$; * $p < 0.05$.

4. Discussion

4.1. Influence of Reservoir Water Level Rhythm on the Survival and Growth of Plants in the Riparian Zones

The periodic fluctuation of water levels in the Three Gorges Reservoir disrupts the natural flood–drought pattern of rivers and creates a specific reservoir water level rhythm. The impact of water level elevation on plant communities may be related to resource differentiation and vegetation ecological adaptation differences [1]. In the lower part of the riparian zone, where flooding stress was intense, the establishment of vegetation was hindered, and intolerant plant species perished due to the lack of organismal structures and functions that adapted to extreme environments, resulting in simplified community composition. The upper part of the riparian zone, where microhabitat conditions were more complex and resource combinations were optimal, was conducive to the establishment and growth of vegetation species with a wider ecological niche, leading to a higher species diversity in the community [19,20]. The results of this study indicated that there was a correlation among the main influencing factors (duration of flooding, elevation of water level, and soil moisture content) that affected the spatial distribution of plant communities, and they are all related to the hydrological characteristics of the reservoir. As the elevation of the water level increased, flooding duration, frequency, and depth decreased (Figure 3). Soil moisture content also showed a strong correlation with hydrological factors such as flooding duration, frequency, and depth. For example, soil moisture content was significantly negatively correlated with water level elevation ($R = -0.61$, $p < 0.001$) and significantly positively correlated with the number of days of inundation ($R = 0.61$, $p < 0.001$). These results are consistent with previous studies. Capon [21] and Su et al. [22] considered the duration of flooding as the main factor influencing plant community composition and diversity. In their study on the species richness patterns of riparian plant communities in the Pengxi River, Tong et al. [23] found that flooding duration, soil moisture content, and substrate heterogeneity had important effects on the distribution patterns of plant communities. Wang and Hong [24] found in their study on the effects of the Three Gorges Dam on vegetation coverage at different elevations in the riparian zone that an increase in water level had a negative impact on vegetation coverage below an elevation of 175 m.

4.2. Impact of Multiple Environmental Stressors on the Vegetation Biomass in the Riparian Zone

Previous studies have shown various forms of relationship between riparian plant biomass and elevation of water level [25,26]. It is generally believed that they exhibit a negative correlation, meaning that species diversity decreases with increasing water level [27]. Another form is the "mid-domain bulge", where biomass initially increases and then decreases with increasing water level [28]. The conclusions drawn in this study align with the "mid-domain bulge" theory. Influenced by the fluctuation of water levels in the Three Gorges Reservoir (Figure 4), the diversity index of vegetation in the riparian zone reached its lowest value in the low elevation zone (145~155 m). After experiencing long-term annual water level fluctuations, vegetation biomass in the reservoir riparian zone showed a trend of initially increasing and then decreasing with increasing elevation, reaching its highest point in the mid-elevation zone (155~165 m).

The seasonal inundation–exposure regime in the riparian zone of the Three Gorges Reservoir created specific macro-habitats. However, soil erosion, sediment deposition, and changes in the soil matrix environment have increased habitat fragmentation and vulnerability, making the ecosystem more fragile and sensitive [29]. The periodic rise and fall of water levels disrupted the stability of soil structure through water erosion and sediment deposition, leading to soil nutrient loss and unstable growth substrates [30,31]. The accumulated duration of flooding formed during the periodic rise and fall of water levels (Figure 3) primarily affects soil physical properties such as soil moisture content and porosity. For example, with the increasing duration of flooding, soil moisture content showed a significant positive correlation ($R = 0.61$, $p < 0.01$). It further affected soil chemical properties, such as organic matter and nutrient content. Soil moisture content was significantly negatively correlated with soil pH ($R = -0.49$, $p < 0.05$), and soil bulk density was significantly negatively correlated with total nitrogen content ($R = -0.45$, $p < 0.05$). Therefore, the accumulated duration of flooding and average flooding depth formed by the periodic rise and fall of water levels were the primary stressors determining the differentiation of vegetation biomass along the elevation gradient. The intense and prolonged flooding and delayed exposure in the lower part of the riparian zone hindered the photosynthesis and metabolic processes of vegetation. To reduce energy consumption, plants adopted strategies such as reducing population density and allocating more resources to reproduction, thereby slowing down plant growth [30]. In the upper part of the riparian zone, vegetation was greatly affected by land-based infrastructure and human factors, and the low moisture content was unfavorable for plant growth and nutrient uptake, resulting in a distribution trend of low biomass at both ends and high biomass in the middle.

Soil, as an important component of material and energy cycling in the riparian zone, plays a crucial role in coordinating plant growth and supplying nutrients, thus determining the productivity of plant communities. The periodic rise and fall of water levels lead to the breakdown of large-sized soil aggregates into microaggregates, accelerating the release, transport, and diffusion of soil nutrients and resulting in nutrient-poor soil conditions in the riparian zone. The positive correlation between vegetation biomass and duration of flooding and total nitrogen content indicated that these factors were the main soil limiting factors determining the differentiation of vegetation biomass along the elevation gradient. High concentrations of total nitrogen can stimulate seed germination [32], promote root absorption to maintain plant nitrogen stoichiometry balance, and enhance chlorophyll synthesis to increase ecosystem productivity. Thus, soil total nitrogen is in line with changes in vegetation biomass. On the other hand, the duration of flooding can increase soil moisture content. As an important carrier of energy cycling, soil moisture content affects the transformation and transport of nutrients in the soil, thereby determining the efficiency of vegetation in utilizing soil water and nutrients and promoting biomass accumulation.

4.3. Adaptation Mechanisms of C. dactylon Morphology in the Reservoir Riparian Zones

The morphological mechanisms by which plants in the riparian zone adapt to flooding stress involve various processes that facilitate gas transport to avoid hypoxia. These mechanisms include the development of adventitious roots, the formation of root and leaf aerenchyma, and the formation of leaf gas films, all of which enhance oxygen and carbon dioxide exchange in plants and maintain root aeration status [33]. During the flooding period, the root aerenchyma tissue of *C. dactylon* develops [34] and root biomass increases [35], indicating that the roots remain vital, enabling quick sprouting after the riparian zone is exposed. Additionally, during flooding, the aboveground stolons of *C. dactylon* quickly die off, while the underground rhizomes firmly anchor in the soil, absorbing soil nutrients and storing energy. When the aboveground stems of *C. dactylon* emerge from the water surface, photosynthesis and aerobic respiration can take place, facilitating rapid plant growth.

Plant height, upright stem length, primary stolon length, and root length of *C. dactylon* in the riparian zone were positively correlated with elevation and negatively correlated with flooding duration. The dominance of *C. dactylon* in the 145~155 m elevation of the Three Gorges Reservoir riparian zone may be related to its internal aerenchyma tissue [36–38]. In the low elevation zone, the plant height and average internode length of *C. dactylon* decreased, whereas the number of tillers increased. This facilitated the rapid diffusion of gases from stems to roots, providing oxygen for root respiration and promoting root growth. Consequently, it increased tissue porosity and oxygen leakage, enhancing the porosity of primary and adventitious roots and the waterlogging tolerance of the plant. In the high elevation zone, as the duration of waterlogging decreased, the time available for photosynthesis and aerobic respiration increased, leading to more pronounced growth. Considering the growth pattern of *C. dactylon* under flooding conditions, shorter flooding durations allowed for a longer period of photosynthesis in the aboveground stems, resulting in longer vegetation growth. This aligns with the findings of this study.

5. Conclusions

This study focused on the riparian zone of the Daning River, a typical tributary in the Three Gorges Reservoir area. Through field investigations and laboratory analysis, the distribution characteristics of riparian vegetation under different soil types and elevations of water level were revealed. *C. dactylon*, as the dominant species, was selected for studying its plant characteristics under different soil types and elevations, and the environmental factors influencing its growth were explored. The main conclusions are as follows:

(1) There was no significant difference in plant biomass and morphological indicators between the yellow loam and purple soil riparian zones along the same elevation gradient.
(2) Plant biomass showed a trend of initial increase followed by a decrease with increasing elevation of water level, reaching a higher value in the riparian zone at an elevation of 155–165 m.
(3) Soil total nitrogen content was identified as a key limiting factor for plant biomass.
(4) The plant height and stolon length of *C. dactylon* decreased with increasing flooding duration, facilitating the rapid diffusion of gases from stems to roots, providing oxygen for root respiration, and adapting to long-term flooding in the riparian zone. Consequently, *C. dactylon* became the dominant species in the 145–155 m elevation range in the riparian zone.

Author Contributions: Conceptualization, X.L., S.L. and Z.L.; methodology, X.L. and S.L.; software, X.L., S.L. and Z.W.; validation, Y.X. and Z.L.; formal analysis, X.L., S.L. and Z.W.; investigation, X.L. and S.L.; resources, S.L., Z.L. and Y.X.; data curation, X.L. and S.L.; writing—original draft preparation, X.L., S.L. and Z.W.; writing—review and editing, X.L., S.L., Y.X., Z.W. and Z.L.; visualization, X.L., S.L. and Z.W.; supervision, Z.L.; project administration, S.L. and Z.L.; funding acquisition, S.L., Y.X. and Z.L. All authors have read and agreed to the published version of the manuscript.

Funding: This research was funded by the National Natural Science Foundation of China project (No. 51809287, No. 42202269), the Young Elite Scientists Sponsorship Program by CAST (No. 2022QNRC001), and the Follow-up Work of the Three Gorges Project (No. 2136902).

Data Availability Statement: The data that support the findings of this study are available from the corresponding author upon reasonable request.

Acknowledgments: We thank Aimin Cai, Yijie Liu and Pengcheng Du for their help with sample collection. We also thank Xiaoru Su for the English translation.

Conflicts of Interest: The authors declare no conflict of interest.

References

1. Li, S.; Wang, Y.; Hu, L.; Zhao, J.; Liao, X.; Xie, T.; Wen, J.; Bao, Y.; Li, L. Nitrogen fixation of *Cyndon dactylon*: A possible strategy coping with long-term flooding in the Three Gorges Reservoir. *Sci. Total Environ.* **2023**, *866*, 161422. [CrossRef] [PubMed]
2. Wang, Y.; Yeh, T.C.J.; Wen, J.C.; Gao, X.; Zhang, Z.; Huang, S.Y. Resolution and ergodicity issues of river stage tomography with different excitations. *Water Resour. Res.* **2019**, *55*, 4974–4993. [CrossRef]
3. Chen, S.; Li, S.; Liu, L.; Wang, Y.; Zeng, X.; Long, S.; Zhou, H.; Yang, J.; Li, F.; Luo, H. Seasonal variations alter the effect of an invasive plant on the decomposition of a native plant in a subtropical eutrophic lake, China. *Hydrobiologia* **2022**, *849*, 4153–4165. [CrossRef]
4. Ye, C.; Butler, O.M.; Chen, C.; Liu, W.; Du, M.; Zhang, Q. Shifts in characteristics of the plant-soil system associated with flooding and revegetation in the riparian zone of Three Gorges Reservoir, China. *Geoderma* **2020**, *361*, 114015. [CrossRef]
5. Huang, Y.; Wang, J.; Yang, M. Unexpected sedimentation patterns upstream and downstream of the Three Gorges Reservoir: Future risks. *Int. J. Sediment Res.* **2019**, *34*, 27–36. [CrossRef]
6. Ren, Q.; Li, C.; Yang, W.; Hong, S.; Ma, P.; Wang, C.; Schneider, R.L.; Morreale, S.J. Revegetation of the riparian zone of the Three Gorges Dam Reservoir leads to increased soil bacterial diversity. *Environ. Sci. Pollut. Res.* **2018**, *25*, 23748–23763. [CrossRef]
7. Wen, Z.; Ma, M.; Zhang, C.; Yi, X.; Chen, J.; Wu, S. Estimating seasonal aboveground biomass of a riparian pioneer plant community: An exploratory analysis by canopy structural data. *Ecol. Indic.* **2017**, *83*, 441–450. [CrossRef]
8. Al-Snafi, A.E. Chemical constituents and pharmacological effects of *Cynodon dactylon*—A review. *J. Pharm.* **2016**, *6*, 17–31. [CrossRef]
9. Valencia-Gredilla, F.; Royo-Esnal, A.; Juárez-Escario, A.; Recasens, J. Different ground vegetation cover management systems to manage *Cynodon dactylon* in an irrigated vineyard. *Agronomy* **2020**, *10*, 908. [CrossRef]
10. Ye, C.; Zhang, K.; Deng, Q.; Zhang, Q. Plant communities in relation to flooding and soil characteristics in the water level fluctuation zone of the Three Gorges Reservoir, China. *Environ. Sci. Pollut. Res.* **2013**, *20*, 1794–1802. [CrossRef]
11. Xiao, L.; Zhu, B.; Kumwimba, M.N.; Jiang, S. Plant soaking decomposition as well as nitrogen and phosphorous release in the water-level fluctuation zone of the Three Gorges Reservoir. *Sci. Total Environ.* **2017**, *592*, 527–534. [CrossRef] [PubMed]
12. Schneider, F.D.; Morsdorf, F.; Schmid, B.; Petchey, O.L.; Hueni, A.; Schimel, D.S.; Schaepman, M.E. Mapping functional diversity from remotely sensed morphological and physiological forest traits. *Nat. Commun.* **2017**, *8*, 1441. [CrossRef]
13. Herms, C.H.; Hennessy, R.C.; Bak, F.; Dresbøll, D.B.; Nicolaisen, M.H. Back to our roots: Exploring the role of root morphology as a mediator of beneficial plant–microbe interactions. *Environ. Microbiol.* **2022**, *24*, 3264–3272. [CrossRef]
14. Li, S.; Cui, B.; Bai, J.; Xie, T.; Yan, J.; Wang, Q.; Zhang, S. Effects of soil abiotic factors on the plant morphology in an intertidal salt marsh, Yellow River Delta, China. *Phys. Chem. Earth Parts A/B/Cb* **2018**, *103*, 75–80. [CrossRef]
15. Wang, H.; Wang, R.; Harrison, S.P.; Prentice, I.C. Leaf morphological traits as adaptations to multiple climate gradients. *J. Ecol.* **2022**, *110*, 1344–1355. [CrossRef]
16. Peng, X.; Shi, D.; Jiang, D.; Wang, S.; Li, Y. Runoff erosion process on different underlying surfaces from disturbed soils in the Three Gorges Reservoir Area, China. *Catena* **2014**, *123*, 215–224. [CrossRef]
17. Li, S.; Cui, B.; Xie, T.; Zhang, K. Diversity pattern of macrobenthos associated with different stages of wetland restoration in the yellow river delta. *Wetlands* **2016**, *36*, 57–67. [CrossRef]
18. Li, S.; Cui, B.; Xie, T.; Bai, J.; Wang, Q.; Shi, W. What drives the distribution of crab burrows in different habitats of intertidal salt marshes, Yellow River Delta, China. *Ecol. Indic.* **2018**, *92*, 99–106. [CrossRef]
19. Guo, Y.; Yang, S.; Shen, Y.F.; Xiao, W.F.; Cheng, R.M. Study on the natural distribution characteristics and community species diversity of existing plants in the Three Gorges Reservoir. *Acta Ecol. Sin.* **2019**, *39*, 4255–4265.
20. Hong, M.; Guo, Q.; Nie, B.; Kang, Y.; Pei, S.; Jin, J.; Wang, X. Responses of *Cynodon dactylon* population in hydro-fluctuation belt of Three Gorges Reservoir area to flooding-drying habitat change. *Chin. J. Appl. Ecol.* **2011**, *22*, 2829–2835.

21. Capon, S.J. Flood variability and spatial variation in plant community composition and structure on a large arid floodplain. *J. Arid. Environ.* **2005**, *60*, 283–302. [CrossRef]
22. Su, X.; Zeng, B.; Qiao, P.; Ayi, Q.; Huang, W. The effects of winter water submergence on flowering phenology and reproductive allocation of *Salix variegata Franch.* in Three Gorges reservoir region. *Acta Ecol. Sin.* **2010**, *30*, 2585–2592.
23. Tong, X.X.; Chen, C.D.; Wu, S.J.; Jia, Z.Y.; Yi, X.M.; Ma, M.H. Spatial distribution pattern of plant community and habitat impact analysis of the drawdown zone of Pengxi River in the Three Gorges Reservoir. *Acta Ecol. Sin.* **2018**, *38*, 571–580.
24. Wang, Y.; Liu, Y.; Liu, S.; Huang, H. Vegetation reconstruction in the water-level-fluctuation zone of the Three Gorges Reservoir. *Chin. Bull. Bot.* **2005**, *22*, 513–522.
25. Lomolino, M.V. Elevation gradients of species-density: Historical and prospective views. *Glob. Ecol. Biogeogr.* **2011**, *10*, 3–13. [CrossRef]
26. Tilman, D.; Reich, P.B.; Knops, J.M. Biodiversity and ecosystem stability in a decade-long grassland experiment. *Nature* **2006**, *441*, 629–632. [CrossRef] [PubMed]
27. Wang, Q.; Yuan, X.; Willison, J.H.M.; Zhang, Y.; Liu, H. Diversity and above-ground biomass patterns of vascular flora induced by flooding in the drawdown area of China's Three Gorges Reservoir. *PLoS ONE* **2014**, *9*, e100889. [CrossRef] [PubMed]
28. Oommen, M.A.; Shanker, K. Elevational species richness patterns emerge from multiple local mechanisms in Himalayan woody plants. *Ecology* **2005**, *86*, 3039–3047. [CrossRef]
29. Bao, Y.; He, X.; Anbang, W.; Gao, P.; Tang, Q. Dynamic changes of soil erosion in a typical disturbance zone of China's Three Gorges Reservoir. *Catena* **2018**, *169*, 128–139. [CrossRef]
30. He, X.; Bao, Y. Research advances on soil erosion and ecological restoration in the riparian zone of the Three Gorges Reservoir. *Sci. Soil Water Conserv.* **2019**, *17*, 160–168.
31. Li, L.; Xie, T.; Zhang, S.; Yuan, Z.; Liu, M.; Li, C. Characteristics of nutrient content and enzyme activity in the rhizosphere and bulk soils of four suitable plant species in the hydro-fluctuation zone of the Three Gorges Reservoir. *Acta Ecol. Sin.* **2020**, *40*, 7611–7620.
32. Li, T.; Zhu, Z.; Shao, Y.; Chen, Z.; Roß-Nickoll, M. Soil seedbank: Importance for revegetation in the water level fluctuation zone of the reservoir area. *Sci. Total Environ.* **2022**, *829*, 154686. [CrossRef]
33. Peng, C.; Zhang, L.; Qin, H.; Li, D. Revegetation in the water level fluctuation zone of a reservoir: An ideal measure to reduce the input of nutrients and sediment. *Ecol. Eng.* **2014**, *71*, 574–577. [CrossRef]
34. Abiko, T.; Kotula, L.; Shiono, K.; Malik, A.I.; Colmer, T.D.; Nakazono, M. Enhanced formation of aerenchyma and induction of a barrier to radial oxygen loss in adventitious roots of *Zea nicaraguensis* contribute to its waterlogging tolerance as compared with maize (*Zea mays* ssp. *mays*). *Plant Cell Environ.* **2012**, *35*, 1618–1630. [CrossRef] [PubMed]
35. Li, Q.; Ding, W.; Wang, S.; Zhu, Q.; Yang, J.; Ke, S.; Qin, L.; Yang, L.; Zheng, J.; Meng, Y. Influence of multi-year high water level running on growth recovery of *Cynodon dactylon* population in water-level-fluctuating zone of the Three Gorges Reservoir. *Acta Ecol. Sin.* **2020**, *40*, 985–992.
36. Ribeiro, C.L.; Silva, C.M.; Drost, D.R.; Novaes, E.; Novaes, C.R.; Dervinis, C.; Kirst, M. Integration of genetic, genomic and transcriptomic information identifies putative regulators of adventitious root formation in Populus. *BMC Plant Biol.* **2016**, *16*, 1–11. [CrossRef]
37. Voesenek, L.; Colmer, T.; Pierik, R.; Millenaar, F.; Peeters, A. How plants cope with complete submergence. *New Phytol.* **2006**, *170*, 213–226. [CrossRef]
38. Yamauchi, T.; Watanabe, K.; Fukazawa, A.; Mori, H.; Abe, F.; Kawaguchi, K.; Oyanagi, A.; Nakazono, M. Ethylene and reactive oxygen species are involved in root aerenchyma formation and adaptation of wheat seedlings to oxygen-deficient conditions. *J. Exp. Bot.* **2014**, *65*, 261–273. [CrossRef]

water

MDPI

Article

Biochar/Clay Composite Particle Immobilized Compound Bacteria: Preparation, Collaborative Degradation Performance and Environmental Tolerance

Pengfei Sun [1,2,3], Jun Wei [4], Yaoyao Gao [4], Zuhao Zhu [1,2,3] and Xiao Huang [4,*]

1 Ministry of Natural Resources, Fourth Institute of Oceanography, Beihai 536000, China; sunpengfei@4io.org.cn (P.S.); zhuzuhao@4io.org.cn (Z.Z.)

2 Key Laboratory of Tropical Marine Ecosystem and Bioresource, Ministry of Natural Resources, Beihai 536000, China

3 Guangxi Beibu Gulf Key Laboratory of Marine Resources, Environment and Sustainable Development, Beihai 536000, China

4 Jiangsu Key Laboratory of Atmospheric Environment Monitoring and Pollution Control, Collaborative Innovation Center of Atmospheric Environment and Equipment Technology, School of Environmental Science and Engineering, Nanjing University of Information Science and Technology, Nanjing 210000, China; 20211248087@nuist.edu.cn (J.W.); 202212480164@nuist.edu.cn (Y.G.)

* Correspondence: 003199@nuist.edu.cn

Abstract: Immobilized microbial materials can effectively remove pollutants from surface water, and a biochar/clay composite particle (BCCP) material is prepared with immobilized *Flavobacterium mizutaii* sp. and *Aquamicrobium* sp. to remove ammonia nitrogen (NH_4^+–N) and petroleum hydrocarbons (PHCs). The results indicated that the optimal ratios of biochar, Na_2SiO_3 and $NaHCO_3$ were 15%, 3%, and 3%, and the adsorption process was found to be better described with the pseudo-second-order kinetic equation. The individual immobilization of *Flavobacterium mizutaii* sp. and *Aquamicrobium* sp. with sodium alginate–polyvinyl alcohol (PVA + SA) achieved 80% and 90% removal efficiencies for NH_4^+–N and PHCs at the 10th d. The composite immobilization of two efficient bacteria could degrade 82.48% NH_4^+–N and 74.62% PHCs. In addition, immobilization relieved the effects of temperature and salinity. This study can provide guidance for the application of immobilized microbial composite materials in natural water environments.

Keywords: biochar; composite particle; immobilization; compound bacteria; environmental tolerance

1. Introduction

Wetland systems are special zones between marine and terrestrial ecosystems that play a crucial role in maintaining ecological balance [1]. The Liaohe Estuarine Wetland (LEW) is an important crab breeding area and also a reused oil extraction area. In recent years, aquaculture and oil exploration have led to excessive emissions of ammonia nitrogen (NH_4^+–N) and severe petroleum pollution in wetland water systems, resulting in damage to the living environment of river crabs and affecting the ecological environment of wetlands [2]. Therefore, it is particularly important to control the concentrations of NH_4^+–N and petroleum hydrocarbons (PHCs) in wetlands.

Secondary pollution during the removal of NH_4^+–N and PHCs from surface water using conventional chemical methods used in wastewater treatment plants renders these methods unsuitable. The adsorption method has a wide application range and good treatment performance. Previous studies have shown that the use of biochar and clay to prepare biochar/clay composite particles (BCCP) to adsorb NH_4^+–N has a significant effect [3]. However, adsorption alone cannot completely degrade petroleum pollutants such as PHCs. With the development of biotechnology, many microbial strains can be screened and applied for the degradation of NH_4^+–N and PHCs. *Flavobacterium mizutaii* sp. was reported as a predominant bacterial genus in the denitrification process and can effectively degrade NH_4^+–N in water [4]. *Aquamicrobium* sp. was found to be effective in degrading alkanes [5]. However, the use of high-efficiency bacterial agents is affected by the persistence of residence in freely flowing water bodies. Using BCCP as a carrier to immobilize efficient bacterial communities on BCCP can effectively maintain the concentration of microorganisms and provide pollutant degradation effects [6]. Meanwhile, environmental factors, such as salinity and temperature, restrict pollutant degradation, and it is necessary to study the degradation of pollutants by immobilized bacteria technology in wetlands [6]. The activity of nitrite oxidizing bacteria (NOB) gradually decreased as the salinity increased from approximately zero to 35.0 g/L [7]. When the temperature was lower than the optimal temperature, it affected the growth rate of bacteria, while higher temperatures reduced protein activity and even lead to cell death [8].

Sodium alginate (SA) and sodium alginate–polyvinyl alcohol (PVA + SA) as crosslinking materials can protect the bacteria against the intrusion of the environment [9]. Reddy and Osborne 2020 immobilized *Pseudomonas guariconensis* in the biocarrier matrix to degrade Reactive red 120, and the degradation efficiency could reach 91% [10]. Yan et al., 2020 found that the degradation efficiency of Ca^{2+} and Mg^{2+} can reach 90% and 70% under the action of immobilized *Lysinibacillus fusiformis DB1-3* bacteria [11]. Hence, immobilization technology has good performance on the microbial degradation of pollutants. The previous research showed that the immobilized ammonia-oxidizing bacteria (AOB) could resist the influence of low temperature and maintained a good degradation efficiency [12]. However, LEW has received severe combined pollution of PHCs and NH_4^+–N, and whether immobilized composite microbial communities still have good effects is unknown. Hence, the degradation performance of combined immobilization of oil degrading bacteria and AOB needs further research.

In addition, the LEW is located in northern China and is affected by low temperature, high salinity and tides. The application of high-efficiency degrading bacteria is a challenge. Therefore, immobilization methods with BCCP do not only resist low temperature and high salinity but also avoid being dispersed by tides in the wetland. Immobilized BCCP is an effective method to solve the above problems, and its tolerance to low temperature and high salt needs to be further explored.

In this study, we investigated the effectiveness of immobilized microbial composite materials in removing NH_4^+–N and PHCs in wetland environments. (1) Orthogonal experiments and adsorption kinetics were studied to explore the optimal formulation of BCCP, and the adsorption effectiveness of BCCP on NH_4^+–N was investigated. (2) The degradation efficiency of NH_4^+–N and PHCs by BCCP immobilized with *Flavobacterium mizutaii* sp. and *Aquamicrobium* sp. was studied. (3) The tolerance to low temperatures and high salinity on the immobilized microorganisms was examined. This experiment can provide guidance for the application of immobilized microbial composite materials in wetlands.

2. Materials and Methods

2.1. Preparation of BCCP

Clay and reed stalk materials were obtained from the LEW. The reed stalks were repeatedly washed with deionized water to remove impurities. Then, they were placed in a crucible and dried at 105 °C for 24 h to eliminate any remaining moisture and impurities. The dried reed stalks were ground into powder using a mini plant grinder (FZ 102, Beijing Weiye, Nanhai, China), and the resulting powder was sieved to obtain particles with a size of 0.85 mm for further use. The reed stalk powder was subjected to carbonization in a pyrolyzer under a constant oxygen-limited condition with a heating rate of 10 °C per minute. The carbonization process was carried out at 600 °C for 3 h. Subsequently, the biochar and clay samples were crushed and sieved to obtain a uniform particle size of 0.15 mm. The biochar samples were dried at 105 °C for 24 h and then sealed in brown containers. Prior to use, they were rinsed multiple times with deionized water to remove any remaining ash content. Detailed information regarding the characteristics of the biochar can be found in [3]. The specific preparation process of the biochar/clay is shown in Figure 1.

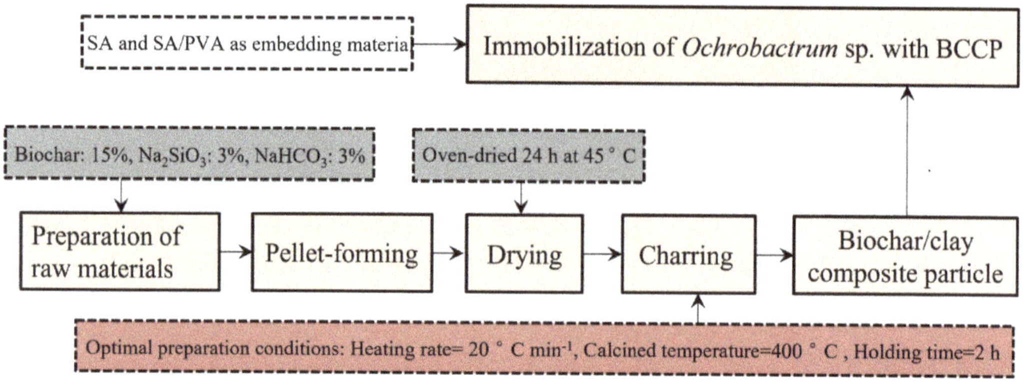

Figure 1. The preparation process of BCCP.

The formulation for the preparation of BCCP includes the base amount of binder, the amount of Na_2SiO_3 and the amount of $NaHCO_3$. An orthogonal experiment, L_9 (3^4), was conducted to optimize the best preparation method. The levels of the orthogonal experiment for the preparation formulation are shown in Table 1.

Table 1. The orthogonal test table level of BCCP preparation factors.

Number.	Biochar Dosage (%)	Na_2SiO_3 Dosage (%)	$NaHCO_3$ Dosage (%)
1	5	1	1
2	5	2	2
3	5	3	3
4	10	1	2
5	10	2	3
6	10	3	1
7	15	1	3
8	15	2	1
9	15	3	2

2.2. *Flavobacterium mizutaii sp. and Aquamicrobium sp.*

The screening methods for *Flavobacterium mizutaii* sp. and *Aquamicrobium* sp. can be found in Huang et al., 2017 and Huang et al., 2022 [13,14]. The 16S rDNA gene of strain HXN-2 has been cloned and sequenced using the SeqMatch program in RDP (http://rdp.cme.msu.edu/ (accessed on 10 July 2023)). In Figure 2, it has been classified that HXN-2 shows a 95% similarity to the 16S rDNA gene of *Aquamicrobium* sp. genus [13].

The 16S rDNA gene of strain SY-I has been cloned and sequenced, and the gene sequence has been submitted to GenBank. Using the SeqMatch program in RDP (http://rdp.cme.msu.edu/ (accessed on 10 July 2023)) and conducting a BLAST analysis against the online database, it was found that SY-I shows a 94% similarity to the 16S rDNA gene of *Flavobacterium mizutaii* sp. Through searching for other closely related strains to SY-I and using software such as MNGA, a 16S rDNA phylogenetic tree has been constructed. Table 2 shows the physiological and chemical reactions of high-efficiency degrading bacteria.

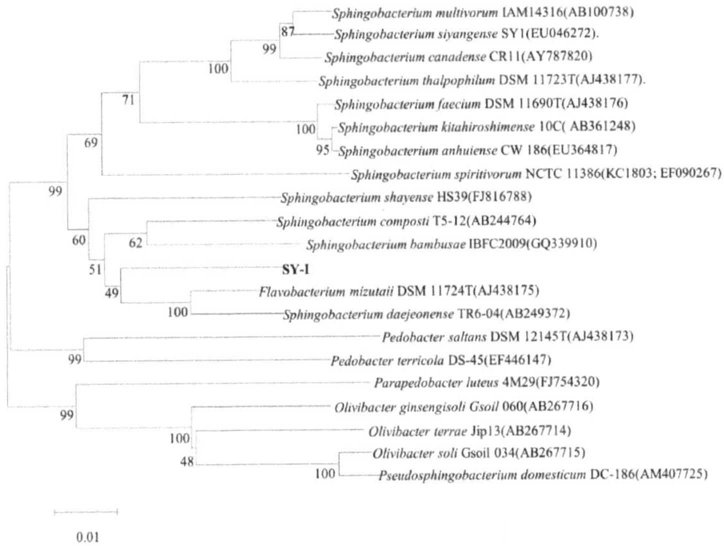

Figure 2. Phylogenetic tree of strain SY-I based on the complete sequences of the 16SrDNA gene.

Table 2. Physiological and biochemical responses of HXN-2 and SY-1.

No. of Strains	Catalase Test	Starch Hydrolysis Test	Citrate Test	MR Test	Glucose Oxidation Fermentation Test	VP Test	Indole Test
HXN-2	+	+	−	+	−	−	+
SY-I	−	−	−	+	−	+	+

2.3. Adsorption kinetics of $NH_4{}^+$–N by BCCP

In order to comprehensively understand the adsorption kinetics characteristics of biochar spheres on $NH_4{}^+$–N, this study used the pseudo-first-order kinetic Equation (1), pseudo-second-order kinetic Equation (2) and intra-particle diffusion model (3) to fit the experimental data.

$$q_t = q_e\left(1 - e^{-k_1 t}\right) \quad (1)$$

$$\frac{t}{q_t} = \frac{1}{k_2 {q_e}^2} + \frac{t}{q_e} \quad (2)$$

$$q_t = k_p\sqrt{t} + C \quad (3)$$

In the equations, k_1 represents the rate constant of the pseudo-first-order kinetic equation, min^{-1}; k_2 represents the rate constant of the pseudo-second-order kinetic equation, g/mg min; k_p represents the rate constant of intra-particle diffusion, mg/g $min^{0.5}$; q_e represents the adsorption capacity of the biochar spheres, expressed in mg/g; and q_t represents the adsorption capacity of the biochar spheres at time t, mg/g.

2.4. Preparation of Immobilized Compound Bacteria

A 2% SA solution and a 12% PVA solution were prepared with deionized water. Then, a certain amount of $CaCl_2$ was weighed and dissolved in deionized water to prepare a 2% $CaCl_2$ solution. Both solutions dissolved in a constant-temperature water bath at 100 °C, and the solution was then sterilized at 121 °C and high pressure for 30 min. *Flavobacterium mizutaii* sp. and *Aquamicrobium* sp. were concentrated using a centrifuge at 4000 rpm, 20 °C for 10 min once they were cultured in logarithmic growth phase (OD600≈0.6). The supernatant was discarded and rinsed with sterile water to remove surface nutrients, and this process was repeated 2–3 times. The cultured and concentrated bacterial strains were mixed with the embedding material in a 1:2 ratio to obtain an embedding mixture. In addition, the BCCP was added into the embedding mixture and then removed into a 2% $CaCl_2$ solution, where gel particles formed. The prepared gel particles were placed in the sterilized $CaCl_2$ solution and then crosslinked in a refrigerator at 4 °C for 24 h.

2.5. Degradation of NH_4^+–N and PHCs by Immobilized Compound Bacteria

The experiment involved the addition of four different treatments to a 100 mL solution containing NH_4^+–N at a concentration of 50 mg L^{-1} and PHCs at a concentration of 1000 mg L^{-1}. The treatments included the following: (1) Control group: addition of BCCP alone; (2) FB group: addition of free bacteria; (3) P-B group: addition of BCCP followed by the adsorption of bacterial species; and (4) P-B-SA + PVA group: addition of BCCP with SA+PVA encapsulated bacterial species. The experiment was run continuously for 10 d to investigate the removal efficiency of NH_4^+–N and PHCs. The concentration of NH_4^+–N in the effluent was detected every day, and a sample of the effluent was taken every 3 days to test for PHCs.

2.6. Tolerances of Low Temperature and High Salinity

The preparation of high-efficiency bacteria with SA+PVA immobilized BCCP particles was based on the above immobilization method. They were placed separately under temperature conditions of 10, 15, 20, 25, 30 and 35 °C as well as salinity conditions of 10‰, 15‰, 20‰, 25‰, 30‰ and 35‰ They were run continuously to investigate the performance of temperature and salinity on NH_4^+–N and PHCs degradation.

2.7. Analysis Methods

The determination method for NH_4^+–N involves using Nessler's reagent and spectrophotometry. The analysis method for PHCs is as follows. Transfer the test water sample along with 2.0 g of anhydrous sulfuric acid to a 250 mL separating funnel and mix well. Add 10 mL of n-hexane and rinse the sample bottle twice with 10 mL of n-hexane; then, transfer all the rinsing solution to the separating funnel. Shake the separating funnel for 5 min (release any trapped gas), and let it stand for 10 min. After sufficient phase separation between the extract and the water sample, transfer the lower aqueous layer back to the original water sample bottle. Use filter paper to remove any moisture from the neck of the separating funnel. Transfer the n-hexane extract to a 50 mL stoppered colorimetric tube. Repeat the process two more times to ensure complete extraction of PHCs from the test sample. Transfer the extract to a 1 cm quartz cuvette and measure the absorbance (A) at a wavelength of 225 nm using n-hexane as a reference. Record the measured data and calculate the concentration of PHCs in the water sample according to Formula (4):

$$C_{oil} = \frac{QV_1}{V_2} \quad (4)$$

In the formula, C_{oil} represents the concentration of oil in the water sample, mg L^{-1}; Q represents the concentration of oil in the n-hexane extract obtained from the standard curve, mg L^{-1}; V_1 represents the volume of n-hexane extraction solvent, mL; and V_2 represents the volume of the sample, mL.

3. Results

3.1. Preparation and Condition Optimization of BCCP

3.1.1. Effect of Preparation Formula on the Adsorption Characteristics of NH_4^+–N

The control of the preparation formula (biochar dosage, Na_2SiO_3 dosage and $NaHCO_3$ dosage) could change the surface structure of biochar to a certain extent and affect its adsorption performance for NH_4^+–N [15]. Therefore, in this study, the orthogonal experiment method was used to investigate the effect of the BCCP preparation formula on NH_4^+–N adsorption (Table 3). The preparation formula of BCCP had a great influence on the adsorption of NH_4^+–N. The minimum adsorption capacity was 0.452 mg/g; the maximum was 0.538 mg/g. With the dosage increase of the biochar, Na_2SiO_3 and $NaHCO_3$, the adsorption capacity of the prepared biochar spheres showed an increasing trend. When the dosage ratio of biochar, Na_2SiO_3 and $NaHCO_3$ was 15%, 3% and 3%, the adsorption capacity of the biochar spheres was the largest, which was 0.528, 0.500 and 0.506 mg/g, respectively. This may be because the increase in biochar dosage increases the specific surface area of the biochar spheres and the porosity of the biochar spheres, which increases the adsorption capacity of the biochar spheres. The addition of the Na_2SiO_3 crosslinking agent increased the interlayer distance of the clay on the basis of maintaining the original layered structure of the clay. Meanwhile, the addition of Na_2SiO_3 increased the amount of Na_2SiO_3 to generate $NaHCO_3$ with CO_2 in the air [3]. After the decomposition of Na_2CO_3 at high temperature, CO_2 was released, which promoted the formation of a microporous structure of the biochar spheres, thereby increasing the adsorption capacity of the biochar spheres [16]. $NaHCO_3$ is decomposed into CO_2 at high temperature, which further increases the porosity of the material and enhances the adsorption capacity of NH_4^+–N.

Table 3. Orthogonal experiment results with different preparation formulas.

Levels	Biochar Dosage (%)	Na_2SiO_3 Dosage (%)	$NaHCO_3$ Dosage (%)	Result
	A	B	C	Adsorption Capacity (mg/g)
1	5	1	1	0.452
2	5	2	2	0.463
3	5	3	3	0.484
4	10	1	2	0.487
5	10	2	3	0.495
6	10	3	1	0.491
7	15	1	3	0.538
8	15	2	1	0.521
9	15	3	2	0.526
Average value 1	0.466	0.492	0.488	
Average value 2	0.491	0.493	0.492	0.506
Average value 3	0.528	0.500	0.506	

Table 4 shows the range analysis of NH_4^+–N adsorption by BCCP preparation formula, and the primary and secondary relationships of the influence of three factors on NH_4^+–N adsorption were judged by the range. Through the experimental results $T_3 > T_1 > T_2$, it showed that the primary and secondary relationships affecting the adsorption capacity of the biochar spheres were CAB ($NaHCO_3$ dosage > biochar dosage > Na_2SiO_3 dosage). Through the analysis of the influence of each factor, it was concluded that $A_3B_3C_3$ was the best biochar configuration formula, and the corresponding parameters were biochar 15%, Na_2SiO_3 3% and $NaHCO_3$ 3%.

Table 4. The range analysis of NH_4^+–N adsorption with different biochar preparation formulas.

Levels	Factor				
	δ1	δ2	δ3	R	T
1	δ11 = −0.029	δ21 = −0.003	δ31 = −0.007	R_{01} = −0.003 R_{11} = −0.029	0.026
2	δ12 = −0.004	δ22 = −0.002	δ32 = −0.003	R_{02} = −0.002 R_{12} = −0.004	0.002
3	δ13 = 0.033	δ23 = 0.005	δ33 = 0.011	R_{03} = 0.033 R_{13} = 0.005	0.028
Primary relation	CAB				
Optimal scheme	$A_3B_3C_3$ (15% − 3% − 3%)				

3.1.2. Effect of Raw Material Ratio on NH_4^+–N Adsorption

By fitting the experimental data of different biochar dosages and different $NaHCO_3$ dosages, the quasi-first-order kinetic equation, quasi-second-order kinetic equation and intra-particle diffusion model of NH_4^+–N adsorption by the biochar spheres were obtained. The kinetic model parameters and correlation coefficients are shown in Tables 5 and 6, respectively. By comparing the correlation coefficients, the quasi-second-order kinetic equation is more suitable for the adsorption of NH_4^+–N by BCCP than the quasi-first-order kinetic equation and the intra-particle diffusion equation under the conditions of different biochar dosages and different $NaHCO_3$ dosages. Intra-particle diffusion is not the control step to control the adsorption rate.

Table 5. The adsorption kinetic parameters of BCCP with different biochar dosing quantities.

$\frac{Biochar}{Clay}$		Pseudo-First-Order			Pseudo-Second-Order			Intra-Particle Diffusion	
	q_e (mg/g)	K^{-1} (min^{-1})	q_{eq} (mg/g)	R^2	K_2 (g/mg min)	q_{eq} (mg/g)	R^2	k_p (g/mg $min^{0.5}$)	R^2
5%	0.450	0.036	0.455	0.929	0.093	0.513	0.986	0.024	0.944
10%	0.482	0.036	0.469	0.939	0.088	0.530	0.991	0.025	0.944
15%	0.518	0.039	0.495	0.916	0.092	0.557	0.983	0.026	0.945

Table 6. The adsorption kinetic parameters of BCCP with different $NaHCO_3$ dosing quantities.

$\frac{Na_2CO_3}{Clay}$		Pseudo-First-Order			Pseudo-Second-Order			Intra-Particle Diffusion	
	q_e (mg/g)	K^{-1} (min^{-1})	q_{eq} (mg/g)	R^2	K_2 (g/mg min)	q_{eq} (mg/g)	R^2	k_p (g/mg $min^{0.5}$)	R^2
1.0%	0.395	0.035	0.370	0.877	0.117	0.414	0.959	0.019	0.960
2.0%	0.404	0.029	0.385	0.899	0.091	0.435	0.953	0.021	0.961
3.0%	0.411	0.039	0.394	0.934	0.121	0.440	0.985	0.020	0.928

The pseudo-second-order kinetic equation can better describe the adsorption process of NH_4^+–N by the biochar spheres [17]. The model includes all the processes of biochar sphere adsorption, namely external membrane diffusion, surface adsorption and intra-particle diffusion. Compared with the pseudo-first-order kinetic model, the formation of chemical bonds affects the pseudo-second-order kinetic adsorption and is the main reason [18], indicating that the adsorption process of BCCP on NH_4^+–N is mainly chemical adsorption.

The adsorption process of BCCP can be divided into three stages, namely external diffusion, internal diffusion (pore internal diffusion) and adsorption reaction stage, and finally, the adsorption equilibrium is reached [19]. In the internal diffusion model, it indicates that the internal diffusion is the rate-determining step of the reaction during the adsorption process. However, the internal diffusion fitting curve did not pass the coordinate origin in this study, which indicated that the internal diffusion was not the only rate-determining step in the adsorption process of NH_4^+–N by BCCP.

3.2. Performances of Individual AOB and Petroleum-Degrading Bacteria Immobilization

As a carrier for microbial immobilization, BCCP can avoid the dispersion of free bacteria and reduce its pollutant degradation performance, which is a common and effective method [20]. However, whether immobilization becomes the main controlling factor limiting the conversion of pollutants needs further research. In this study, the individual immobilization of AOB and petroleum-degrading bacteria were studied to discuss the degradation efficiency of NH_4^+–N and PHCs, and the results are shown in Figure 3A,B.

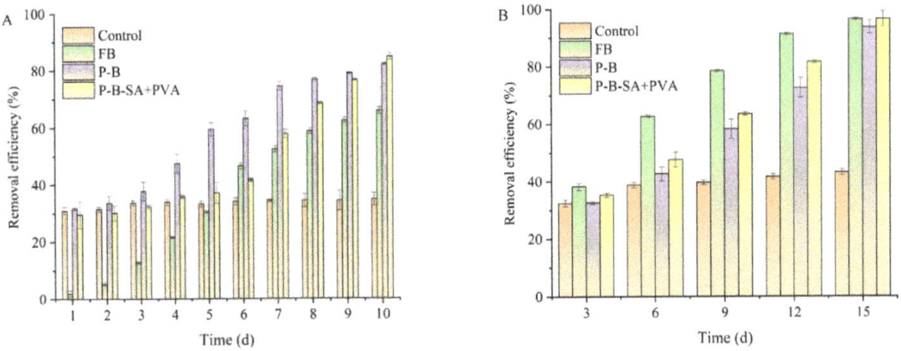

Figure 3. Effect of BCCP-immobilized AOB (**A**) and petroleum-degrading bacteria (**B**).

3.2.1. NH_4^+–N Removal Performance with Immobilized *Flavobacterium mizutaii* sp.

Figure 3A demonstrates the degradation performance of BCCP-immobilized AOB on NH_4^+–N. The removal efficiency of NH_4^+–N in the control group (only added BCCP) was stable at approximately 30%. The control group did not add bacteria but had a certain adsorption effect, which relied on the adsorption of the BCCP themselves. In the experimental group of (FB) (only free bacteria and no composite particles), the effect was little in the first 2 d, and the degradation efficiency was less than 10%. Then, with the time extending, the degradation efficiency continued to increase and reached 65.74% at 10 d.

The experimental groups of P-B (BCCP adsorbed the bacteria) and the SA + PVA combined embedding bacteria increased with the time extending and reached more than 80% at 10 d. The NH_4^+–N rapid adsorption performance of the P-B and P-B-SA + PVA groups at the beginning resulted in a higher removal performance than free bacteria. However, the effect of the P-B-SA + PVA group is lower than that of the P-B group after a period of time, and then the removal efficiency of the P-B-SA + PVA group restored and equivalented to the P-B group with the further extension of the reaction time. It indicated that the embedding first had a certain inhibitory effect on the experiment, and then the permeability became stronger, and the degradation effect was significantly improved with the increase in enrichment [6]. The results showed that BCCP could improve the degradation efficiency of NH_4^+–N as the carrier of immobilized AOB. Due to the porous structure of the biochar sphere that provides a larger surface area and more pores, it can store more substrates and promote the growth of microorganisms [21].

3.2.2. The Effect of Immobilized *Aquamicrobium* sp. on the Removal of PHCs

Figure 3B shows the removal efficiency of PHCs under different conditions. From the figure, it can be observed that the PHC degradation in the control group with the addition of BCCP exhibited a gradual increase from 32.45% on the 3rd d to 43.14% on the 15th d, which indicated the process was only adsorption efficiency. However, upon the addition of bacterial strains, the removal efficiency of PHCs significantly improved. FB, P-B and P-B-SA + PVA demonstrated removal efficiencies of 96.48%, 93.69% and 96.53% for PHCs at the 15th d, respectively. Notably, P-B-SA + PVA showed superior removal performance compared to PB, primarily due to its porous structure, which enhances PHC adsorption while providing a larger substrate reservoir to promote microbial growth [22].

3.3. Performance of Immobilized Compound Bacteria

Individual immobilization of *Aquamicrobium* sp. and *Flavobacterium mizutaii* sp. showed better degradation effects for NH_4^+–N and PHCs. However, whether the composite embedding of the two bacterial communities still has the same effect for NH_4^+–N and PHCs removal needs to be studied and is demonstrated in Figure 4. At the 1st d, the degradation effects of NH_4^+–N and PHCs by the two bacteria were unsatisfactory at only 28.74% and

32.5%, respectively. With the time extending, the adaptability of the flora to the environment was enhanced, and the abundance of microorganisms was enriched [23], resulting in increased degradation of NH_4^+–N and PHCs. At the 10th d, the degradation efficiency reached 74.62% and 82.48%, respectively. Compared with the individual immobilized *Aquamicrobium* sp. and *Flavobacterium mizutaii* sp. (the removal efficiencies of NH_4^+–N and PHCs were 84.69% and 96.53%), the removal efficiencies of NH_4^+–N and PHCs with the immobilized compound bacteria were slightly reduced (individual immobilized bacteria was 10.07% and 14.05% higher than those by the composite embedding).

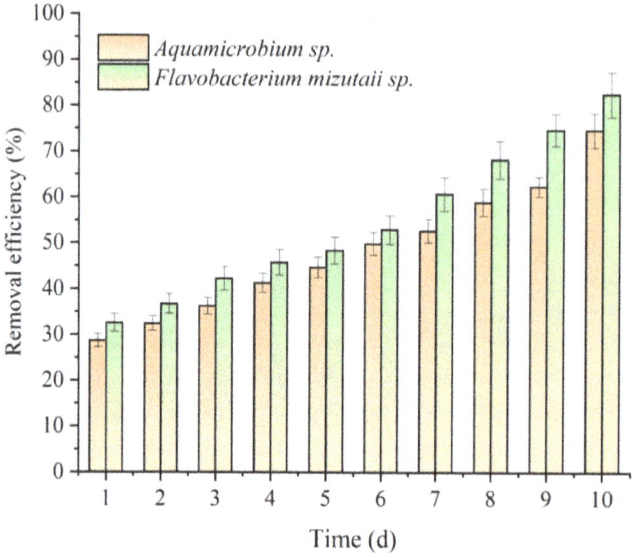

Figure 4. Removal performance of immobilized compound bacteria on NH_4^+–N and PHCs.

3.4. *Tolerance of Immobilized Compound Bacteria to Low Temperature and High Salinity*

3.4.1. Tolerance to Temperature

Temperature is an important environmental factor that affects microbial growth. Figure 5A demonstrates the effect of temperature on the degradation of NH_4^+–N and PHCs by immobilized *Flavobacterium mizutaii* sp. and *Aquamicrobium* sp. As the temperature gradually increased, the degradation efficiency of NH_4^+–N and PHCs also increased. At 10 °C, the removal efficiencies of NH_4^+–N and PHCs were 52.5% and 41.25%, respectively. However, as the temperature rose to 35 °C, the removal efficiencies of NH_4^+–N and PHCs increased to 92.76% and 89.92%, respectively. When the temperature was lower than the optimal growth temperature for *Flavobacterium mizutaii* sp. and *Aquamicrobium* sp., the intracellular activity within the microbial cells decreased, which led to a slower growth rate of bacteria [24]. The previous study by Huang et al. (2017) confirmed that the *Aquamicrobium* sp. was a cold-tolerant AOB and could degrade 53.48% NH_4^+–N when the temperature was 15 °C [13]. The high removal efficiency reflected by SA and PVA immobilization with BCCP is owed to cold-resistant function. Meanwhile, Huang et al. (2022) utilized SA and PVA immobilized petroleum degrading bacteria and found it to have certain salt- and cold-resistance performances [14].

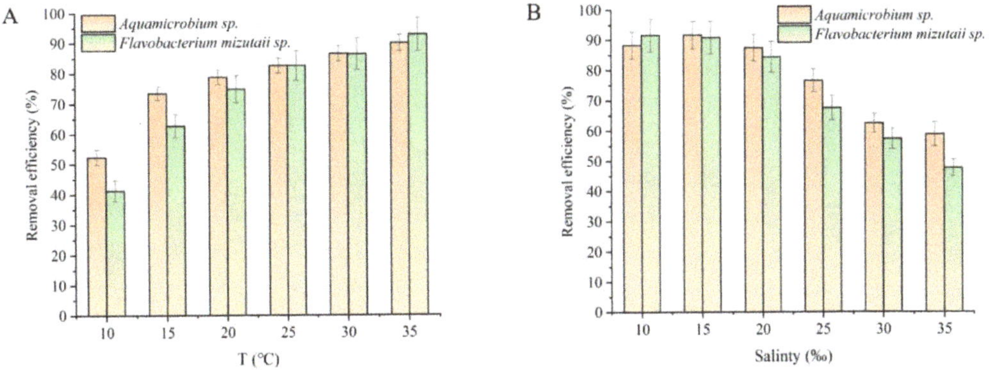

Figure 5. Effects of temperature (**A**) and salinity (**B**) on immobilized compound bacteria.

3.4.2. Tolerance to Salinity

Salinity is another important environmental factor that affects microbial growth. As shown in Figure 5B, the optimal salinity for the degradation of NH_4^+–N and PHCs is 15‰ with removal efficiencies of 88.27% and 91.64%, respectively. However, as salinity increases, the removal efficiencies of PHCs and NH_4^+–N continuously decrease. At a salinity of 35‰, the removal efficiencies of NH_4^+–N and PHCs decrease to 58.74% and 47.68%, respectively. This is mainly due to the impact of high salt environments on cell osmotic pressure, leading to a decrease in microbial abundance [25]. Gao et al., 2020 found that immobilization materials provide protection to bacteria in high-salinity environments [25]. When salinity exceeds 15%, bacterial growth is inhibited, which is consistent with the findings of this study. The previous study by Huang et al., 2017 found that the *Aquamicrobium* sp. was a salt-tolerant AOB, and the NH_4^+–N removal efficiency could reach 60% when the salinity was 20‰ [13]. In this study, the removal efficiencies were 58.74% and 47.68% under the conditions of 35‰ and 20‰, which indicated the SA and PVA immobilization with BCCP can protect bacteria and slow down the impact of high salinity.

3.5. NH_4^+–N and PHCs Degradation Mechanism with BCCP Immobilization

Based on the above research, a preliminary discussion of the mechanisms underlying the degradation of NH_4^+–N by immobilized *Flavobacterium mizutaii* sp. and the degradation of PHCs by *Aquamicrobium* sp. has been conducted and is illustrated in Figure 6. The BCCP exhibits physical adsorption, which facilitates the rapid adsorption and accumulation of NH_4^+–N and PHCs, providing better conditions for subsequent microbial degradation. Additionally, the encapsulation of bacterial populations within the biochar spheres allows for the degradation of high concentrations of NH_4^+–N. The porous nature of biochar provides a habitat for the bacterial community, and its adsorption capacity promotes microbial transformation, leading to the degradation of NH_4^+–N and PHCs [26]. In summary, the BCCP contribute to the effective removal of NH_4^+–N and PHCs through physical adsorption, while the encapsulation of bacterial populations within the biochar spheres and the porous nature of biochar provide favorable conditions for bacterial degradation and transformation processes, ultimately leading to the degradation of NH_4^+–N and PHCs.

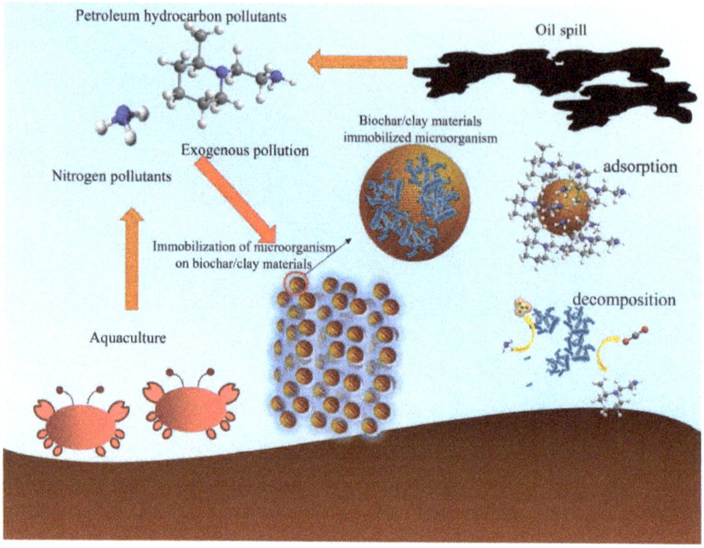

Figure 6. NH_4^+–N and PHCs removal mechanism by immobilization of compound bacteria on BCCP.

4. Conclusions

The results showed that the optimal composition of BCCP was 15% biochar, 3% Na_2SiO_3 and 3% $NaHCO_3$. The primary and secondary factors influencing the adsorption capacity were $NaHCO_3$ dosage > biochar dosage > Na_2SiO_3 dosage. The adsorption process of BCCP for NH_4^+–N fitted the pseudo-second-order kinetic equation. Individual immobilization of AOB (*Flavobacterium mizutaii* sp.) and petroleum-degrading bacteria (*Aquamicrobium* sp.) owned higher removal efficiencies (84.69% and 96.53%) than free bacteria. Compared with the individual immobilized *Aquamicrobium* sp. and *Flavobacterium mizutaii* sp., the removal efficiencies of NH_4^+–N and PHCs with immobilizing compound bacteria were reduced by 10.07% and 14.05%. In addition, immobilization relieved the effects of low temperature and high salinity.

Author Contributions: P.S.: Data curation, Visualization, Methodology, Formal analysis, Writing—original draft. J.W.: Conceptualization, Supervision, Validation, Writing—review and editing. Y.G.: Conceptualization, Formal analysis, Writing—review and editing. Z.Z.: Writing—review and editing. X.H.: Funding acquisition, Writing—review and editing. All authors have read and agreed to the published version of the manuscript.

Funding: This study was funded by the National Natural Science Foundation of China (No. U20A20103), the Special Project of Guangxi Science and Technology Base and Talent (No. GUIKE AD20297065), the Scientific Research Fund of the Fourth Institute of Oceanography (No. 202003), the Science and Technology Planning Projects of Beihai, Guang xi, China (No. 202082042 and 202082031) and the National Major Project of Water Pollution Control and Management Technology in China (No. 2013ZX07202-007).

Data Availability Statement: The data that support the findings of this study are available on request from the corresponding author H. X. upon reasonable request.

Conflicts of Interest: The authors declare no conflict of interest.

References

1. Saha, T.K.; Pal, S. Exploring physical wetland vulnerability of Atreyee river basin in India and Bangladesh using logistic regression and fuzzy logic approaches. *Ecol. Indic.* **2018**, *98*, 251–265. [CrossRef]
2. Duan, M.; Li, C.; Wang, X.; Fang, S.; Xiong, Y.; Shi, P. Solid separation from the heavy oil sludge produced from Liaohe Oilfield. *J. Pet. Sci. Eng.* **2018**, *172*, 1112–1119. [CrossRef]

3. Huang, X.; Bai, J.; Li, K.; Zhao, Y.; Tian, W.; Hu, C. Preparation of Clay/Biochar Composite Adsorption Particle and Performance for Ammonia Nitrogen Removal from Aqueous Solution. *J. Ocean Univ. China* **2020**, *19*, 729–739. [CrossRef]
4. Chen, Y.; Wang, X.; Wang, X.; Cheng, T.; Fu, K.; Qin, Z.; Feng, K. Biofilm Structural and Functional Features on Microplastic Surfaces in Greenhouse Agricultural Soil. *Sustainability* **2022**, *14*, 7024. [CrossRef]
5. Shahi, A.; Aydin, S.; Ince, B.; Ince, O. Evaluation of microbial population and functional genes during the bioremediation of petroleum-contaminated soil as an effective monitoring approach. *Ecotoxicol. Environ. Saf.* **2016**, *125*, 153–160. [CrossRef] [PubMed]
6. Sun, P.; Huang, X.; Xing, Y.; Dong, W.; Yu, J.; Bai, J.; Duan, W. Immobilization of Ochrobactrum sp. on Biochar/Clay Composite Particle: Optimization of Preparation and Performance for Nitrogen Removal. *Front. Microbiol.* **2022**, *13*, 838836. [CrossRef]
7. He, H.; Chen, Y.; Li, X.; Cheng, Y.; Yang, C.; Zeng, G. Influence of salinity on microorganisms in activated sludge processes: A review. *Int. Biodeterior. Biodegradation* **2016**, *119*, 520–527. [CrossRef]
8. Nwoba, E.G.; Parlevliet, D.A.; Laird, D.W.; Alameh, K.; Moheimani, N.R. Light management technologies for increasing algal photobioreactor efficiency. *Algal Res.* **2019**, *39*, 101433. [CrossRef]
9. Angelim, A.L.; Costa, S.P.; Farias, B.C.S.; Aquino, L.F.; Melo, V.M.M. An innovative bioremediation strategy using a bacterial consortium entrapped in chitosan beads. *J. Environ. Manag.* **2013**, *127*, 10–17. [CrossRef]
10. Reddy, S.; Osborne, J.W. Biodegradation and biosorption of Reactive Red 120 dye by immobilized *Pseudomonas guariconensis:* Kinetic and toxicity study. *Water Environ. Res.* **2020**, *92*, 1230–1241. [CrossRef]
11. Yan, H.; Han, Z.; Zhao, H.; Pan, J.; Zhao, Y.; Tucker, M.E.; Zhou, J.; Yan, X.; Yang, H.; Fan, D. The bio-precipitation of calcium and magnesium ions by free and immobilized Lysinibacillus fusiformis DB1-3 in the wastewater. *J. Clean. Prod.* **2019**, *252*, 119826. [CrossRef]
12. Huang, X.; Bai, J.; Li, K.-R.; Zhao, Y.-G.; Tian, W.-J.; Dang, J.-J. Characteristics of two novel cold- and salt-tolerant ammonia-oxidizing bacteria from Liaohe Estuarine Wetland. *Mar. Pollut. Bull.* **2017**, *114*, 192–200. [CrossRef]
13. Huang, X.; Zhou, T.; Chen, X.; Bai, J.; Zhao, Y. Enhanced Biodegradation of High-Salinity and Low-Temperature Crude-Oil Wastewater by Immobilized Crude-Oil Biodegrading Microbiota. *J. Ocean Univ. China* **2022**, *21*, 141–151. [CrossRef]
14. Mahdi, Z.; Yu, Q.J.; El Hanandeh, A. Removal of lead(II) from aqueous solution using date seed-derived biochar: Batch and column studies. *Appl. Water Sci.* **2018**, *8*, 181. [CrossRef]
15. Ojeda-López, R.; Ramos-Sánchez, G.; García-Mendoza, C.; Azevedo, D.C.S.; Guzmán-Vargas, A.; Felipe, C. Effect of Calcination Temperature and Chemical Composition of PAN-Derived Carbon Microfibers on N_2, CO_2, and CH_4 Adsorption. *Materials* **2021**, *14*, 3914. [CrossRef] [PubMed]
16. Xue, S.; Zhang, X.; Ngo, H.H.; Guo, W.; Wen, H.; Li, C.; Zhang, Y.; Ma, C. Food waste based biochars for ammonia nitrogen removal from aqueous solutions. *Bioresour. Technol.* **2019**, *292*, 121927. [CrossRef]
17. Yao, Y.; Gao, B.; Fang, J.; Zhang, M.; Chen, H.; Zhou, Y.; Creamer, A.E.; Sun, Y.; Yang, L. Characterization and environmental applications of clay–biochar composites. *Chem. Eng. J.* **2014**, *242*, 136–143. [CrossRef]
18. Ömeroğlu, A.; Erdoğan, Y.; Özcan, A.S. Modification of bentonite with a cationic surfactant: An adsorption study of textile dye Reactive Blue 19. *J. Hazard. Mater.* **2007**, *140*, 173–179. [CrossRef]
19. Zhang, D.; Zhang, Y.; Shen, F.; Wang, J.; Li, W.; Li, E.; Falandysz, J. Removal of cadmium and lead from heavy metals loaded PVA–SA immobilized *Lentinus edodes*. *Desalination Water Treat.* **2013**, *52*, 4792–4801. [CrossRef]
20. Chen, Y.; Yu, B.; Lin, J.; Naidu, R.; Chen, Z. Simultaneous adsorption and biodegradation (SAB) of diesel oil using immobilized Acinetobacter venetianus on porous material. *Chem. Eng. J.* **2016**, *289*, 463–470. [CrossRef]
21. Zhao, Y.-G.; Zheng, Y.; Tian, W.; Bai, J.; Feng, G.; Guo, L.; Gao, M. Enrichment and immobilization of sulfide removal microbiota applied for environmental biological remediation of aquaculture area. *Environ. Pollut.* **2016**, *214*, 307–313. [CrossRef] [PubMed]
22. Nie, H.; Nie, M.; Diwu, Z.; Wang, L.; Yan, H.; Lin, Y.; Zhang, B.; Wang, Y. Biological treatment of high salinity and low pH produced water in oilfield with immobilized cells of P. aeruginosa NY3 in a pilot-scale. *J. Hazard. Mater.* **2019**, *381*, 121232. [CrossRef] [PubMed]
23. Binnal, P.; Babu, P.N. Optimization of environmental factors affecting tertiary treatment of municipal wastewater by Chlorella protothecoides in a lab scale photobioreactor. *J. Water Process. Eng.* **2017**, *17*, 290–298. [CrossRef]
24. Ge, C.-H.; Dong, Y.; Li, H.; Li, Q.; Ni, S.-Q.; Gao, B.; Xu, S.; Qiao, Z.; Ding, S. Nitritation-anammox process–A realizable and satisfactory way to remove nitrogen from high saline wastewater. *Bioresour. Technol.* **2018**, *275*, 86–93. [CrossRef]
25. Gao, Y.; Wang, X.; Li, J.; Lee, C.T.; Ong, P.Y.; Zhang, Z.; Li, C. Effect of aquaculture salinity on nitrification and microbial community in moving bed bioreactors with immobilized microbial granules. *Bioresour. Technol.* **2020**, *297*, 122427. [CrossRef]
26. Li, L.; Wang, T.L.; Li, H.F.; Chen, J.Q.; Su, F.L. Effects of the Liao RiverWetland on removal nitrogen. *J. Irrig. Drain.* **2012**, *31*, 137–139. [CrossRef]

water

MDPI

Article

Quantifying the Impact of Changes in Sinuosity on River Ecosystems

Zicheng Yu [1,2], Yicheng Fu [3,*], Ye Zhang [1,2], Zhe Liu [4] and Yixuan Liu [3]

1 School of Water Conservancy and Hydroelectric Power, Hebei University of Engineering, Handan 056002, China; skycx666666@163.com (Z.Y.)
2 Hebei Key Laboratory of Smart Water Conservancy, Hebei University of Engineering, Handan 056002, China
3 China Institute of Water Resources and Hydropower Research, Beijing 100038, China
4 HeBei Water Conservancy Engineering Bureau Group Limited, Shijiazhuang 050021, China
* Correspondence: swfyc@126.com

Abstract: To quantitatively study the hydrodynamic changes in different river morphologies and clarify the impact of morphological changes on river ecosystems, this study examined a section of the Nansha River near Laoniuwan in the Haidian District, Beijing, and characterized different river morphologies by river sinuosity. The River 2D model was used for simulation and analysis, and the depth and velocity diversity indices were introduced to quantify the distribution of depth and velocity under different sinuosities. *Cyprinus carpio* was selected as the target fish in this study, and its suitability curve was determined using literature and field surveys. Combined with the simulation results, a weighted usable area curve was established to identify its inflection point and maximum value and determine the ecological flow in the river under different sinuosities, that is, to clarify the relationship between sinuosity and ecological flow. The results showed that the lower the sinuosity, the worse the depth and velocity diversity, but a greater sinuosity did not lead to better depth and velocity diversity. The depth and velocity diversity of a sinuosity of 1.5 were better than those of 1.89 in general, except for low flow conditions ($Q = 5\ m^3/s$). For rivers with water use restricted by nature and society and where ecological needs exist, ecological engineering that appropriately changes the planform of rivers can be considered to increase the diversity of river/channel geometry and provide a basis for the ecological restoration of rivers.

Keywords: sinuosity; depth and velocity diversity; ecological flow; habitat

1. Introduction

Globally, more than two-thirds of rivers are substantially affected by human activity [1]. In recent years, several urban rivers have been channelized and linearized in the planning stages because of the convenience of land use and the safety of flood discharge; thus, their natural form and direction are often drastically changed [2]. For example, only approximately 2% of the rivers and lakes in the United States and Germany are in a natural state [3,4]. Between 1929 and 1942, the U.S. Army Corps of Engineers constructed 14 cut-offs in the Lower Mississippi River between Memphis, Tennessee, and Red River Landing, Louisiana. Engineered cut-offs, in combination with two natural cut-offs, resulted in a net shortening of the river by approximately 235 km [5]. In the mid-to-late 19th century, systematic river channelization engineering was applied in Europe and elsewhere, leading to notable changes in the landscape, from multi-to single-line channel configurations such as the Danube, Rhone, Rhine, Italian rivers, and braided rivers in the French Alps [6]. River channelization in China began in the 1940s, but its rapid development was concentrated from the late 1950s to the mid-1970s. From 1958 to 1962, channelization or segmented channelization projects were conducted nationwide with a total length of approximately 1500 km [7].

Narrowing of the river channel and hardening of its banks accelerate the flow of river water, reducing the ratio of infiltration and interception of runoff during rainfall, increasing the peak volume, advancing it, and reducing the river water quality accordingly [8]. These canalized rivers often have regular and geometric cross sections owing to land occupation and engineering volumes. However, the channelization of curved rivers has not yet been fully demonstrated [9]. When the curved shape of a river changes in its natural state, it also changes its hydrological and hydraulic characteristics [9]. In addition to changes in river scouring and siltation, which have an impact on flood safety, it also leads to the disappearance of geomorphological diversity within the river; that is, it results in changes in the patterns of mainstreams, tributaries, shoals, and jets in natural rivers [10].

In the 1980s, countries such as Germany and Switzerland proposed the concept of "re-naturalization" to restore rivers to a state close to that of nature [11,12]. The United Kingdom has adopted near-natural river management techniques that emphasize the need to give priority to the ecological functions of rivers when restoring and maintaining them [13]. The Netherlands has emphasized the combination of ecological restoration of rivers and flood control and has put forward the concept of "giving space to the river" [14]. Guaranteeing ecological flows is one of the most important initiatives for restoring the naturalization of rivers [9]. With the increasing destruction of ecological and environmental resources, the ecological flow of rivers has received increasing attention. In recent years, research related to the ecological flow of rivers has developed rapidly and has become a major global concern in the 20th century [15–17]. Ecological flow is the amount of water, time, and water quality required to maintain the natural ecosystem of rivers and estuaries and to sustain the ecosystem on which human survival and development depend. The scientific determination of ecological flow is essential for high-quality regional ecological development [18].

After years of research and development, ecological flow calculation methods have been divided into four major categories: hydrological, hydraulic, habitat, and integrated. Both hydraulic and habitat methods determine the ecological flow in a study area by analyzing the relationship between the flow and the type and quantity of habitat provided by the water flowing through a river [19]. Changes in river morphology inevitably cause changes in the diversity of water depths and velocities in a river, affecting the type and quantity of habitats available to certain organisms and causing a considerable change in the ecological flow that maintains that section of the river [9,20,21].

To address the above problems, this study selected a river section near Laoniuwan in the Nansha River Basin of the Haidian District, Beijing, China, as the research object. Based on local historical and field research data, the River 2D model was used to simulate and analyze the results of the simulation, introduce a weighted usable area (WUA) diversity index, and quantitatively analyze the influence of changes in river morphology on WUA diversity. Using *Cyprinus carpio* as the target species, we statistically analyzed the changes in the WUA under different sinuosities and summarized the relationship between sinuosity and ecological flow to provide a basis for the ecological restoration of a river.

2. Materials and Methods

2.1. Study Area and Indicators

Haidian District is located in Northwest Beijing at the junction of the North China Plain and Taihang Mountains remnants (of a mountain range). The total area of Haidian District is 430.73 km^2, lying between latitudes 39°53′–40°09′ north and longitudes 116°03′–116°23′ east.

The Nansha River is located in the northwestern part of Haidian District, in a floodplain behind mountains. The total length of the main channel and the river basin is approximately 30.6 km and 263 km^2, respectively (Figure 1). The basin belongs to the warm belt of semi-humid monsoon continental climate, with average precipitation and surface evaporation for many years of up to 619 and 1883 mm, respectively. The river depth, upper mouth width, and bottom of the river longitudinal slope of the basin are approximately

4.0–5.0 m, 80.0–130.0 m, and 1–2.6‰, respectively, belonging to the Wenyu river system. A U-shaped bend is present on the west side of the Beijing–Bao Expressway near Laoniuwan, with a small turning radius and poor flow conditions, which are not conducive to flooding and regional drainage safety. The original Nansha River was scheduled to be straightened in the local comprehensive improvement plan of 2008; however, this plan has not yet been implemented. Moreover, there has always been controversy over whether the reach should remain in its current state or transform into a straight state. In recent years, as the concept of river management has advanced, the importance of maintaining the river's existing form in terms of flood control and ecology has been recognized. However, there has been no quantitative comparison of the impact on river ecology between the two maintenance cases of the existing planar meander form and its planned straightening. In this context, discussions and analyses in this area have practical importance for the development of locally related work.

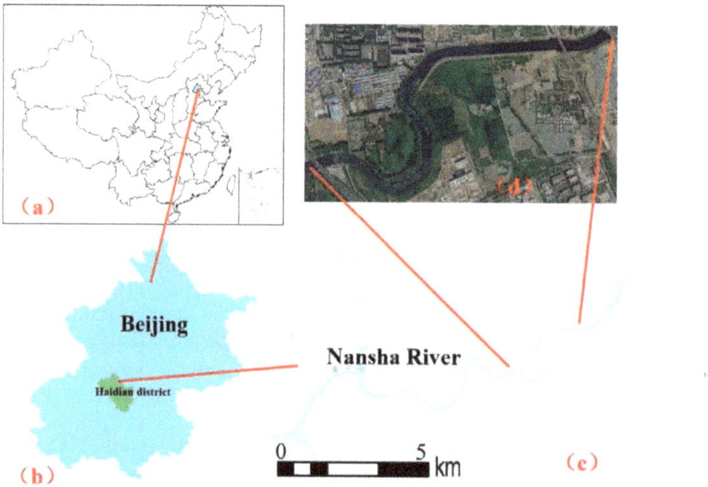

Figure 1. Study area. (**a**) Map of China; (**b**) scope of Beijing; (**c**) the main stream of the Nansha River and some of its tributaries; and (**d**) schematic of the study reach.

Fish are the top predators in river ecosystems, play an important role in the energy flow and material cycle of the entire ecosystem, and are important indicators of regional water conservation and environmental safety [22]. Based on the principles of representation, accessibility, and feasibility, a species with rich historical research data and a definitive research foundation that fully reflects the changes in habitat, that is, the changes in habitat suitability caused by changes in different planar forms, was selected. In addition, *C. carpio* is the dominant species in this area owing to its large population. Based on these factors, *C. carpio* was selected as the indicator species through a combination of field research and data mining.

Fish are the top predators in river ecosystems and play an important role in the energy flow and material cycle of the whole ecosystem, as well as being an important indicator of regional water conservation and environmental safety [22]. Based on the principles of representation, accessibility, and feasibility, species with rich historical research data and a definitive research foundation were selected, fully reflecting the changes in habitat—that is, the changes in habitat suitability caused by changes in different planar forms. In addition, *Cyprinus carpio* is the dominant species in this area, with a large number. Based on the above factors, *Cyprinus carpio* was selected as the indicator species through a combination of field research and data mining.

2.2. Hydraulic Model and WUA

In this paper, the ecological flow was determined using a habitat simulation method based on the principle of the in-channel flow increase method (IFIM)—which is the earliest and most widely used method in habitat simulation [23]—to establish the relationship between the quantity and quality of a suitable fish habitat and flow through a hydraulic ecological model, evaluate the impact of flow changes on the fish habitat, and adjust the flow to improve the ecological environment. The model used was River 2D, which is an unsteady two-dimensional depth-averaged finite element hydrodynamic model written by Professor Peter Steffler of the University of Alberta [24]. Its hydrodynamic simulation is based on a two-dimensional Saint-Venant set of equations consisting of mass conservation and momentum conservation equations in the x- and y-directions [25,26].

The weighted usable area of the selected target species is abbreviated as *WUA*:

$$WUA = \sum_{i=1}^{n} CSF(V_i, C_i, D_i) \times A_i \tag{1}$$

In Equation (1), *WUA* represents the WUA of the selected study reach. $CSF(V_i, D_i, C_i)$ is the overall suitability value of each cell. Among them, i represents the number of cells; V, C, and D represent the flow velocity, bed substrate, and depth suitability index, respectively; and A_i represents the area of each cell level. After the field survey, the substrate in the study section is known to be more uniform, so its suitability index can be considered to be 1.

2.3. Data Processing

Sinuosity is defined as the ratio of the length of the curved arc along the river channel between the two endpoints of a river segment to the length of a straight line between the two endpoints [27]. According to the river classification and the actual situation of the studied river section, based on the current elevation data of the Nansha River, six planar meander forms were constructed using the replication and interpolation methods as follows.

First, based on the actual situation and the reasons for the selection of the above-mentioned study river section, the range of sinuosities was determined to be 1.0 (planning) to 1.89 (current situation).

Second, based on the current situation, we identified and analyzed the elevation data of the meander section; the elevation data of the river section without the sinuosity changing was guaranteed to be consistent, and the data of the changed river section was replicated based on the location correspondence.

Thirdly, when the sinuosity changed, the length of the river section would change. Based on the replication method, the River 2D model was used to perform model interpolation during data preprocessing. Finally, the six determined planar meander morphology maps are shown below in Figure 2.

2.4. Determination of the Suitability Curve

A key factor that determines the accuracy of habitat simulation results is the habitat suitability index (HSI), which is used to quantitatively describe the suitability of a species to a habitat, with values ranging from 0 to 1, with 0 being completely unsuitable and 1 being completely suitable; the larger the value, the better the suitability [28].

Currently, there is a lack of information on the suitability of the hydraulic characteristics for fish distributed in the Nansha River. In this study, we refer to the existing literature [29] and actual local information to obtain a preliminary suitable flow and water depth for the target fish and derive the suitability curve for *C. carpio*. The substrate size and water quality in the study area were homogeneous and were ignored during this estimation. The suitable and optimal flow velocities for *C. carpio* were 0.1–1.1 m/s and 0.2–0.6 m/s, respectively. The suitable water depths for *C. carpio* were 0.2–2.5 m, and the optimum water depths were 1.0–1.5 m (Figure 3).

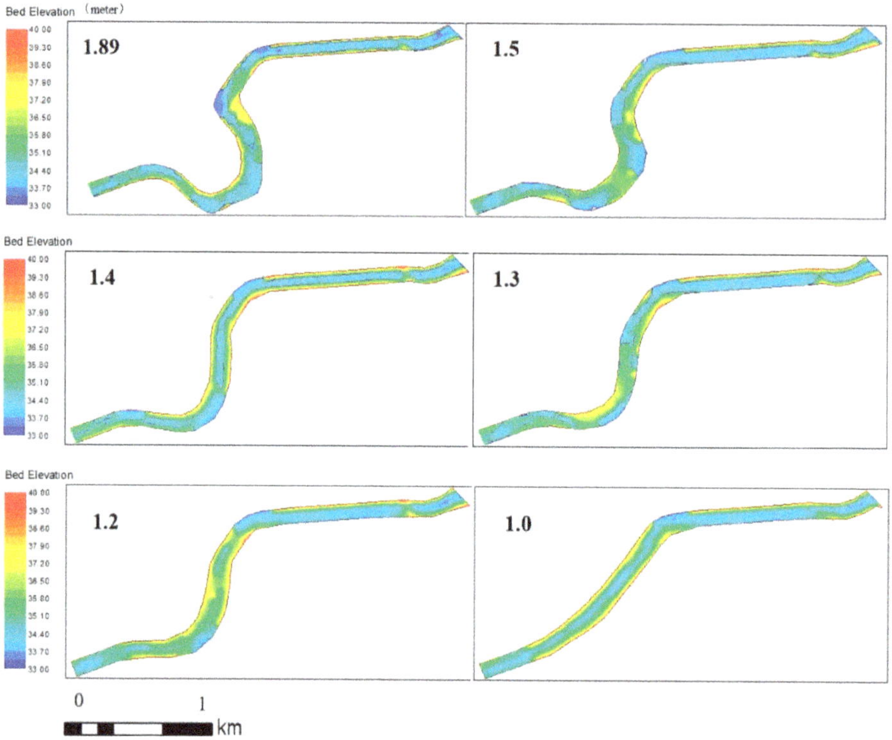

Figure 2. Different sinuosities (1.89–1.0).

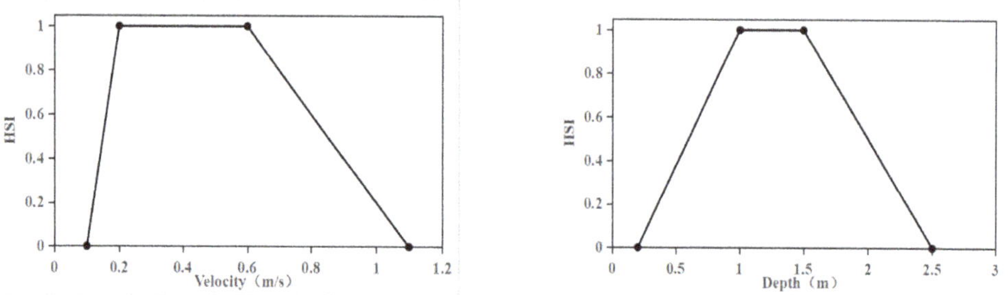

Figure 3. Velocity and depth suitability curves for *Cyprinus carpio*. (Drawings based on the research of Yang et al. [29]).

2.5. Model Building and Validation

Based on the local hydrological data, six flow conditions were simulated—5, 10, 15, 20, 25, and 30 m^3/s—and the roughness was taken as 0.035. River sections 1–6 were selected (see Figure 4). Comparing the simulated water level of the model—with an initial flow of 520 m^3/s—to the design water level of the "Haidian District Nansha River (Shangzhuang gate district boundary) Desilting Control Project" (the planning report of which has been approved and is being implemented) in 2016, it can be seen that the simulated value of the water level of the constructed model matched well with the design water level value of the desilting control project being implemented. The relative error was small (see Table 1), and the model established in this study exhibited good simulation ability for the hydrodynamic conditions of the Nansha River, where the spacing between 1 and 6 in each section is 300 m.

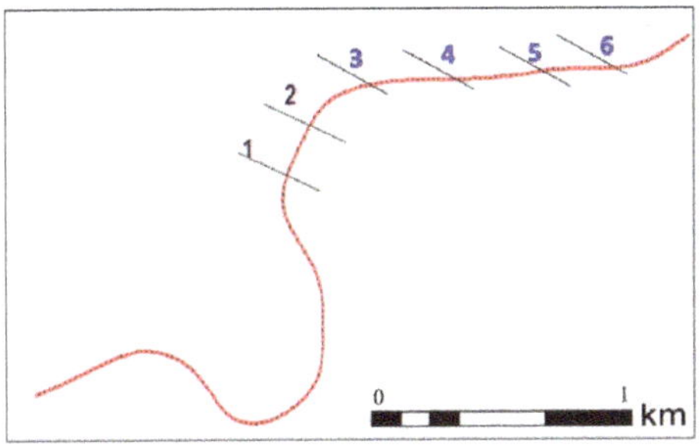

Figure 4. Schematic diagram of the cross section.

Table 1. Sections 1–6 simulation and water level design.

Section Name	1	2	3	4	5	6
Simulated water level (m)	38.55	38.43	38.20	38.08	37.96	37.92
Design water level (approved dredging project) (m)	38.45	38.40	38.40	38.39	38.36	38.28
Relative error (%)	0.26	0.08	0.52	0.80	1.05	0.98

2.6. Depth and Velocity Diversity Index

Diversity indices originated from the quantitative characterization of species diversity, Fisher proposed the concept of the species diversity index in 1943 [30]. Later, it was developed and gradually applied in the fields of environmental science, physical geography, and basic agricultural science—Fatch Paul et al. [31] proposed the overall agricultural diversity index for measuring agricultural diversity. Li Changchao et al. proposed a comprehensive index of microplastic diversity by clarifying the differences between microplastics in different environments and analyzing them retrospectively [32]. Currently, the indices for quantifying depth and velocity diversity are not yet clear. Combining the generalization and application of diversity indices from previous authors, this paper constructs a depth and velocity diversity index based on Shannon's index to better characterize depth and velocity diversity quantitatively.

The Shannon (H) diversity index can be used to reflect the degree of depth and velocity heterogeneity [33]. The higher its value, the better the degree of depth and velocity flow heterogeneity, and the more stable the survival of organisms in the region.

$$H = -\sum_{i=1}^{m} (P_i) \times \log_2(P_i) \tag{2}$$

In Equation (2), H denotes the diversity index; m denotes the number of different types of areas; and P_i denotes the proportion of the study area occupied by the ith type.

In order to consider the heterogeneous results of the combination of water depth and velocity, this paper introduces the depth and velocity diversity index based on the above analysis, referring to the research of related experts and scholars. We use H_e to express it, which is calculated as follows:

$$H_e = \sum_{i=1}^{m} (P_i) \times \log_2(P_i) \sum_{j=1}^{n} (P_j) \times \log_2(P_j) \tag{3}$$

In Equation (3), H_e represents the depth and velocity diversity index; m represents the number of different depth regions, n is the number of different velocity regions; P_i is the proportion of the study area occupied by the ith depth interval, and P_j is the proportion of the study area occupied by the jth velocity interval.

3. Results

3.1. Analysis of Water Depth and Flow Velocity Diversity at Each Sinuosity

The simulation results were processed and analyzed, and the water depth and velocity distribution of each sinuosity at different flows were calculated. The water depth and flow velocity scatter distribution maps and box plots are shown in Figures 5 and 6. Notably, most scatter points fall in an area with a water depth of 0.5–2.5 m and a flow velocity of 0–0.5 m/s. The studied river section was suitable for the survival of the target fish. Under the same sinuosity, the scattering points showed a more dispersed trend with increasing flow, and the maximum depth and velocity diversity values gradually increased. That is, depth and velocity diversity exhibited better trends, and the distribution of depth and velocity was more dispersed for sinuosities 1.3, 1.5, and 1.89. Under the same flow, as the sinuosity increased, the water depth exhibited a steady upward trend, and the flow velocity initially increased before falling.

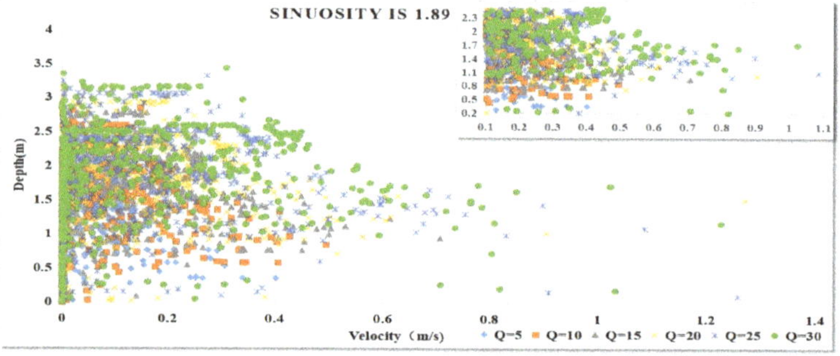

(**a**) Sinuosity is 1.89

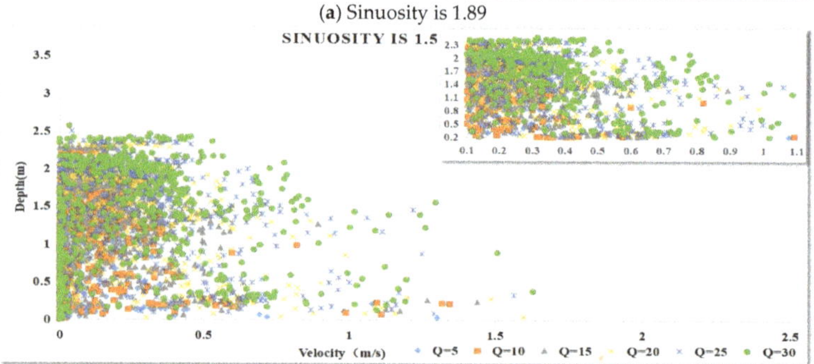

(**b**) Sinuosity is 1.5

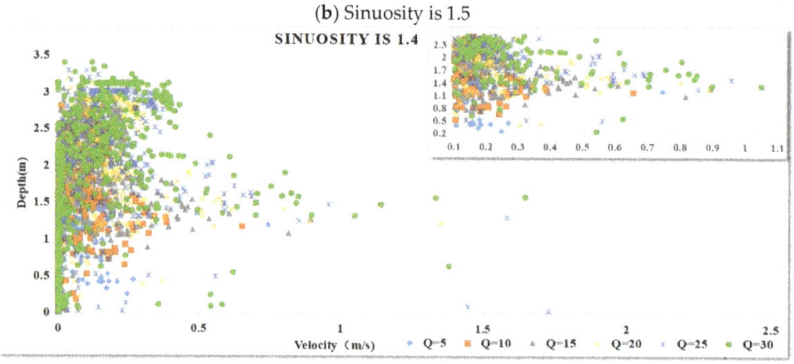

(**c**) Sinuosity is 1.4

Figure 5. *Cont.*

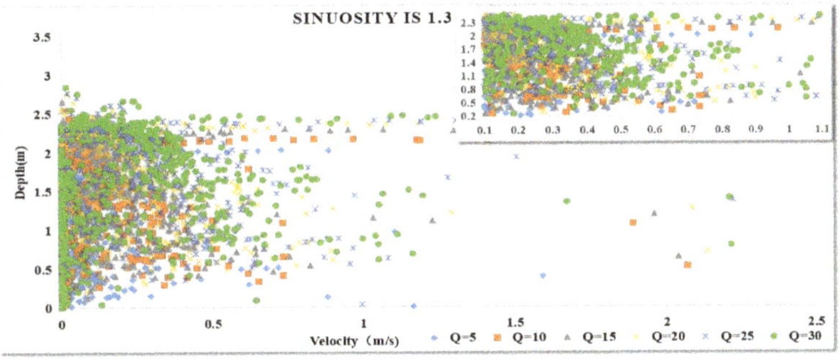

(d) Sinuosity is 1.3

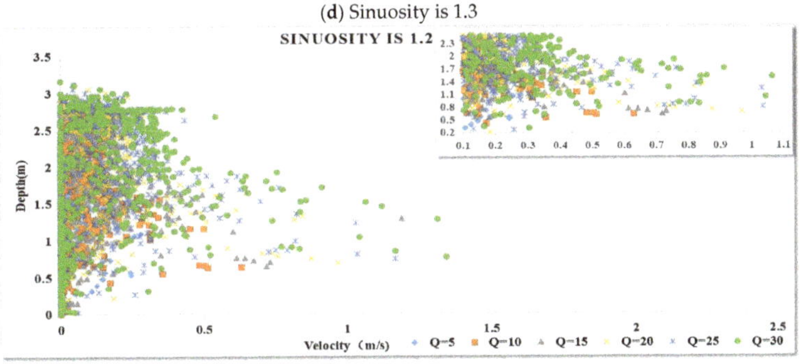

(e) Sinuosity is 1.2

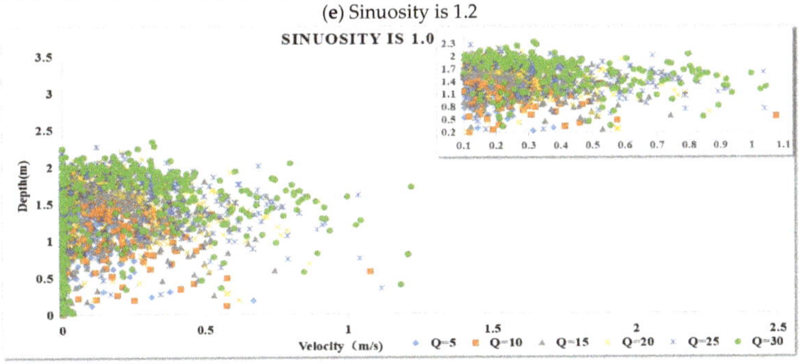

(f) Sinuosity is 1.0

Figure 5. Scatter diagrams of depth and velocity distribution under different sinuosities with different flows.

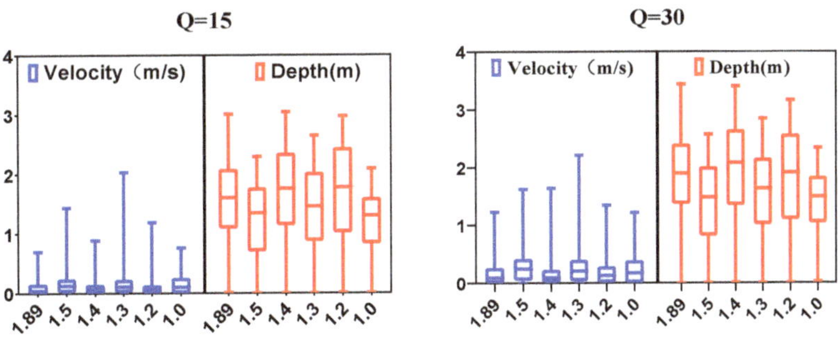

Figure 6. Velocity and depth distribution under different sinuosities.

To quantify the relationship between sinuosity and diversity of water depth and velocity, the velocity was divided into nine intervals, 0–0.1, 0.1–0.2, 0.2–0.4, 0.4–0.6, 0.6–0.8, 0.8–1.0, 1.0–1.2, 1.2–1.5, and 1.5 m/s or more, based on the actual conditions of the studied river and the suitable survival threshold of the target fish. The water depth was divided into six intervals: 0.5–1.0, 1.0–1.5, 1.5–2.0, 2.0–2.5, 2.5–3.0, and 3.0 m or more. Based on the simulation results, the percentage of each interval in the water area of the river section was derived and used as the P_{ij} quantity to calculate the water depth flow diversity index and quantitatively describe the relationship between sinuosity, depth, and velocity diversity under different flows (Figure 7 and Table 2).

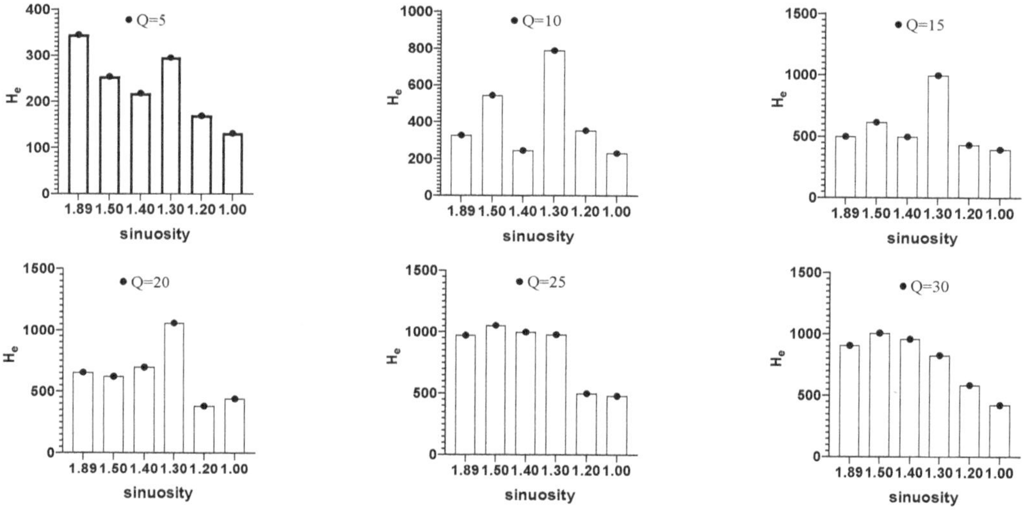

Figure 7. Diversity index of water depth and rate at different flows.

Table 2. Diversity index values of depth and velocity at different sinuosities.

Q (m³/s) \ S	1.89	1.5	1.4	1.3	1.2	1.0
5	345	254	218	296	170	132
10	329	544	246	789	355	232
15	499	616	498	998	431	396
20	653	621	697	1059	381	441
25	971	1053	998	978	499	480
30	906	1007	958	827	585	425

The overall trends in depth and velocity diversity increased with increasing sinuosity. This does not necessarily imply that greater sinuosity results in better depth and velocity diversity. For the index calculation results, the maximum value of the index was 1059 for the river form with a sinuosity of 1.3. However, lower sinuosity resulted in worse depth and velocity diversity. The sinuosities of 1.2 and 1.01 exhibited a substantially lower depth and velocity diversity index than the other forms under all flow conditions (Figure 8).

With increasing flow, most plane forms with different sinuosities showed an increasing trend, but this did not imply that greater flow resulted in better depth and velocity diversity. Furthermore, the maximum value of the index of sinuosity of 1.89–1.0 is obtained in the following order: Q = 25, 25, 25, 20, 30, and 25 m^3/s. In the case of a low flow of Q = 5 m^3/s, the greater the sinuosity, the better the depth and velocity diversity within a certain range (Figure 9).

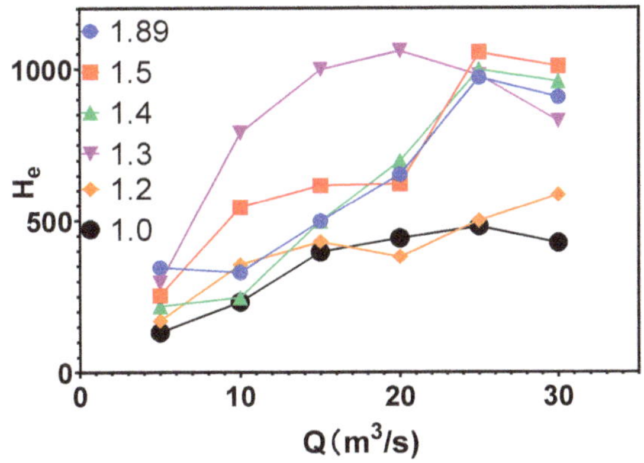

Figure 8. Trend of depth and velocity diversity index.

3.2. Sinuosity and Ecological Flow

Based on the suitability curve of *C. carpio*, the WUA distribution of the species at six sinuosities under different flow conditions was determined by simulation analysis (Figure 10). With an increasing flow rate, the WUA exhibited an overall increasing trend. At the same flow, river sections with greater sinuosity were relatively better; these more suitable areas were mostly at river bends, and the areas more suitable for *C. carpio* in the straight section were substantially smaller. The WUA distribution was organized, and the Q-WUA relationship curves were plotted.

As seen from the curves, a peak sinuosity of 1.5 was the largest, and that of 1.4 was the smallest. Under low flow, the suitable survival area of the river section with high sinuosity was substantially better than that of the river section with low sinuosity. At higher flows, the meanders were also better; however, there were more suitable areas for smaller meanders. Combined with the habitat simulation method, the highest point of the curve was the ecological flow of the river, and we could identify the ecological flow of meanders from 1.0–1.89 as being 20, 25, 20, 15, 22.5, and 15 m^3/s. When determining the ecological flow of meanders at a sinuosity of 1.5, the flows of 20 and 25 m^3/s did not exhibit monotonic increasing or decreasing trends. Thus, using a trial calculation of 22.5 m^3/s WUA, we found that the WUA corresponding to Q = 22.5 was the maximum value on the curve.

With increasing sinuosity, the overall ecological flow required by the target fish in the studied river section exhibited a decreasing trend, and sinuosity and ecological flow exhibited a negative correlation (Figure 11). Meanders with high sinuosity have a greater diversity of water depths and velocities, more geomorphological unit variety, and a strong regulation ability, which makes the rivers more suitable for the survival and reproduction of fish at lower flows. The smallest ecological flow of 15 m^3/s was required with sinuosities of 1.89 and 1.4. Comparing the WUA under the two morphologies, 1.89 was much larger than that of 1.4. In this study, considering the relationship between the social economy and ecology, a sinuosity of 1.89 was considered the appropriate morphology.

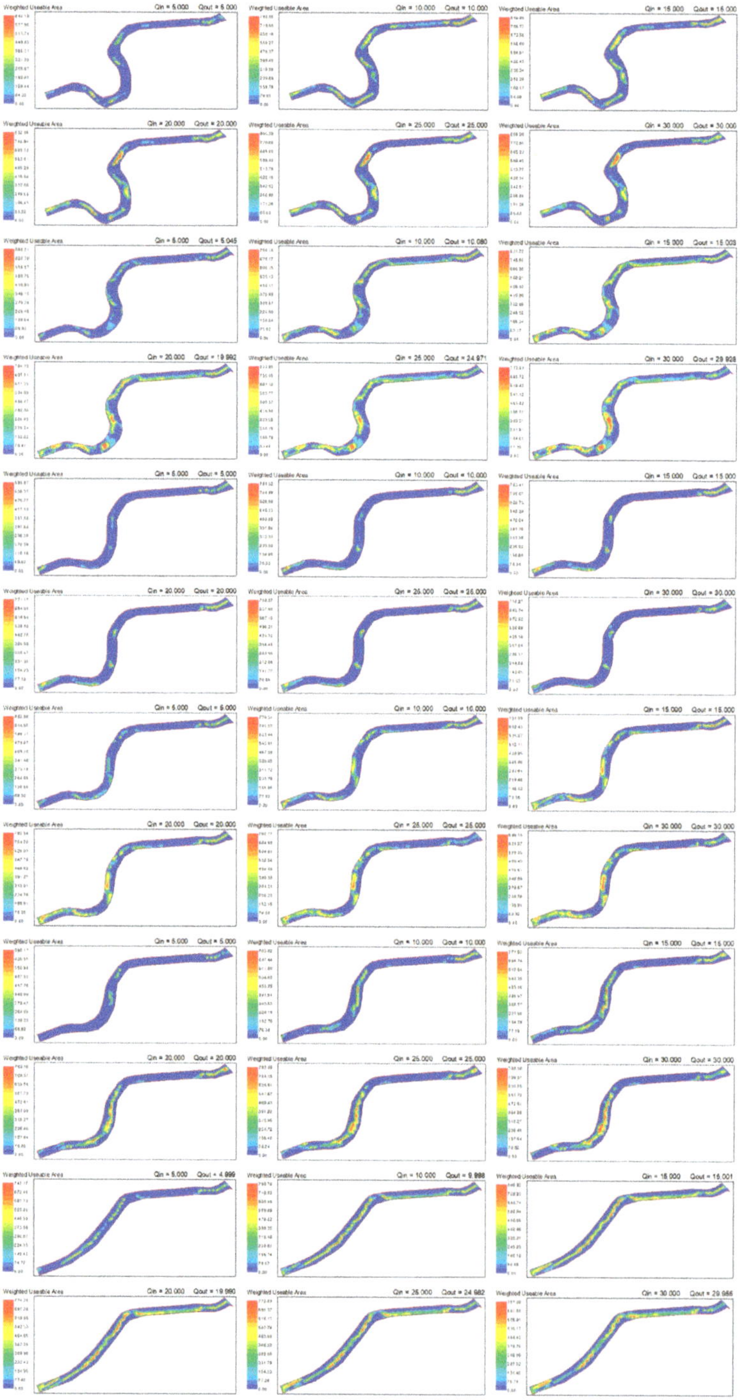

Figure 9. Distribution of WUA at different flows (with sinuosities of 1.89–1.0).

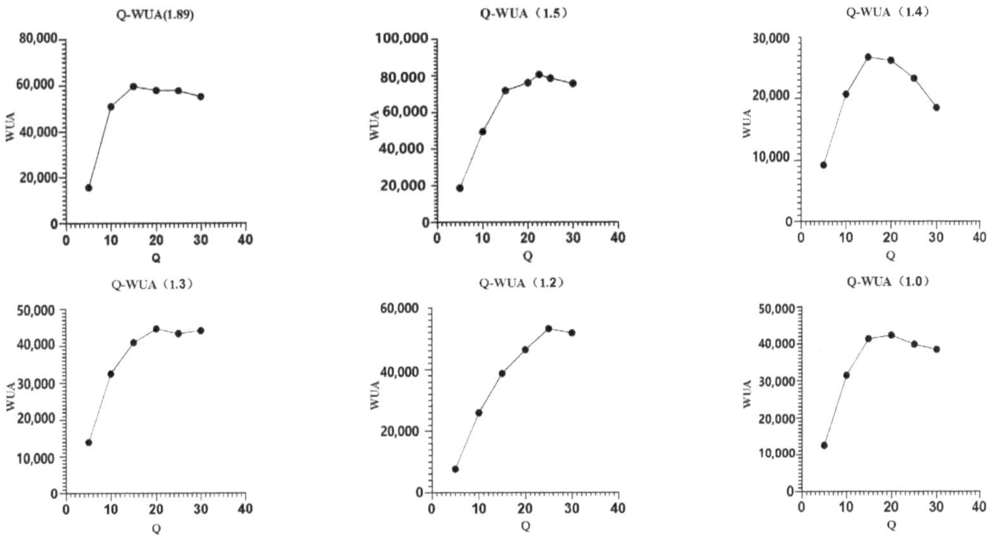

Figure 10. Q-WUA curves at different sinuosities.

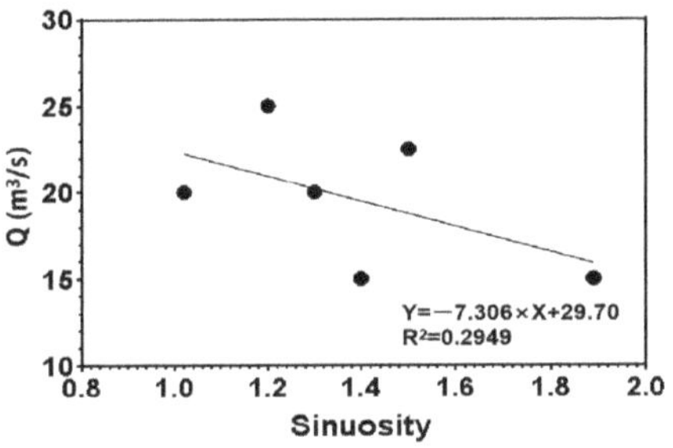

Figure 11. Relationship between sinuosity and ecological flow.

4. Discussion

4.1. Quantification of River Depth and Velocity Diversity

Many experts and scholars have conducted studies on the depth and velocity diversity. Scholars have introduced the habitat depth and velocity diversity index to quantitatively analyze the effects of artificial step-deep pool systems on aquatic habitats and river ecology [34]. Stähly et al. quantified the spatial variation in aquatic habitats in river segments using the hydroform diversity index [35]. The establishment of related indices allows a more intuitive quantitative study of the relationship between geomorphology, hydrology, and river ecosystems, and the use of hydrodynamic models to quantitatively simulate and analyze the spatial distribution and trend changes in water depth and flow velocity can provide technical support for the layout of river ecological restoration measures and the assessment of restoration effects [36–38].

4.2. Habitat Modeling Methodology and Ecological Flows

The habitat simulation method is used to determine the ecological water demand of rivers according to the physical habitat conditions required by the indicator species through field monitoring or water environment modeling to obtain the spatial and temporal distribution of habitat factors and habitat suitability evaluation indexes. Simulation of the quantitative relationship between the flow and distribution of suitable habitats is performed to obtain the ecological flow of aquatic organisms for the protection of the indicator species, mostly fish, and to provide a basis for the rational development and utilization of water resources [39–41]. Habitat simulation methods more adequately consider a single or multiple species, such as this study, which considers the dominant species in the study area, *C. carpio*, which reflects specific ecological needs, although the entire river ecosystem is ignored. Our results show that, as the flow changed, the inflection point of the Q-He curve did not coincide with the maximum value and Q-WUA curve, which initially showed the limitations of the ecological flow calculated by habitat simulation methods alone. In general, the better the diversity of the habitat, the better the biodiversity and the healthier the river ecosystem [42,43]. Further research is needed to determine if the flow required to maintain the stability of the entire river ecosystem should be considered.

4.3. Relationships between Changes in Sinuosity, River Ecology, and Socioeconomics

Changes in river morphology affect not only the physical form and dynamic river processes of individual reaches but also longer reaches and even entire river systems, including some tributaries [44]. Human intervention in river environments always needs to consider the unintended side effects and potential long-term legacies that may create new problems upstream or downstream [45], thereby affecting habitat availability and ecological status in longer reaches [9]. Furthermore, it is particularly important to clarify the relationship between sinuosity and the ecological and regional socioeconomic needs of rivers when carrying out comprehensive river sinuosity adjustment-oriented improvement projects.

Scholars have studied the mechanism of sinuosity in the self-purification of water bodies and proposed increasing the sinuosity to improve the ability of rivers to remove pollutants from water, which was thought to increase the growth rate of dissolved oxygen. However, the excessive meandering of rivers can affect flooding, sand drainage, and the safety of riverbanks [46]. Changes in the meandering of rivers must be justified, as there is a corresponding relationship between the meandering of rivers, society, and ecology; that is, no blind remediation is possible. Moreover, meanders of rivers that do not require remediation and whose meanders have remained unchanged for a long time should be left unchanged [47]. Meandering river ecosystems are sensitive and have difficulty recovering from damage [48]. Consequently, unnecessary human interference should be reduced to prevent the destruction of river habitats, which affect the stability of river ecosystems [49]. River morphology characterized by sinuosity is linked to the abundance, evenness, and diversity of organisms in the structure of river ecosystems, and there are large differences in organisms living at different geomorphic units of rivers [50,51]. In general, the complex sinuosity of water supports the diversity of river organisms [52,53]. Against the risk of riparian soil erosion along dammed rivers, the configuration of river morphology should be considered as one of the potential management strategies for offsetting the negative impacts of damming [54]. Moreover, after changing the sinuosity of rivers, monitoring of river habitat types and the corresponding biological changes should continue [55]. In this study, we quantified the relationship between sinuosity, depth, velocity diversity, and ecological flow and argued for an appropriate sinuosity, which has practical implications for the multi-objective ecological restoration of rivers.

For some water-scarce rivers affected by the natural environment, geographic location, socioeconomic factors, and human factors, there may be much less in-channel runoff than the ecological flow calculated through hydraulics and habitat simulation. In such cases where water diversion and replenishment cannot be guaranteed, research such as this study should be considered to determine the appropriate sinuosity through quantitative methods to maintain the stability of river ecosystems at relatively low flow rates through ecological engineering to adjust the river sinuosity.

5. Conclusions

In this study, a River 2D model was used to simulate the hydrodynamics and habitats of six morphologies of the studied river, which, combined with the depth and velocity diversity index and the Q-WUA curve, were used to determine the relationship between sinuosity, depth and velocity diversity, and the ecological flow required by the target species. Quantitative research showed that, as the sinuosity increased, the depth and velocity diversity of the water increased, and the two were positively correlated; however, this did not imply that the greater the sinuosity, the better the depth and velocity diversity, except under low flow conditions. Our results showed that the ecological flow required for the target species in the reach exhibited a downward trend and was negatively correlated with sinuosity.

Author Contributions: Investigation, data curation, writing—original draft, Z.Y.; methodology, data curation, writing—original draft, Y.F.; writing—review and editing, Y.Z.; writing—review and editing, Z.L.; investigation and methodology, Y.L. All authors have read and agreed to the published version of the manuscript.

Funding: This research received no external funding.

Data Availability Statement: The datasets generated and analyzed during the current study are not publicly available due to the fact that basic and analytical data involve local planning, but are available from the corresponding author on reasonable request.

Acknowledgments: Thank you to all authors for their contributions to the article.

Conflicts of Interest: The authors declare no conflict of interest.

References

1. Giller, P.S. River restoration: Seeking ecological standards. Editor's introduction. *J. Appl. Ecol.* **2005**, *42*, 201–207. [CrossRef]
2. Walteros, J.M.; Ramírez, A. Urban streams in Latin America: Current conditions and research needs. *Rev. Biol. Trop.* **2020**, *68*, 13–28. [CrossRef]
3. Jähnig, S.; Hering, D.; Sommerhäuser, M. *Fließgewässer-Renaturierung Heute und Morgen: EG-Wasserrahmenrichtlinie, Maßnahmen und Effizienzkontrolle; mit 55 Tabellen*; Schweizerbart: Stuttgart, Germany, 2011.
4. Lytle, D.A.; Poff, N. Adaptation to natural flow regimes. *Trends Ecol. Evol.* **2004**, *19*, 94–100. [CrossRef] [PubMed]
5. Winkley, B.R. *Man-Made Cutoffs on the Lower Mississippi River, Conception, Construction and River Response*; Army Engineer District: Vicksburg, MS, USA, 1977.
6. Arnaud, F.; Schmitt, L.; Johnstone, K.; Rollet, A.-J.; Piégay, H. Engineering impacts on the Upper Rhine channel and floodplain over two centuries. *Geomorphology* **2019**, *330*, 13–27. [CrossRef]
7. Tao, G. Development and Prospects of Canalization Engineering in China. *Adv. Sci. Technol. Water Resour.* **1995**, *4*, 4–8+35. (In Chinese)
8. Kadoya, M. A review of the study on runoff changes due to urbanization. *Proc. JSCE* **1985**, 363. [CrossRef]
9. Dong, Z.; Sun, D.; Zhao, J.; Jing, Z. *River Restoration*; China Water Conservancy and Hydropower Publishing: Beijing, China, 2013. (In Chinese)
10. Dutta, V.; Sharma, U.; Iqbal, K.; Adeeba; Kumar, R.; Pathak, A.K. Impact of river channelization and riverfront development on fluvial habitat: Evidence from Gomti River, a tributary of Ganges, India. *Environ. Sustain.* **2018**, *1*, 167–184.
11. Larsen, P. Restoration of River Corridors: German Experiences. In *The Rivers Handbook: Hydrological and Ecological Principles*; Wiley: Hoboken, NJ, USA, 1994; pp. 419–438.
12. Li, P.; Li, D.; Sun, X.; Chu, Z.; Xia, T.; Zheng, B. Application of Ecological Restoration Technologies for the Improvement of Biodiversity and Ecosystem in the River. *Water* **2022**, *14*, 1402. [CrossRef]
13. Tijdeman, E.; Hannaford, J.; Stahl, K. Human influences on streamflow drought characteristics in England and Wales. *Hydrol. Earth Syst. Sci.* **2018**, *22*, 1051–1064. [CrossRef]
14. WarWarner, J.F.; van Buuren, A.; Edelenbos, J. *Making Space for the River: Governance Experiences with Multifunctional River Flood Management in the US and Europe*; IWA Publishing: London, UK, 2012.
15. Wineland, S.M.; Başağaoğlu, H.; Fleming, J.; Laura, G.-D.; Kellogg, W.; Koch, J.; Lane, B.A.; Mirchi, A.; Nava, L.F.; Neeson, Y.M.; et al. The environmental flows implementation challenge: Insights and recommendations across water-limited systems. *Wiley Interdiscip. Rev. Water* **2022**, *9*, e1565. [CrossRef]
16. Xu, Z.; Wu, W.; Yu, S. Ecological baseflow: Progress and challenge. *J. Hydroelectr. Eng.* **2016**, *35*, 1–11. (In Chinese)
17. Palmer, M.A.; Bernhardt, E.S. Hydroecology and river restoration: Ripe for research and synthesis. *Water Resour. Res.* **2006**, *42*, W03S07.
18. Acreman, M.; Arthington, A.H.; Colloff, M.J.; Couch, C.; Crossman, N.D.; Dyer, F.J.; Overton, I.; Pollino, C.A.; Stewardson, M.J.; Young, W. Environmental flows for natural, hybrid, and novel riverine ecosystems in a changing world. *Front. Ecol. Environ.* **2014**, *12*, 466–473. [CrossRef] [PubMed]

19. Ying, C.; Zhang, Q.; Chen, X.A.; Qiang, T. Advances in the theories and calculation methods of ecological water requirement. *J. Lake Sci.* **2010**, *4*, 465–480. (In Chinese)
20. Brookes, A. River channel change. *The Rivers Handbook: Hydrological and Ecological Principles*; Wiley-Blackwell: Hoboken, NJ, USA, 1994; Volume 2, pp. 55–75.
21. Wang, Y.; Xia, J.; Cai, W.; Liu, Z.; Li, J.; Yin, J.; Zu, J.; Dou, C. Response of Fish Habitat Quality to Weir Distribution Change in Mountainous River Based on the Two-Dimensional Habitat Suitability Model. *Sustainability* **2023**, *15*, 8698. [CrossRef]
22. Chunguang, Z.; Yahui, Z.; Yingchun, X.; Ruilu, G.; Qing, Z.; Yun, F.; Enyuan, F. Fish species diversity and conservation in Beijing and adjacent areas. *Biodivers. Sci.* **2011**, *19*, 597–604. [CrossRef]
23. McManamay, R.A.; Brewer, S.K.; Jager, H.I.; Troia, M.J. Organizing Environmental Flow Frameworks to Meet Hydropower Mitigation Needs. *Environ. Manag.* **2016**, *58*, 365–385. [CrossRef]
24. Steffler, P.; Blackburn, J. Two-Dimensional Depth Averaged Model of River Hydrodynamics and Fish Habitat. In *River2D User's Manual*; University of Alberta: Edmonton, AB, Canada, 2002.
25. Naderi, M.H.; Zakerinia, M.; Salarijazi, M. Improvement of habitat simulation method using River2D hydrodynamic model to determine ecological regime of river. *Proc. R. Soc. Edinb.* **2019**, *72*, 263–277.
26. Bum, R.K.; Chun, P.S.; Hoon, J.Y.; Ok, P.M. Study on ecological instream flow estimation using River2D model in the Seomjin River. *J. Korean Soc. Water Environ.* **2011**, 822–829.
27. Leopold, L.B.; Wolman, M.G. *River Channel Patterns: Braided, Meandering, and Straight*; US Government Printing Office: Washinton, DC, USA, 1957.
28. Yi, Y.; Wang, Z.; Yang, Z. Two-dimensional habitat modeling of Chinese sturgeon spawning sites. *Ecol. Model.* **2010**, *221*, 864–875. [CrossRef]
29. Yang, Z.; Yu, S.; Chen, H.; She, D. Model for defining environmental flow thresholds of spring flood period using abrupt habitat change analysis. *Adv. Water Sci.* **2010**, *21*, 567–574. (In Chinese)
30. Williams, C.B. Area and number of species. *Nature* **1943**, *152*, 264–267. [CrossRef]
31. Fatch, P.; Masangano, C.; Hilger, T.; Jordan, I.; Mambo, I.; Kamoto, J.F.M.; Kalimbira, A.; Nuppenau, E.-A. Holistic agricultural diversity index as a measure of agricultural diversity: A cross-sectional study of smallholder farmers in Lilongwe district of Malawi. *Agric. Syst.* **2020**, *187*, 102991. [CrossRef]
32. Li, C.; Gan, Y.; Chao, Z.; Huan, H.; Fang, J.; Wang, L.; Yan, W.; Jian, L. "Microplastic communities" in different environments: Differences, links, and role of diversity index in source analysis. *Water Res.* **2020**, *188*, 116574. [CrossRef] [PubMed]
33. Zhang, A.; Dong, Z.; Zhao, J.; Sun, D. Progress of analysis and research on landscape pattern of river basin. *Water Resour. Hydropower Eng.* **2012**, *7*, 5. (In Chinese)
34. Yu, G.A.; Wang, Z.Y.; Zhang, K.; Wang, F.X. Effect of artificial step-pools on improving aquatic habitats and stream ecological in incised river channel. *J. Hydraul. Eng.* **2008**, *2*, 162–167. (In Chinese)
35. Stähly, S.; Gostner, W.; Franca, M.J.; Robinson, C.T.; Schleiss, A.J. Sampling sufficiency for determining hydraulic habitat diversity. *J. Ecohydraulics* **2018**, *3*, 130–144. [CrossRef]
36. Bockelmann, B.N.; Fenrich, E.K.; Lin, B.; Falconer, R.A. Development of an ecohydraulics model for stream and river restoration. *Ecol. Eng.* **2004**, *22*, 227–235. [CrossRef]
37. Roni, P.; Hall, J.E.; Drenner, S.M.; Arterburn, D. Monitoring the effectiveness of floodplain habitat restoration: A review of methods and recommendations for future monitoring. *WIREs Water* **2019**, *6*, e1355. [CrossRef]
38. Xing, Y.; Lv, B.; Ma, D.; Zhang, H.; Yu, G. Hydrodynamic modeling to evaluate the influence of inland navigation channel training works on fish habitats in the Upper Yellow River. *Ecol. Eng.* **2021**, *169*, 106289.
39. Maddock, I. The importance of physical habitat assessment for evaluating river health. *Freshw. Biol.* **1999**, *41*, 373–391. [CrossRef]
40. Poff, N.L.; Zimmerman, J.K.H. Ecological responses to altered flow regimes: A literature review to inform the science and management of environmental flows. *Freshw. Biol.* **2010**, *55*, 194–205. [CrossRef]
41. Alsterberg, C.; Roger, F.; Sundbäck, K.; Juhanson, J.; Hulth, S.; Hallin, S.; Gamfeldt, L. Habitat diversity and ecosystem multifunctionality—The importance of direct and indirect effects. *Sci. Adv.* **2017**, *3*, e1601475.
42. Palmer, M.A.; Menninger, H.L.; Bernhardt, E. River restoration, habitat heterogeneity and biodiversity: A failure of theory or practice? *Freshw. Biol.* **2010**, *55*, 205–222. [CrossRef]
43. Geist, J.; Hawkins, S.J. Habitat recovery and restoration in aquatic ecosystems: Current progress and future challenges. *Aquat. Conserv. Mar. Freshw. Ecosyst.* **2016**, *26*, 942–962. [CrossRef]
44. Hohensinner, S.; Hauer, C.; Muhar, S. River morphology, channelization, and habitat restoration. *Riverine Ecosystem Management: Science for Governing towards a Sustainable Future*; Springer: Cham, Switzerland, 2018; pp. 41–65.
45. Gregory, K. The human role in changing river channels. *Geomorphology* **2006**, *79*, 172–191.
46. Wanjie, Z.; Linjuan, X. Review on Artificial Bending Cutoff and Regulation of Distorted River Bends. In Proceedings of the 2018 3rd International Conference on Smart City and Systems Engineering (ICSCSE), Xiamen, China, 29–30 December 2018; pp. 905–907.
47. Murniningsih, S. Analysis of the relation between meander river transformation and soil classification in urban area. *IOP Conf. Ser. Earth Environ. Sci.* **2020**, *612*, 012073.
48. Graf, W.; Leitner, P.; Hanetseder, I.; Ittner, L.D.; Dossi, F.; Hauer, C. Ecological degradation of a meandering river by local channelization effects: A case study in an Austrian lowland river. *Hydrobiologia* **2016**, *772*, 145–160. [CrossRef]

49. Saha, P.; Islam, M.; Oyshi, J.T.; Khanum, R.; Nishat, A. A sustainability analysis on the trends and frequency of the channel flow of a carp breeding river against human interventions and governing public–private partnership (PPP) as adaptation. *SN Appl. Sci.* **2020**, *2*, 969. [CrossRef]
50. Gray, D.P.; Harding, J.S. *Braided River Ecology: A Literature Review of Physical Habitats and Aquatic Invertebrate Communities*; Science & Technical Pub., Department of Conservation: Setúbal, Portugal, 2007.
51. Blettler, M.C.M.; Amsler, M.L.; Eberle, E.G.; Szupiany, R.; Latosinski, F.G.; Abrial, E.; Oberholster, P.J.; Espinola, L.A.; Paira, A.; Poza, A.; et al. Linking hydro-morphology with invertebrate ecology in diverse morphological units of a large river-floodplain system. *Water Resour. Res.* **2016**, *52*, 9495–9510. [CrossRef]
52. Garcia, X.-F.; Schnauder, I.; Pusch, M.T. Complex hydromorphology of meanders can support benthic invertebrate diversity in rivers. *Hydrobiologia* **2011**, *685*, 49–68. [CrossRef]
53. Lorenz, A.W.; Jähnig, S.C.; Hering, D. Re-meandering German lowland streams: Qualitative and quantitative effects of restoration measures on hydromorphology and macroinvertebrates. *Environ. Manag.* **2009**, *44*, 745–754. [CrossRef] [PubMed]
54. Ran, Y.; Liu, Y.; Wu, S.; Li, W.; Zhu, K.; Ji, Y.; Mir, Y.; Ma, M.; Huang, P. A higher river sinuosity increased riparian soil structural stability on the downstream of a dammed river. *Sci. Total. Environ.* **2021**, *802*, 149886. [CrossRef] [PubMed]
55. Nagayama, S.; Nakamura, F. The significance of meandering channel to habitat diversity and fish assemblage: A case study in the Shibetsu River, northern Japan. *Limnology* **2018**, *19*, 7–20. [CrossRef]

water

MDPI

Brief Report

Study on Ferritin Gene Expression to Evaluate the Health of White Leg Shrimp (*Litopenaeus vannamei*) Postlarvae Due to Changes in Water Temperature, Salinity, and pH

Chul-Won Kim [1], Ju-Wook Lee [2], Seung-Won Kang [3] and Han-Seung Kang [4,*]

[1] Major of Aquaculture, Korea National University of Agriculture and Fisheries, Jeonju 54874, Republic of Korea; aquaworld@korea.kr
[2] Incheon Regional Office of National Fishery Products Quality Management Service, Incheon 22346, Republic of Korea; leejuwook84@gmail.com
[3] Daesang Aquaculture Trout Association Corporation, Taean 32158, Republic of Korea; rtt666@hanmail.net
[4] Department of Marine Environment, MS BioLab, Daejeon 34576, Republic of Korea
* Correspondence: hanseungkang66@gmail.com; Tel.: +82-42-632-9753

Abstract: The growth and survival of marine organisms are influenced by environmental factors such as water temperature, salinity, and pH. Unsuitable environmental conditions may negatively impact marine organisms. The white leg shrimp (*Litopenaeus vannamei*), a euryhaline organism highly adapted to salinity, is a valuable species for aquaculture. This study examined the effects of water temperature, salinity, and pH on the health of postlarvae *L. vannamei*. Stress levels within the organisms were analyzed through the expression of a biomarker gene. Ferritin was selected as the biomarker gene for analysis. The experimental animal samples used were the hepatopancreas of *L. vannamei* postlarvae. The analysis was performed by qRT-PCR. The results showed that the adaptation of *L. vannamei* postlarvae to temperature was dependent on salinity. Under low-salinity conditions (5 psu), ferritin expression increased at 25 °C and 30 °C after 48 h of exposure; however, it decreased after 72 h of exposure. Under normal salinity conditions (27 psu), ferritin expression increased from 24 h to 72 h at water temperatures of 25 °C and 30 °C. These results indicate that low-salinity conditions may enable *L. vannamei* postlarvae to rapidly adapt to high temperatures. In conclusion, *L. vannamei* postlarvae adapt more efficiently to high temperatures under low-salinity conditions than that under high-salinity conditions. The results of this study could beneficially impact *L. vannamei* farming.

Keywords: marine organisms; euryhaline organism; aquaculture; stress; biomarker gene

1. Introduction

In marine ecosystems, environmental factors such as water temperature, salinity, and pH affect physiological processes, such as reproduction, physiology, metabolism, growth, and regulation of osmotic pressure, as well as the habitat of marine animals [1–4]. Marine animals are sensitive to temperature change, and may be affected by seasonal or long-term climate change [5]. These changes in temperature affect reproduction, growth, physiology, metabolism, energy balance, and diseases in marine organisms [1,6–9]. Changes in temperature may also disrupt the homeostasis of marine animals, to which the animals may adapt by changing their habitat [10–12]. Increased water temperature may cause hypoxia through a decrease in oxygen solubility and weakening of the binding force of hemoglobin [13]. Even for species targeted for aquaculture, water temperature conditions are crucial for seeding production, cultivation, and early development [7,14–16]. Temperature stress may affect animal immunity and metabolism through the generation of intracellular reactive oxygen species (ROS), which may trigger oxidative stress [17].

Salinity is another crucial environmental factor that affects marine animals, impacting their ecological tolerance, stress levels, and distribution [18]. The standard salinity concentration of the ocean is 33–34 psu. To recover from the imbalance in homeostasis caused by stress, marine animals may attempt to regain homeostasis through metabolic activities such as osmosis control, control of body fluid concentration, and oxygen consumption. However, during extreme changes in salinity, adaptation may not be possible, which may lead to fatal consequences [19–24]. pH represents the hydrogen ion concentration in water and is an indicator of acidity or basicity; it varies according to various dissolved substances and the assimilation and respiration of organisms and, therefore, affects the physiological processes of organisms.

Dissolved oxygen (DO) determines optimal aquaculture stocking density and production, as oxygen affects the growth and reproduction of aquaculture animals by influencing food intake, metabolism, and rearing conditions. The concentration of oxygen in water is limited by its solubility in water, which decreases with increasing temperature and salt concentration. DO levels are related to the pH; therefore, oxygen levels in water affect organisms variably based on the pH. The stress caused by environmental change may lead to the production of reactive oxygen species (ROS) in organisms, thereby influencing intracellular stress and cell and tissue damage. It may also lead to cell proliferation, decreased metabolism, and even death. Genes related to antioxidant activity may be activated as a defense mechanism against ROS.

Ferritin is a protein present in the cytoplasm and mitochondria of organisms that plays a role in preparing cells for damage caused by excess iron and in replenishing iron if it is lacking [25–28]. This replenishing process regulates intracellular iron homeostasis and helps remove heavy metals when their concentration in the cell is high [29]. Iron is an essential nutrient for organisms; however, high concentrations may cause oxidative stress and increase the expression of the ferritin gene, accompanied by an increase in antioxidant enzyme activity. Accordingly, iron may cause oxidative stress and ferritin may be associated with antioxidant activity, which protects the cells from oxidative damage. Ferritin plays a crucial role in marine animals, which is similar to its role in terrestrial organisms. In marine animals, ferritin primarily functions as an iron storage protein, helping to regulate iron levels in the body [30]. Ferritin in *L. vannamei* is also known to play a role in enhancing immunity, physiological responses, and survival [31]. Overall, the function of ferritin in marine animals is essential for maintaining iron homeostasis, supporting vital physiological processes, and protecting against oxidative stress, ultimately contributing to their overall health and well-being in aquatic environments [32]. Ferritin has been implicated in the immune response of shrimp, particularly in defense against pathogens and oxidative stress [33]. In aquaculture, shrimp are exposed to various environmental stressors, including temperature fluctuations, salinity changes, and pollutants. In the presence of these stressors, ferritin may play a role in mitigating the effects of oxidative stress and preventing oxidative damage.

L. vannamei can adapt to high temperatures, is euryhalinous, and is able to survive in a salinity range of 1–40 psu, which enables its adaptation to various ecological conditions. In Korea, indigenous shrimp farming varieties include *Penaeus japonicus* and *Fenneropenaeus chinensis*; however, these species are available in limited amounts due to their vulnerability to white spot disease caused by the white spot syndrome virus. Accordingly, *L. vannamei* is becoming the major species for shrimp farming in Korea and its demand is also increasing in other countries due to its high stocking density and environmental adaptability, especially to variable levels of salinity. However, several studies have reported that the ability of *L. vannamei* to adapt to a various extreme environmental conditions can have a negative impact on the organism if the environmental conditions such as water temperature, salinity, and pH are not appropriate [34–38]. However, reports on the postlarvae, which have weak immune systems, are rare.

In this study, we aimed to investigate the effects of changes in water temperature, salinity, and pH on *L. vannamei* postlarvae through the expression patterns of the ferritin gene. The expression pattern of the ferritin gene is believed to be able to serve as a biomarker gene that can indirectly indicate the level of stress in the body of *L. vannamei* postlarvae. Also, we hope that these gene expression patterns will reveal the level of stress in the body and how this organism may overcome stress through environmental adaptation.

2. Materials and Methods

2.1. Preparation of Experimental Animals

The *L. vannamei* postlarvae used in this study were obtained from the breeding farm of the National University of Agriculture and Fisheries of Korea. Seawater sterilized by autoclave was used as the breeding water. A 5 L glass beaker was used as the breeding container. A total of 30 postlarvae (length: 3.0 ± 0.2 cm) per experimental group were assigned; they were incubated in a multiroom incubator.

2.2. Water Temperature, Salinity, and pH

Water temperature was set at 15, 20, 25, and 30 °C. Salinity was set to 5 or 27 psu in each water temperature. The pH was set at 6.5 or 7.5 within each water temperature and salinity concentration combination. The experimental period was 72 h. Five experimental animals were collected from each experimental group at different time points (6, 24, 48, and 72 h) for sampling. Experimental animals housed at a water temperature of 15 °C were used as the control group.

2.3. Total RNA Extraction

Hepatopancreatic tissues were used for total RNA extraction. The tissues were washed with a saline solution, then placed in liquid nitrogen, and triturated. Total RNA was extracted using the RNAiso Plus reagent (TaKaRa Bio, Otsu, Japan) according to the manufacturer's instructions. The extracted total RNA was quantified using a spectrophotometer (NanoVue, GE Healthcare, Amersham, UK), and RNA quality was confirmed based on the standard A260/280 ratio of 1.8–2.0.

2.4. Reverse Transcription–Quantitative Polymerase Chain Reaction (RT-qPCR)

One microgram of total RNA extracted from hepatopancreatic tissue was used as the template. Reverse transcription (RT) was performed using an oligo(dT)18 (0.5 μg) primer and the AccuPower RT premix kit (Bioneer Co., Daejeon, Korea) according to the manufacturer's instructions. RT was carried out in a 20 μL reaction mixture at 42 °C for 1 h to synthesize cDNA. For RT-qPCR analysis, 1 μL of cDNA was used. The primers used for ferritin gene expression analysis were as follows: F (5′-CAAGTCCGCCAGAACTAC-3′) and R (5′-TGGCAAATCCAGGTAGAG-3′). The nucleotide sequences of the β-actin was used as an internal standard using the following primers: F (5′-CCACGAGACCACCTACAAC-3′) and R (5′-AGGCGAGGGCAGTGATTTC-3′). Each primer was used at a concentration of 20 pmol for RT-qPCR. RT-qPCR was performed after adjusting the final volume of the reaction solution to 20 μL using the SFCgreen Fast qPCR Master Mix (2×) kit (SFCprobes Co., Chungju, Korea) according to the manufacturer's instructions. The cycling conditions for RT-qPCR were 94 °C for 5 min, followed by 35 cycles of 20 s at 94 °C, 20 s at 55 °C, and 20 s at 72 °C. To confirm the specificity and purity of all PCR products, melt curve analysis was carried out after amplification, under the following conditions: 94 °C for 15 s and 55 °C for 20 s. Fluorescence data were acquired during each annealing phase. Gene expression was assessed using the $2^{-\Delta\Delta CT}$ method as described by Livak [39].

2.5. Statistical Analysis

Significant differences between the control and the experimental groups were analyzed using Student's *t*-test and only those with $p < 0.05$ were considered significant.

3. Results and Discussion

Published research indicates that *L. vannamei* is subject to stress upon being exposed to varying water salinity and temperature [34–36]. In this study, we investigated this possibility by comparing the effects of water temperature, salinity, and pH on the health of *L. vannamei* postlarvae.

The expression of the ferritin in response to changes in water temperature over time was compared and analyzed in low- (5 psu; pH 6.5, 7.5) and high-salinity conditions (27 psu; pH 6.5, 7.5). The highest expression of ferritin at a salinity of 5 psu and pH 6.5 was observed in shrimp reared at 30 °C. There was a gradual and significant increase in ferritin expression through the experimental period in shrimp reared at 20 °C. With water temperatures of 25 and 30 °C, ferritin expression was the highest after 48 h (compared to that at 15 °C) and significantly decreased after 72 h (Figure 1). With salinity concentrations of 5 psu and a pH of 7.5, the expression of ferritin increased with increasing water temperatures and was higher than that observed at 15 °C. Ferritin expression increased over time in shrimp reared at 20, 25, and 30 °C. When the water temperature was 25 or 30 °C, ferritin expression was highest after 48 h; there was a trend for ferritin expression to decrease after 72 h (Figure 2). Under low-salinity conditions (5 psu), ferritin expression increased as the water temperature increased. This increased ferritin expression with increase in rearing water temperature under low salinity suggests that *L. vannamei* postlarvae can adapt to increasing water temperatures. In addition, the expression of ferritin increased with the increase in rearing water temperature and over time with no significant difference at pH 6.5 or 7.5. Ferritin expression was the highest at 48 h at rearing water temperatures of 25 and 30 °C, with a decrease at 72 h.

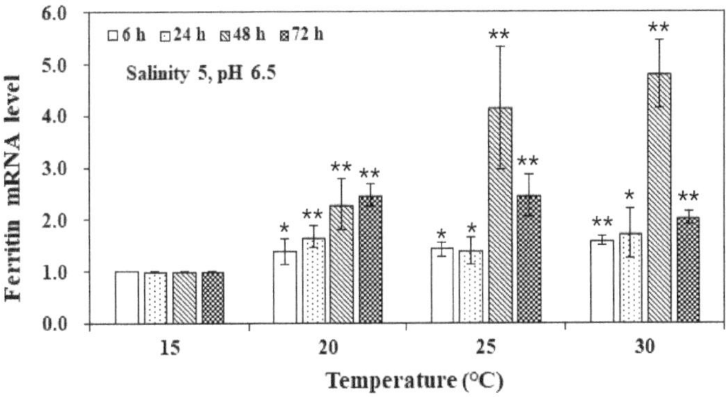

Figure 1. Expression levels of ferritin mRNA in *L. vannamei* postlarvae reared at different temperatures at 5 psu and pH 6.5 over time. * Significantly different from the control by Student's *t*-test ($p < 0.05$), ** $p < 0.01$.

At a salinity concentration of 27 psu and pH of 6.5, ferritin gene expression increased over time with elevated rearing temperatures (20, 25, and 30 °C) compared to that at 15 °C. The expression patterns of ferritin with respect to temperature and time exhibited a gradual increase at 20, 25 °C, and 30 °C, showing elevated expression levels at 24 and 72 h. At 72 h, the expression decreased compared to that at 48 h, although this was not significant (Figure 3). At a salinity of 27 psu and pH of 7.5, ferritin expression exhibited patterns similar to those at a salinity of 27 psu and pH of 6.5 with respect to rearing temperature and time. Ferritin expression was higher at temperatures over 15 °C. Ferritin gene expression at 20 and 30 °C showed a gradual increase with respect to temperature and time, reaching the highest expression at 48 h and showing a significant decrease at 72 h. Similarly, at 25 °C, a gradual increase was observed in ferritin expression up to 48 h when it reached its highest expression, followed by an insignificant decrease at 72 h

(Figure 4). Under high-salinity conditions (27 psu), ferritin gene expression increased with increasing water temperature. Similar to the results obtained at low-salinity conditions, these results suggest that an increase in rearing temperature imposes physiological stress on white shrimp. Furthermore, the expression of the ferritin gene, influenced by increases in rearing temperature and time, exhibited a significant increase at pH 6.5, maintaining high expression levels at 48 and 72 h. This suggests a state of stress owing to temperature adaptation for 72 h. At a pH of 7.5, a decreasing tendency was observed for the expression after 72 h.

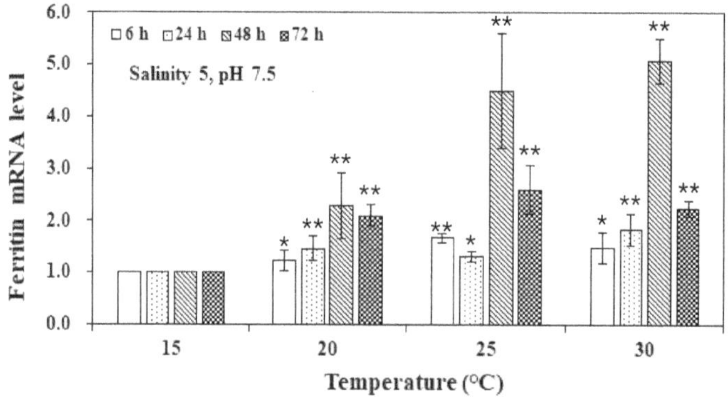

Figure 2. Expression levels of ferritin mRNA in *L. vannamei* postlarvae reared at different temperatures at 5 psu and pH 7.5 over time. * Significantly different from the control by Student's *t*-test ($p < 0.05$), ** $p < 0.01$.

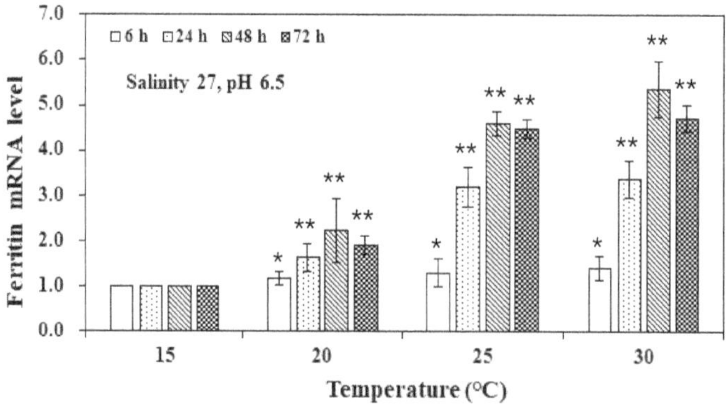

Figure 3. Expression levels of ferritin mRNA in *L. vannamei* postlarvae at 27 psu and pH 6.5 across different rearing temperatures over time. * Significantly different from the control by Student's *t*-test ($p < 0.05$), ** $p < 0.01$.

The results of comparing ferritin expression in the low-salinity (5 psu; pH 6.5 and 7.5) and high-salinity (27 psu; pH 6.5 and 7.5) experimental groups were as follows: In the low-salinity (5 psu; pH 6.5 and 7.5) experimental groups, the highest ferritin expression was observed after 48 h at water temperatures of 25 and 30 °C, and decreased after 72 h. This suggests that under low-salinity conditions, environmental adaptation of *L. vannamei* occurs after 48 h and stress begins to decrease. However, in the high-salinity experimental groups, ferritin expression increased from 24 h to 72 h at water temperatures of 25 and 30 °C. This suggests that under high-salinity conditions (27 psu; pH 6.5 and 7.5), environmental adaptation of *L. vannamei* postlarvae was not achieved and that they were in a state of stress.

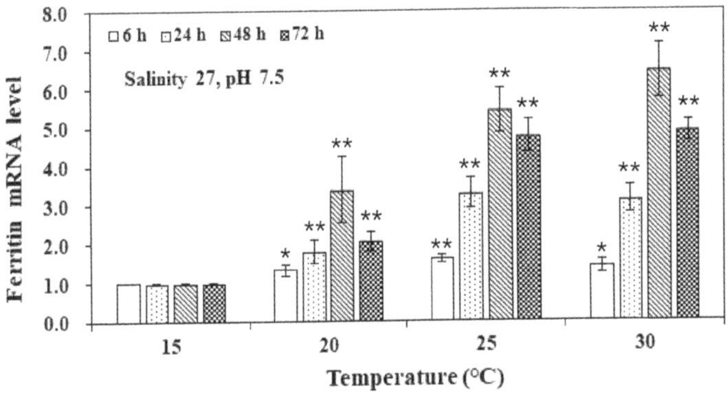

Figure 4. Expression levels of ferritin mRNA in *L. vannamei* postlarvae at 27 psu and pH 7.5 across different rearing temperatures over time. * Significantly different from the control by Student's t-test ($p < 0.05$), ** $p < 0.01$.

In summary, we found that an increase in water temperature under both low- and high-salinity conditions induced physiological stress in *L. vannamei* postlarvae. However, the transition to a stable state after temperature adaptation varied depending on salinity. Under low-salinity conditions (5 psu; pH 6.5 and 7.5), the recovery period appeared to be faster than that under high-salinity conditions (27 psu; pH 6.5 and 7.5). This conclusion is supported by the significant and rapid decrease in the expression pattern of ferritin at 72 h under low-salinity conditions (5 psu; pH 6.5 and 7.5) compared to that under high-salinity conditions (27 psu; pH 6.5 and 7.5).

In addition, the expression of ferritin at both low-salinity (5 psu; pH 6.5 and 7.5) and high-salinity conditions (27 psu; pH 6.5 and 7.5) at 6 h was consistently higher at 20, 25, and 30 °C than that at 15 °C. Upon comparing the expression of the groups exposed to different temperatures (20, 25, and 30 °C), no significant differences were observed (Figure 5). The expression of ferritin at 6 h, influenced by changes in temperature, salinity, and pH, showed no significant differences among the temperature groups (20, 25, and 30 °C).

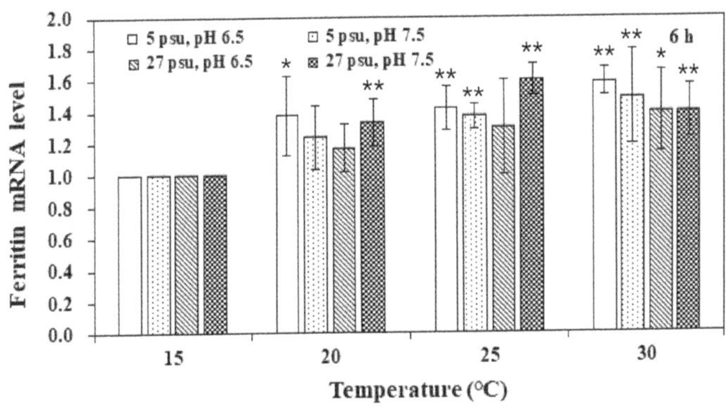

Figure 5. Ferritin mRNA expression levels in *L. vannamei* postlarvae exposed to different salinity (5, 27 psu) and pH (6.5, 7.5) levels for 6 h. * Significantly different from the control by Student's t-test ($p < 0.05$), ** $p < 0.01$.

At 24 h, ferritin expression under low-salinity conditions (5 psu; pH 6.5 and 7.5) was significantly higher at 20, 25 and 30 °C than that at 15 °C. No significant differences were observed in ferritin expression when comparing the groups exposed to 5 psu (pH 6.5) and 5 psu (pH 7.5) at each temperature (20, 25, and 30 °C). Under high-salinity conditions

(27 psu; pH 6.5 and 7.5), a significant increase was observed in ferritin expression with an increase in temperature compared to that observed in the 15 °C group. Particularly, 27 psu (pH 6.5 and 7.5) led to a significant increase in ferritin expression at both 25 and 30 °C compared to the low-salinity conditions (5 psu; pH 6.5 and 7.5). No significant differences were observed between the results of the groups exposed to 27 psu (pH 6.5) and 27 psu (pH 7.5) at any temperature (20, 25, and 30 °C) (Figure 6). At 24 h, ferritin expression, influenced by changes in rearing temperature, salinity, and pH, exhibited significantly higher expression under high-salinity conditions (27 psu; pH 6.5 and 7.5) at 25 and 30 °C. Given these results, it is likely that a salinity concentration of 27 psu (pH 6.5 and 7.5) induced potent stress in *L. vannamei* postlarvae during the initial 24 h of rearing at temperatures of 25 °C and 30 °C, whereas low-salinity conditions (5 psu; pH 6.5 and 7.5) had no significant impact during the first 24 h.

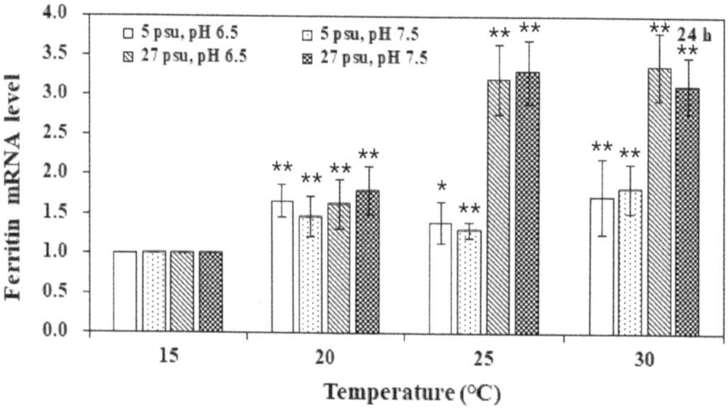

Figure 6. Ferritin RNA expression levels in *L. vannamei* postlarvae exposed to different salinity (5, 27 psu) and pH (6.5, 7.5) levels for 24 h. * Significantly different from the control by Student's *t*-test ($p < 0.05$), ** $p < 0.01$.

Ferritin expression at 48 h increased significantly with increased water temperatures (20, 25, and 30 °C) compared to that at 15 °C for both the low-salinity (5 psu; pH 6.5 and 7.5) and high-salinity (27 psu; pH 6.5 and 7.5) conditions. In particular, low-salinity conditions (5 psu; pH 6.5 and 7.5) did not cause any significant changes in ferritin expression with increased temperatures up to 24 h of rearing; however, a gradual increase in ferritin expression was observed from 48 h onward. Similarly, high salinity (27 psu; pH 6.5 and 7.5) led to a progressive increase in ferritin gene expression with an increase in water temperature; ferritin expression levels were also elevated compared to those at 24 h. Comparing ferritin gene expression in response to increasing water temperature in the low-salinity (5 psu, pH 6.5, 7.5) and high-salinity (27 psu, pH 6.5, 7.5) groups, the high-salinity (27 psu, pH 7.5) group exhibited relatively higher expression than the other experimental groups (Figure 7). The characteristics of ferritin gene expression in response to changes in rearing temperature, salinity, and pH at 48 h revealed a significant increase in gene expression with an increase in water temperature under low-salinity conditions (5 psu; pH 6.5 and 7.5). Therefore, it is likely that a salinity concentration of 5 psu (pH 6.5 and 7.5) induces stress in *L. vannamei* postlarvae after 48 h of rearing at temperatures of 25 and 30 °C. Additionally, a high salinity (27 psu; pH 6.5 and 7.5) resulted in increased ferritin expression compared to that at 24 h.

At 72 h, the expression of ferritin under low-salinity conditions (5 psu; pH 6.5 and 7.5) sharply decreased at both 25 and 30 °C compared to at 48 h. Contrarily, the high-salinity conditions (27 psu; pH 6.5 and 7.5) showed a trend for increasing ferritin expression with increasing temperature. At 25 and 30 °C, high-salinity conditions (27 psu; pH 6.5 and 7.5) led to a significant increase in ferritin expression compared to low-salinity conditions (5 psu; pH 6.5 and 7.5; Figure 8). Ferritin expression abruptly decreased in response to changes in rearing temperature, salinity, and pH at 72 h with a rise in water temperature under low salinity (5 psu; pH 6.5 and 7.5). Therefore, it is possible that a salinity of 5 psu (pH 6.5 and 7.5) alleviates internal stress through temperature adaptation at rearing temperatures of 25 and 30 °C after 72 h. Conversely, high-salinity conditions (27 psu, pH 6.5, 7.5) maintained elevated levels of ferritin expression from 24 h to 72 h. These results suggest that in high-salinity conditions (27 psu; pH 6.5 and 7.5), shrimp still experience the stress of adaptation at 72 h, particularly at rearing temperatures of 20 and 25 °C.

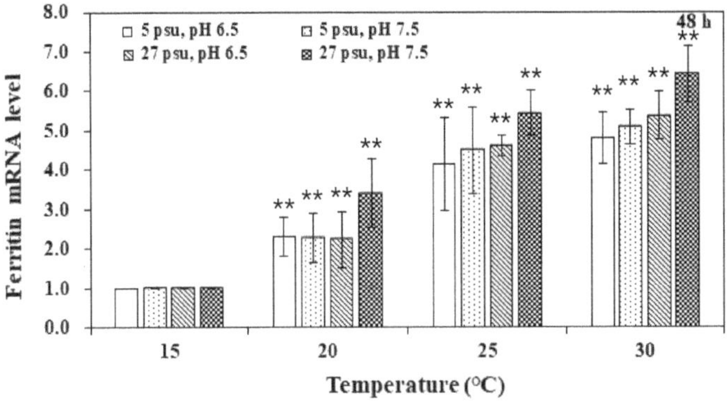

Figure 7. Ferritin mRNA expression levels in *L. vannamei* postlarvae reared at different temperatures exposed to changes in salinity (5 and 27 psu) and pH (6.5 and 7.5) for 48 h. ** Significantly different from the control by Student's *t*-test ($p < 0.01$).

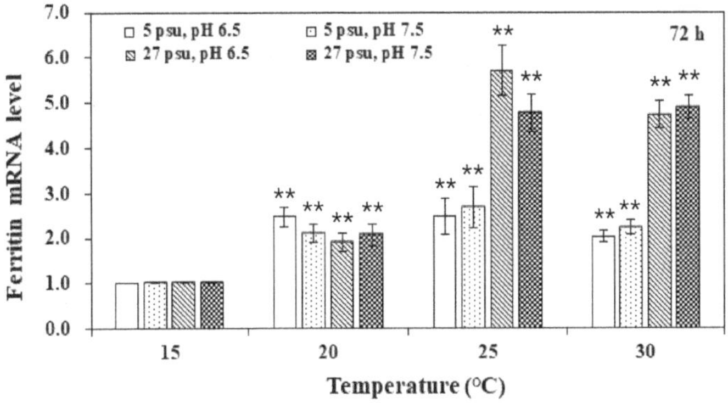

Figure 8. Ferritin mRNA expression levels in *L. vannamei* postlarvae reared at different temperatures exposed to changes in salinity (5 and 27 psu) and pH (6.5 and 7.5) for 72 h. ** Significantly different from the control by Student's *t*-test ($p < 0.01$).

The results of the study summarized the expression patterns of ferritin over time under low- (5 psu; pH 6.5 and 7.5) and high-salinity conditions (27 psu; pH 6.5 and 7.5) across different rearing temperatures, using the expression at 15 °C as a reference. At 20 °C, the expression patterns under both low- (5 psu; pH 6.5 and 7.5) and high-salinity (27 psu; pH 6.5 and 7.5) conditions showed an increasing trend over time (Figure 9). At 25 °C, the expression pattern of ferritin under low-salinity conditions (5 psu, pH 6.5, 7.5) showed no significant changes up to 24 h of rearing; however, expression had increased significantly at 48 h and then decreased significantly after 72 h. Under high-salinity conditions (27 psu; pH 6.5 and 7.5), ferritin expression gradually increased over time and remained significantly increased through 72 h of rearing. Comparing ferritin expression at 25 and 20 °C, all treatment groups had increased expression (Figure 10). At a rearing temperature of 30 °C, the expression patterns of ferritin in all treatment groups were similar to those at 25 °C. Under low-salinity conditions (5 psu; pH 6.5 and 7.5), no significant changes were observed in ferritin expression up to 24 h of rearing; however, expression significantly increased at 48 h and significantly and rapidly decreased at 72 h. Under high-salinity conditions (27 psu; pH 6.5 and 7.5), ferritin expression gradually increased from 24 h to 48 h. At 72 h, ferritin expression in the 27 psu (pH 6.5) group was similar to that at 48 h, whereas in the 27 psu (pH 7.5) group, ferritin expression at 72 h was lower than that at 48 h (Figure 11). These results suggest that salinity and rearing temperature are highly sensitive factors affecting the adaptation of *L. vannamei* postlarvae. At a temperature of approximately 20 °C, the impact of salinity on the occurrence of internal stress in *L. vannamei* postlarvae can be considered to be low. In summary, the optimal conditions for *L. vannamei* postlarvae rearing observed in this study were a rearing temperature of 25 or 30 °C at a low-salinity concentration (5 psu).

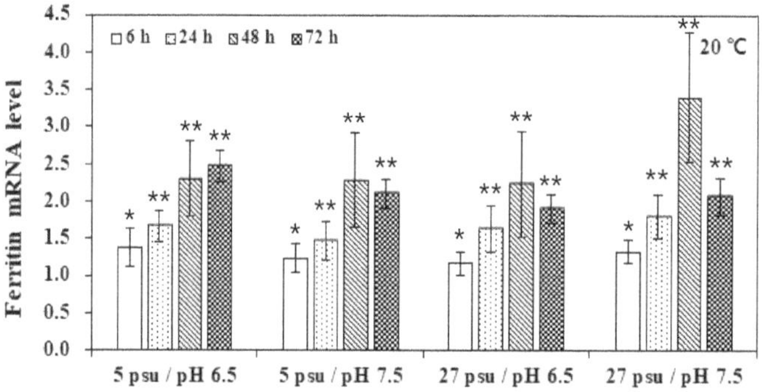

Figure 9. Ferritin mRNA expression levels in *L. vannamei* postlarvae at 20 °C in response to varying salinity and pH conditions. * Significantly different from the control by Student's *t*-test ($p < 0.05$), ** $p < 0.01$.

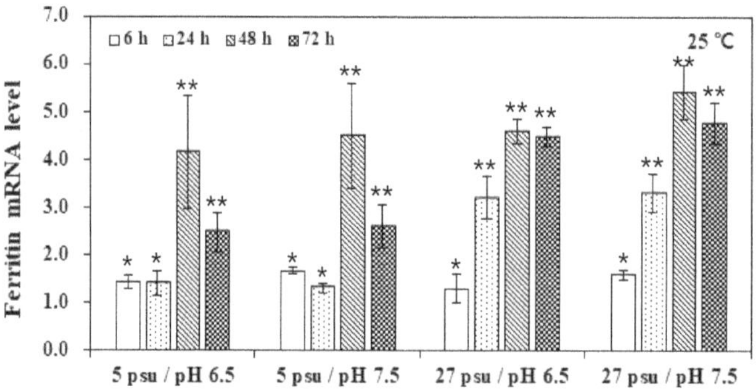

Figure 10. Ferritin mRNA expression levels in *L. vannamei* postlarvae at 25 °C in response to varying salinity and pH conditions. * Significantly different from the control by Student's *t*-test ($p < 0.05$), ** $p < 0.01$.

Cultured organisms are significantly influenced by environmental parameters. In the case of *L. vannamei* cultivation, the crucial parameters include water temperature and salinity, which affect cultivation. The optimal water temperature for *L. vannamei* cultivation has been shown to be 26–33 °C, considering aspects of growth and survival [40]. Research suggests that as the shrimp size increases, low water temperatures are preferable. For postlarvae, temperatures above 30 °C are recommended, while for adults, 27 °C is considered suitable; temperatures higher than this adversely affect growth [41,42]. Additionally, the metabolic rate and growth of *L. vannamei* decrease at rearing temperatures lower than 23 °C [41]. The acceptable range of salinity for *L. vannamei* is broad, ranging from 0.5 to 45 psu [43]. *L. vannamei* is a typical euryhaline crustacean that inhabits different locations throughout its life cycle. The larvae develop in the ocean, whereas postlarvae, juveniles, and adults inhabit estuaries and brackish waters [40]. Therefore, lower salinity is more suitable for *L. vannamei* postlarvae cultivation, lower than that of typical seawater. However, some reports suggest that shrimp growth rates are higher at 25 psu than at low-salinity conditions [44,45]. Low salinity is considered beneficial for habitats, whereas high salinity is considered better for growth. Failure to maintain optimal water temperature and salinity may lead to environmental stress, resulting in weakened immune responses and reduced productivity [40]. For *L. vannamei* cultivation, the recommended pH range for rearing water is 6.4–9.1; alkalinity is associated with low deformity rates [46]. In this study, the optimal conditions for *L. vannamei* postlarvae, considering water temperature, salinity, and pH, were characterized by a short and mild stress adaptation period at rearing temperatures of 25 and 30 °C under low salinity (5 psu). These conditions are suitable for survival, growth, and immunity of postlarvae of *L. vannamei*. The results of this research may be important for the production and cultivation of *L. vannamei* postlarvae.

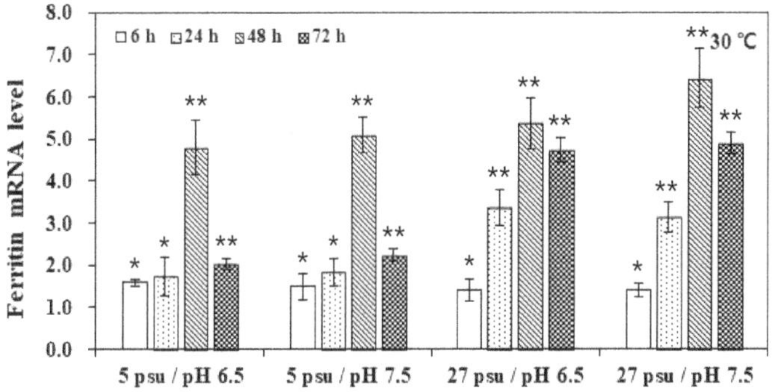

Figure 11. Ferritin mRNA expression levels in *L. vannamei* postlarvae at 30 °C in response to varying salinity and pH conditions. * Significantly different from the control by Student's *t*-test ($p < 0.05$), ** $p < 0.01$.

4. Conclusions

If the environmental factors (temperature, salinity, and pH) of a marine organism's habitat are not appropriate, their health can be adversely affected due to stress. This poor health can have adverse effects on reproduction, survival, and growth. When living organisms are under stress, ROS are typically produced. These ROS attack cells and deteriorate the health of the body. ROS are generated under prolonged stress, including oxidative stress, causing adverse effects on the immune and endocrine systems. Ferritin serves as a marker of antioxidant stress and can be used as an indicator of oxidative stress. We analyzed ferritin gene expression to determine the health status of *L. vannamai* postlarvae. Gene expression analysis is an efficient method for rapid and accurate evaluation of the health status of biological entities. From this perspective, the analysis of health status based on ferritin expression, considering environmental factors, such as water temperature, salinity, and pH, could be valuable. In this study, the appropriate water temperature and salinity for rearing *L. vannamei* postlarvae were presented through gene expression patterns. It was confirmed that low salinity (5 psu) under high temperature conditions for growth of *L. vannamei* postlarvae was positive. In the future, it is necessary to identify the conditions of appropriate environmental factors for *L. vannamei* postlarvae through research on the expression of various genes. In this study, analysis of ferritin expression suggested that low salinity (5 psu) and water temperatures of 25 and 30 °C are the most suitable environmental conditions for *L. vannamei* postlarvae.

Author Contributions: Conceptualization, H.-S.K.; methodology, H.-S.K., C.-W.K. and J.-W.L.; validation, H.-S.K.; investigation, J.-W.L. and S.-W.K.; writing—original draft preparation, C.-W.K. and H.-S.K.; writing—review and editing, H.-S.K. and J.-W.L.; project administration, H.-S.K. All authors have read and agreed to the published version of the manuscript.

Funding: This research was supported by the Ministry of Oceans and Fisheries (Korea Institute of Marine Science and Technology Promotion) in the development of an Autonomous Integrated Cultivation Management System for Shrimp Festival-Type Aquaculture Facilities (RS-2021-KS211474).

Institutional Review Board Statement: This study was conducted in accordance with the guidelines of the "Development of an Autonomous Integrated Cultivation Management System for Shrimp Festival-Type Aquaculture Facilities (RS-2021-KS211474)" project approved by the Korea Institute of Marine Science and Technology Promotion.

Data Availability Statement: The original contributions presented in the study are included in the article, further inquiries can be directed to the corresponding author/s.

Acknowledgments: This research was supported by the Ministry of Oceans and Fisheries (Korea Institute of Marine Science and Technology Promotion) in the development of an Autonomous Integrated Cultivation Management System for Shrimp Festival-Type Aquaculture Facilities (RS-2021-KS211474).

Conflicts of Interest: Author Seung-Won Kang was employed by Daesang Aquaculture Trout Association Corporation and author Han-Seung Kang was employed by MS BioLab. The remaining authors declare no conflicts of interest.

References

1. Chapple, J.P.; Smerdon, G.R.; Berry, R.J.; Hawkins, A.J.S. Seasonal changes in stress-70 protein levels reflect thermal tolerance in the marine bivalve *Mytilus edulis* L. *J. Exp. Mar. Biol. Ecol.* **1998**, *229*, 53–68. [CrossRef]
2. Norris, A.L.; Houser, D.S.; Crocker, D.E. Environment and activity affect skin temperature in breeding adult male elephant seals (*Mirounga angustirostris*). *J. Exp. Biol.* **2010**, *213*, 4205–4212. [CrossRef]
3. Hvas, M.; Folkedal, O.; Imsland, A.; Oppedal, F. Metabolic rates; swimming capabilities; thermal niche and stress response of the lumpfish; *Cyclopterus lumpus*. *Biol. Open.* **2018**, *7*, bio036079. [CrossRef]
4. Pavlosky, K.K.; Yamaguchi, Y.; Lerner, D.T.; Seale, A.P. The effects of transfer from steady-state to tidally-changing salinities on plasma and branchial osmoregulatory variables in adult *Mozambique tilapia*. *Comp. Biochem. Physiol. A Mol. Integr. Physiol.* **2019**, *227*, 134–145. [CrossRef]
5. Sampaio, E.; Lopes, A.R.; Francisco, S.; Paula, J.R.; Pimentel, M.; Maulvault, A.L.; Repolho, T.; Grilo, T.F.; Pousão-Ferreira, P.; Marques, A.; et al. Ocean acidification dampens physiological stress response to warming and contamination in a commercially-important fish (*Argyrosomus regius*). *Sci. Total Environ.* **2018**, *618*, 388–398. [CrossRef]
6. Newell, R.C.; Kofoed, L.H. Adjustment of the components of energy balance in the gastropod *Crepidula fornicate* in response to thermal acclimation. *Mar. Biol.* **1977**, *44*, 275–286. [CrossRef]
7. Logue, J.; Tiku, P.; Cossins, A.R. Heat injury and resistance adaptation in fish. *J. Therm. Biol.* **1995**, *20*, 191–197. [CrossRef]
8. Loomis, S.H.; Ansell, A.D.; Gibson, R.N.; Barnes, M. Freezing tolerance of marine invertebrates. *Oceanogr. Mar. Biol. Annu. Rev.* **1995**, *33*, 337–350.
9. Somero, G.N. Adaptation of enzymes to temperature: Searching for basic "strategies". *Comp. Biochem. Physiol. B Biochem. Mol. Biol.* **2004**, *139*, 321–333. [CrossRef]
10. Caissie, D. The thermal regime of rivers: A review. *Freshwater Biol.* **2006**, *51*, 1389–1406. [CrossRef]
11. Crossin, G.T.; Hinch, S.G.; Cooke, S.J.; Welsh, D.W.; Patterson, D.A.; Jones, S.R.M.; Lotto, A.G.; Leggatt, R.A.; Mathes, M.T.; Shrimpton, J.M.; et al. Exposure to high temperature influences the behaviour; physiology; and survival of sockeye salmon during spawning migration. *Can. J. Zool.* **2008**, *86*, 127–140. [CrossRef]
12. Wang, Y.; Li, L.; Hu, M.; Lu, W. Physiological energetics of the thick shell mussel *Mytilus coruscus* exposed to seawater acidification and thermal stress. *Sci. Total Environ.* **2015**, *514*, 261–272. [CrossRef]
13. Quinn, N.L.; McGowan, C.R.; Cooper, G.A.; Koop, B.F.; Davidson, W.S. Identification of genes associated with heat tolerance in Arctic charr exposed to acute thermal stress. *Physiol. Genom.* **2011**, *43*, 685–696. [CrossRef]
14. Burt, J.M.; Hinch, S.G.; Patterson, D.A. Parental identity influences progeny responses to incubation thermal stress in sockeye salmon *Onchorhynchus nerka*. *J. Fish Biol.* **2012**, *80*, 444–462. [CrossRef]
15. Hevrøy, E.M.; Waagbo, R.; Torstensen, B.E.; Takle, H.; Stubhaug, I.; Jorgensen, S.M.; Torgersen, T.; Tvenning, L.; Susort, S.; Breck, O.; et al. Ghrelin is involved in voluntary anorexia in Atlantic salmon raised at elevated sea temperatures. *Gen. Comp. Endocrinol.* **2012**, *175*, 118–134. [CrossRef]
16. Morita, K.; Nakashima, A. Temperature seasonality during fry out-migration influences the survival of hatchery-reared chum salmon *Oncorhynchus keta*. *J. Fish Biol.* **2015**, *87*, 1111–1117. [CrossRef]

17. Tsuji, Y.; Ayaki, H.; Whitman, S.P.; Morrow, C.S.; Torti, S.V.; Torti, F.M. Coordinate transcriptional and translational regulation of ferritin in response to oxidative stress. *Mol. Cell. Biol.* **2000**, *20*, 5818–5827. [CrossRef]
18. Dame, R.F. *Ecology of Marine Bivalves: An Ecosystem Approach*; CRC Marine Science Series; CRC Press: Boca Raton, FL, USA, 1996; p. 254.
19. Hammond, L.S. Experimental studies of salinity tolerance; burrowing behavior and pedicle regeneration in *Lingula anatina* (Brachiopoda; Inarticulata). *J. Paleontol.* **1983**, *57*, 1311–1316.
20. Beckmann, R.P.; Mizzen, L.E.; Welch, W.J. Interaction of HSP70 with newly synthesized proteins: Implications for protein folding and assembly. *Science* **1990**, *248*, 850–854. [CrossRef]
21. Weirich, C.R.; Tomasso, J.R. Confinement and transport induced stress on red drum juveniles: Effect of salinity. *Prog. Fish Cult.* **1991**, *53*, 146–149. [CrossRef]
22. Boeuf, G.; Payan, P. How should salinity influence fish growth? *Comp. Biochem. Physiol. Part C* **2001**, *130*, 411–423.
23. Marshall, W.S. Na^+, Cl^-, Ca^{2+} and Zn^{2+} transport by fish gills: Retrospective review and prospective synthesis. *J. Exp. Zool.* **2002**, *293*, 264–283. [CrossRef] [PubMed]
24. Matthews, Y.T.G.; Fairweather, P.G. Effect of lowered salinity on the survival, condition and reburial of *Soletellina alba* (Lamarch, 1818) (Bivalvia: Psammobiidae). *Aust. Ecol.* **2004**, *29*, 250–257. [CrossRef]
25. Konijn, A.M.; Glickstein, H.; Vaisman, B.; Meyron-Holtz, E.G.; Slotki, I.N.; Cabantchik, Z.I. The cellular labile iron pool and intracellular ferritin in K562 cells. *Blood* **1999**, *94*, 2128–2134. [CrossRef] [PubMed]
26. Wei, J.Z.; Theil, E.C. Identification and characterization of the iron regulatory element in the ferritin gene of a plant (soybean). *J. Biol. Chem.* **2000**, *275*, 17488–17493. [CrossRef] [PubMed]
27. Kakhlon, O.; Gruenbaum, Y.; Cabantchik, Z.I. Repression of the heavy ferritin chain increases the labile iron pool of human K562 cells. *Biochem. J.* **2001**, *356*, 311–316. [CrossRef] [PubMed]
28. Briat, J.F.; Ravet, K.; Arnaud, N.; Duc, C.; Boucherez, J.; Touraine, B.; Cellier, F.; Gaymard, F. New insights into ferritin synthesis and function highlight a link between iron homeostasis and oxidative stress in plants. *Ann. Bot.* **2009**, *105*, 811–822. [CrossRef] [PubMed]
29. Harrison, P.M.; Arosio, P. The ferritins: Molecular properties, iron storage function and cellular regulation. *Biochim. Biophys. Acta (BBA)-Bioenerg.* **1996**, *1275*, 161–203. [CrossRef]
30. Wu, Y.; Ming, T.; Huo, C.; Qiu, X.; Su, C.; Lu, C.; Zhou, J.; Li, Y.; Su, X. Crystallographic characterization of a marine invertebrate ferritin from the sea cucumber *Apostichopus japonicas*. *FEBS Open Bio* **2022**, *12*, 664–674. [CrossRef]
31. Ruan, Y.H.; Kuo, C.M.; Lo, C.F.; Lee, M.H.; Lian, J.L.; Hsieh, S.L. Ferritin administration effectively enhances immunity, physiological respnses, and survival of pacific white shrimp (*Litopenaeus vannamei*) challenged with white spot syndrome virus. *Fish Shellfish Immunol.* **2010**, *28*, 542–548. [CrossRef]
32. Galay, R.L.; Umemiya-Shirafuji, R.; Bacolod, E.T.; Maeda, H.; Kusakisako, K.; Koyama, J.; Tsuji, N.; Mochizuki, M.; Fusisaki, K.; Tanaka, T. Two kinds of ferritin protect ixodid ticks from iron overload and consequent oxidative stress. *PLoS ONE* **2014**, *9*, e90661. [CrossRef] [PubMed]
33. Ye, T.; Wu, X.; Wu, W.; Dai, C.; Yuan, J. Ferritin protect shrimp *Litopenaeus vannamei* from WSSV infection by inhibiting virus replication. *Fish Shellfish Immunol.* **2015**, *42*, 138–143. [CrossRef] [PubMed]
34. McGraw, W.J.; Davis, D.A.; Teichert-Coddington, D.; Rouse, D.B. Acclimation of *Litopenaeus vannamei* postlarvae to low salinity: Influence of Age, Salinity endpoint, and rate of salinity reduction. *J. World Aquac. Soc.* **2002**, *33*, 78–84. [CrossRef]
35. McGraw, W.J.; Scarps, J. Minimum Environmental potassium for survival of Pacific white shrimp *Litopenaeus vannamei* (Boone) in freshwater. *J. Shellfish Res.* **2003**, *22*, 263–267.
36. McGraw, W.J.; Scarps, J. Mortality of freshwater-acclimated *Litopenaeus vannamei* associated with acclimation rate, habituation period, and ionic challenge. *Aquaculture* **2004**, *236*, 285–296. [CrossRef]
37. Roy, L.A.; Davis, D.A.; Saoud, I.P.; Henry, R.P. Effects of varying levels of aqueous potassium and magnesium on survival, growth and respiration of the Pacific white shrimp, *Litopenaeus vannamei*, reared in low salinity waters. *Aquaculture* **2007**, *262*, 461–469. [CrossRef]
38. Castillo-Soriano, F.A.; Ibarra-Junquera, V.A.; Olivos-Ortiz, F.J.; Barragán-Vázquez, F.J.; Meyer-Willerer, A.O. Influence of water supply chemistry on white shrimp (*Litopenaeus vannamei*) culture in low-salinity and zero-water exchange ponds. *Panam. J. Aquat. Sci.* **2010**, *5*, 376–386.
39. Livak, K.J.; Schmittgen, T.D. Analysis of relative gene expression data using real-time quantitative PCR and the $2^{-\Delta\Delta CT}$ method. *Methods* **2001**, *25*, 402–408. [CrossRef] [PubMed]
40. Wickins, J.F.; Lee, D.O.C. *Crustacean Farming: Ranching and Culture*, 2nd ed.; Blackwell Science: Ames, IA, USA, 2008.
41. Wyban, J.; Walsh, W.A.; Godin, D.M. Temperature effects on growth, feeding rate and feed 858 conversion of the Pacific white shrimp. *Aquaculture* **1995**, *138*, 267–279. [CrossRef]
42. Brett, J.R. Environmental factors and growth. In *Fish Physiology*; Hoar, W.S., Randall, D.J., Brett, J.R., Eds.; Academic Press: London, UK, 1979; Volume 8, pp. 599–667.
43. Lester, L.; Pante, M.J. Penaeid temperature and salinity responses. *Dev. Aquacult. Fish. Sci.* **1992**, *23*, 515–534.
44. FAO. *Health Management and Biosecurity Maintenance in White Shrimp (Penaeus vannamei) Hatcheries in Latin America*; FAO Fisheries Technical Paper. No. 450; FAO: Rome, Italy, 2003; 58p.

45. Le Moullac, G.; Haffner, P. Environmental factors affecting immune responses in *Crustacea*. *Aquaculture* **2000**, *191*, 121–131. [CrossRef]
46. Chen, J.C.; Chin, T.S. Effect of Ammonia at different pH levels on *Penaeus monodon* postlarvae. *Asia Fish. Sci.* **1989**, *2*, 233–238. [CrossRef]

MDPI AG
Grosspeteranlage 5
4052 Basel
Switzerland
Tel.: +41 61 683 77 34

Water Editorial Office
E-mail: water@mdpi.com
www.mdpi.com/journal/water

www.ingramcontent.com/pod-product-compliance
Lightning Source LLC
Chambersburg PA
CBHW040743250926
51850CB00007BB/1394
9783725875559